Lecture Notes in Bioengineering

Lecture Notes in Bioengineering (LNBE) publishes the latest developments in bioengineering. It covers a wide range of topics, including (but not limited to):

- Bio-inspired Technology & Biomimetics
- Biosensors
- Bionanomaterials
- Biomedical Instrumentation
- Biological Signal Processing
- Medical Robotics and Assistive Technology
- Computational Medicine, Computational Pharmacology and Computational Biology
- Personalized Medicine
- Data Analysis in Bioengineering
- Neuroengineering
- Bioengineering Ethics

Original research reported in proceedings and edited books are at the core of LNBE. Monographs presenting cutting-edge findings, new perspectives on classical fields or reviewing the state-of-the art in a certain subfield of bioengineering may exceptionally be considered for publication. Alternatively, they may be redirected to more specific book series. The series' target audience includes advanced level students, researchers, and industry professionals working at the forefront of their fields.

Indexed by SCOPUS and Springerlink. The books of the series are submitted for indexing to Web of Science.

More information about this series at http://www.springer.com/series/11564

Stefano Piotto · Simona Concilio ·
Lucia Sessa · Federico Rossi
Editors

Advances in Bionanomaterials II

Selected Papers from the 3rd International Conference on Bio and Nanomaterials, BIONAM 2019, September 29 – October 3, 2019

Springer

Editors
Stefano Piotto
Department of Pharmacy
University of Salerno
Fisciano, Italy

Simona Concilio
Department of Industrial Engineering
University of Salerno
Fisciano, Italy

Lucia Sessa
Department of Pharmacy
University of Salerno
Fisciano, Italy

Federico Rossi
Department of Earth, Environmental
and Physical Sciences
University of Siena
Siena, Italy

ISSN 2195-271X ISSN 2195-2728 (electronic)
Lecture Notes in Bioengineering
ISBN 978-3-030-47704-2 ISBN 978-3-030-47705-9 (eBook)
https://doi.org/10.1007/978-3-030-47705-9

This Springer imprint is published by the registered company Springer Nature Switzerland AG
The registered company address is: Gewerbestrasse 11, 6330 Cham, Switzerland

Preface

This volume of the Springer book series Lecture Notes in Bioengineering gathers the proceedings of BIONAM 2019, the 3rd Workshop on Bio-nanomaterials, held on September 29 – October 3, 2019, on an MSC cruise ship navigating the Mediterranean Sea. BIONAM focused on the analysis, synthesis and design of bionanomaterials. The previous editions were held in Salerno, Italy, in 2013 and 2016, respectively. The second edition was published by Springer with the title "Advances in Bionanomaterials."

Starting from well-known biological structures, scientists and engineers have developed design principles for novel nanomaterials with superior properties. This knowledge has permitted to create complex structures with high toughness and high mechanical strength, or having a remarkable variety of advanced properties.

The peculiar location where BIONAM 2019 took place allowed an intense interchange of ideas and favored cross-contamination among different fields. The workshop represented an effort to bring together researchers active in biomaterials modeling with experimentalists, and the connection map built out of the conference shows the network and the interactions developed among the different attendees.

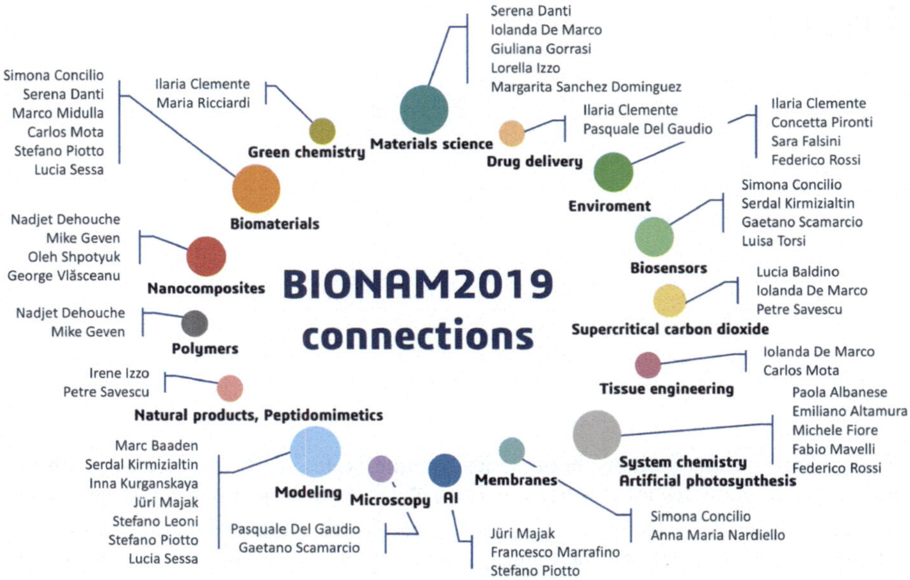

Biophysicists, biochemists and bioengineers presented studies on the fundamental properties of materials suitable for medical use (e.g., implantable devices), for interaction with biological systems and environmental applications. Attendees from therapeutic areas highlighted essential features for developing compatible biomaterials and for the evaluation of such materials in a physiological environment. The computational scientists shared tools to predict the mechanical, physical or biological properties of new biomaterials.

As a whole, the conference provided a comprehensive yet not exhaustive picture of state of the art in the field of bio-nanomaterials. Compared to the previous editions, the 3rd conference gave emphasis to stimuli-responsive and adaptable smart materials. Several researchers coming from nonlinear and far-from-equilibrium chemistry were involved in the scientific committee to cover the new and hot discipline known as dynamical self-assembly, a branch of the material science that try to blend the classical equilibrium self-assembly theory with evolutionary chemical systems and dissipative structures.

As Editors, we wish to express gratitude to the organizing and secretariat committee from the University of Salerno, namely **Iolanda De Marco, Anna Maria Nardiello, Luigi De Biasi** and **Ylenia Miele**, and, of course, to all the attendees of the conference and to the authors who spent time and effort to contribute to this volume. We also acknowledge the precious work of the reviewers and the members of the Program Committee. Special thanks, finally, to the invited speakers for their highly inspiring talks: **Luisa Torsi** from the University of Bari (Italy), **Stefano Leoni** from Cardiff University (UK) and **Marco Midulla** from the University of Bourgogne (France).

The 21 papers presented have been thoroughly reviewed and selected from 54 submissions. The contributions are typical examples of research outcomes in the biomaterial area, and they were grouped into three main categories: “Structure and Properties of Bio and Nanomaterials,” “Modeling of Bio and Nanomaterials” and “Applications of Bio and Nanomaterials.”

They cover the following topics: *nanomaterials, smart and stimuli-responsive materials, applications of nanostructured materials to medicine and biology, supercritical fluids, nanovectors, modeling and simulation of artificial and biological systems, topical controlled release, complex systems, synthetic and systems biology, systems chemistry*, and they represent the most exciting contributions to the 2019 edition of BIONAM.

October 2019

Stefano Piotto
Simona Concilio
Lucia Sessa
Federico Rossi

Organization

BIONAM 2019 has been jointly organized by the University of Salerno (Italy) and the University of Siena (Italy) and held on a cruise in the Mediterranean Sea from September 29 to October 3, 2019.

Chairs

Stefano Piotto	Department of Pharmacy – University of Salerno, Italy piotto@unisa.it
Simona Concilio	Department of Industrial Engineering – University of Salerno, Italy sconcilio@unisa.it
Federico Rossi	Department of Earth, Environmental and Physical Sciences – DEEP Sciences – University of Siena, Italy federico.rossi2@unisi.it
Iolanda De Marco	Department of Industrial Engineering – University of Salerno, Italy idemarco@unisa.it
Lucia Sessa	Department of Pharmacy – University of Salerno, Italy lucsessa@unisa.it

Program Committee

Emiliano Altamura	Università di Bari, Italy
Anna Angela Barba	Università di Salerno, Italy
Marcello Budroni	Université libre de Bruxelles, Belgium
Ugo Caruso	University of Napoli "Federico II", Italy

Supported by

DIPARTIMENTO DI SCIENZE FISICHE, DELLA TERRA E DELL'AMBIENTE

DIFARMA

Università di Salerno
Dipartimento di Ingegneria Industriale

Società Chimica Italiana

Springer

ENRICH
EUROPEAN NETWORK OF RESEARCH AND INNOVATION CENTRES AND HUBS, CHINA

SoftMining

SmartVASE

biomat.net

Contents

List of Contributors

Arti Ahluwalia Research Center “E. Piaggio”, University of Pisa, Pisa, Italy

Paola Albanese Department of Chemistry, University of Bari Aldo Moro, Bari, Italy

Emiliano Altamura Department of Chemistry, University of Bari Aldo Moro, Bari, Italy

Bahareh Azimi Research Unit of DICI-Pisa, Interuniversity Consortium for Materials Science and Technology (INSTM), Florence, Italy

Marc Baaden CNRS, Université de Paris, UPR 9080, Laboratoire de Biochimie Théorique, Paris, France; Institut de Biologie Physico-Chimique-Fondation Edmond de Rotschild, PSL Research University, Paris, France

Lucia Baldino Department of Industrial Engineering, University of Salerno, Fisciano, SA, Italy

Mihail Barboiu Institut Europeen des Membranes, Adaptive Supramolecular Nanosystems Group, Université de Montpellier, ENSCM, CNRS, Montpellier, France

Stefano Berrettini Department of Surgical, Medical, Molecular Pathology and Emergency Medicine, University of Pisa, Pisa, Italy

Ludovica Cacopardo Research Center “E. Piaggio”, University of Pisa, Pisa, Italy

Stefano Cardea Department of Industrial Engineering, University of Salerno, Fisciano, SA, Italy

Angela Casini Department of Chemistry, Technical University of Munich, Garching, Germany

Carolina Chieffo Université de Lyon, Institut de Chimie et Biochimie Moléculaires et Supramoléculaires (ICBMS, UMR 5246) Claude Bernard Lyon 1, Bâtiment Lederer, Villeurbanne Cedex, France

Ilaria Clemente Department of Biotechnology, Chemistry and Pharmacy, University of Siena, Siena, Italy; Department of Chemistry "Ugo Schiff" & CSGI, University of Florence, Sesto Fiorentino, Italy

Simona Concilio Department of Industrial Engineering, University of Salerno, Fisciano, SA, Italy

Serena Danti Research Center "E. Piaggio", University of Pisa, Pisa, Italy; The BioRobotics Institute, Scuola Superiore Sant'Anna, Pontedera, PI, Italy; Department of Civil and Industrial Engineering (DICI), University of Pisa, Pisa, Italy

Nadjet Dehouche Laboratoire des Matériaux Polymères Avancés (LMPA), Faculté de Technologie, Université de Bejaia, Bejaia, Algeria

Iolanda De Marco Department of Industrial Engineering, University of Salerno, Fisciano, SA, Italy

Luigi Di Biasi Department of Pharmacy, University of Salerno, Fisciano, SA, Italy

Miriam Di Martino Department of Pharmacy, University of Salerno, Fisciano, SA, Italy

Rosita Diana Department of Agriculture, University of Napoli Federico II, Portici, NA, Italy

Marta Feula Research Center "E. Piaggio", University of Pisa, Pisa, Italy

Michele Fiore Université de Lyon, Institut de Chimie et Biochimie Moléculaires et Supramoléculaires (ICBMS, UMR 5246) Claude Bernard Lyon 1, Bâtiment Lederer, Villeurbanne Cedex, France

Fabiana Foglia Department of Industrial Engineering, University of Salerno, Fisciano, SA, Italy

Paola Franco Department of Industrial Engineering, University of Salerno, Fisciano, SA, Italy

Giulia Giannone Department of Civil and Industrial Engineering (DICI), University of Pisa, Pisa, Italy

Giuliana Gorrasi Dipartimento di Ingegneria Industriale, Università degli Studi di Salerno, Fisciano, SA, Italy

Arthur Hardiagon CNRS, Université de Paris, UPR 9080, Laboratoire de Biochimie Théorique, Paris, France; Institut de Biologie Physico-Chimique-Fondation Edmond de Rotschild, PSL Research University, Paris, France

Weiwei He Chemistry Program, Science Division, New York University Abu Dhabi, Abu Dhabi, United Arab Emirates; Department of Chemistry, New York University, New York, NY, USA

Fanel Iacobescu Department 29 Horticulture and Food Sciences, University of Craiova, Craiova, Romania

Pio Iannelli Department of Pharmacy, University of Salerno, Fisciano, SA, Italy

Valentina Iozzino Department of Industrial Engineering, University of Salerno, Fisciano, SA, Italy

Lorella Izzo Dipartimento di Biotecnologie e Scienze della Vita, Università degli Studi dell'Insubria, Varese, Italy

Mustapha Kaci Laboratoire des Matériaux Polymères Avancés (LMPA), Faculté de Technologie, Université de Bejaia, Bejaia, Algeria

Salima Kennouche Laboratoire des Matériaux Polymères Avancés (LMPA), Faculté de Technologie, Université de Bejaia, Bejaia, Algeria; Centre des Matériaux des Mines d'Alès, IMT-Mines d'Alès, Alès, Cedex, France

Serdal Kirmizialtin Chemistry Program, Science Division, New York University Abu Dhabi, Abu Dhabi, United Arab Emirates

Andrea Lazzeri Department of Civil and Industrial Engineering (DICI), University of Pisa, Pisa, Italy

Stefano Leoni School of Chemistry, Cardiff University, Cardiff, UK

José-Marie Lopez-Cuesta Centre des Matériaux des Mines d'Alès, IMT-Mines d'Alès, Alès, Cedex, France

Augustin Lopez Université de Lyon, Institut de Chimie et Biochimie Moléculaires et Supramoléculaires (ICBMS, UMR 5246) Claude Bernard Lyon 1, Bâtiment Lederer, Villeurbanne Cedex, France

Xavier Martinez CNRS, Université de Paris, UPR 9080, Laboratoire de Biochimie Théorique, Paris, France; Institut de Biologie Physico-Chimique-Fondation Edmond de Rotschild, PSL Research University, Paris, France

Francesco Marrafino Department of Pharmacy, University of Salerno, Fisciano, SA, Italy

Fabio Mavelli Department of Chemistry, University of Bari Aldo Moro, Bari, Italy; Institute of Nanotechnology, CNR NANOTEC, Italian National Research Council, Bari, Italy

Massimo Mella Dipartimento di Scienza e Alta Tecnologia, Università degli Studi dell'Insubria, Como, Italy

Felicia Menicucci Department of Chemistry "Ugo Schiff" & CSGI, University of Florence, Sesto Fiorentino, Italy

Marco Midulla Department of Diagnostic and Therapeutic Radiology, Center for Mini-Invasive Image-Guided Therapies, Université de Bourgogne Franche-Compté, Centre Hospitalier Universitaire de Dijon, Dijon, France

Ylenia Miele Dipartimento di Chimica e Biologia, Università degli Studi di Salerno, Fisciano, Italy

Francesco Milano CNR-ISPA, Institute of Sciences of Food Production, Lecce Unit, Lecce, Italy

Mario Milazzo The BioRobotics Institute, Scuola Superiore Sant'Anna, Pontedera, PI, Italy

Stefania Moscato Department of Clinical and Experimental Medicine, University of Pisa, Pisa, Italy

Carlos Mota Institute for Technology Inspired Regenerative Medicine (MERLN), Complex Tissue Regeneration Department, Maastricht University, Maastricht, The Netherlands

Samuel Murail Université de Paris, CNRS, INSERM, Biologie Fonctionnelle et Adaptative UMR 8251, Computational Modeling of Protein Ligand Interactions U1133, Paris, France

Anna Maria Nardiello Department of Pharmacy, University of Salerno, Fisciano, SA, Italy

Roberto Pantani Department of Industrial Engineering, University of Salerno, Fisciano, SA, Italy

Stefano Piotto Department of Pharmacy, University of Salerno, Fisciano, SA, Italy

Maria-Magdalena Poenaru Department 29 Horticulture and Food Sciences, University of Craiova, Craiova, Romania

Sandra Ristori Department of Chemistry "Ugo Schiff" & CSGI, University of Florence, Sesto Fiorentino, Italy

Federico Rossi Dipartimento Scienze Fisiche, della Terra e dell'Ambiente, Siena, Italy

Jacopo Santoro Department of Pharmacy, University of Salerno, Fisciano, SA, Italy

Hubert Santuz CNRS, Université de Paris, UPR 9080, Laboratoire de Biochimie Théorique, Paris, France; Institut de Biologie Physico-Chimique-Fondation Edmond de Rotschild, PSL Research University, Paris, France

Giulia Sautariello Department of Chemistry "Ugo Schiff" & CSGI, University of Florence, Sesto Fiorentino, Italy

Petre Savescu Department 32 TAS Food Control and Expertise, University of Craiova, Craiova, Romania

Lucia Sessa Department of Pharmacy, University of Salerno, Fisciano, SA, Italy

Andrea Sorrentino IPCB-CNR, P.le Enrico Fermi, 1, Portici, NA, Italy

Pasquale Stano Department of Biological and Environmental Sciences and Technologies, University of Salento, Lecce, Italy

Fabio Sterpone CNRS, Université de Paris, UPR 9080, Laboratoire de Biochimie Théorique, Paris, France; Institut de Biologie Physico-Chimique-Fondation Edmond de Rotschild, PSL Research University, Paris, France

Andrea Tagliabue Dipartimento di Scienza e Alta Tecnologia, Università degli Studi dell'Insubria, Como, Italy

Luisa Trombi Research Unit of DICI-Pisa, Interuniversity Consortium for Materials Science and Technology (INSTM), Florence, Italy

Massimo Trotta CNR-IPCF, Institute for Physical and Chemical Processes, Bari, Italy

Valentina Volpe Department of Industrial Engineering, University of Salerno, Fisciano, SA, Italy

Darren Wragg School of Chemistry, Cardiff University, Cardiff, UK; Department of Chemistry, Technical University of Munich, Garching, Germany

Structure and Properties of Bio and Nanomaterials

Composition and Microstructure of Biocompatible and pH-Sensitive Copolymers Prepared by a Free Solvent ARGET ATRP

Ylenia Miele[1(✉)], Massimo Mella[2], Lorella Izzo[3], and Federico Rossi[4]

[1] Dipartimento di Chimica e Biologia, Università degli Studi di Salerno, Via Giovanni Paolo II, 132, 84084 Fisciano, Italy
ymiele@unisa.it
[2] Dipartimento di Scienza ed Alta Tecnologia, Università degli Studi dell'Insubria, Via Valleggio 9, 22100 Como, Italy
[3] Dipartimento di Biotecnologie e Scienze della Vita, Università degli Studi dell'Insubria, Via J.H. Dunant 3, 21100 Varese, Italy
[4] Dipartimento Scienze fisiche, della Terra e dell'ambiente, Pian dei Mantellini 44, 53100 Siena, Italy

Abstract. Controlled/living radical polymerizations enable the synthesis of functional polymers with well-defined compositions and architectures. In this paper, we propose the usage of the ARGET ATRP technique to produce copolymers of MMA and DMAEMA. The feed composition was changed systematically to modulate the final composition of the copolymer. The absence of additional organic solvent (bulk polymerization), the use of low amounts of metal catalyst and the reduction of purification steps are the main advantages with respect to a traditional ATRP.

1 Introduction

Stimuli-responsive materials alter their chemical and/or physical properties upon exposure to external stimuli like pH, temperature, redox variations and light [1–5]. Nowadays, engineering of new responsive materials, especially block-copolymers, is a big scientific challenge and involve many researchers in finding new properties for applications that span from environmental monitoring and remediation to biological and medical applications [6–10].

A biologically relevant pH-responsive polymer is poly[2-(dimethylamino)ethyl methacrylate] (DMAEMA), which is both temperature- (lower critical solution temperature around 40 °C) [11] and pH-sensitive (the p*K*a of the amine group in poly-DMAEMA is around 7.3) [12]. Moreover, poly-DMAEMA is able to bind plasmid DNA through electrostatic interactions, yielding polymer/plasmid complexes (also called polyplexes) [13]. As observed for many polycations (see e.g. [14]), poly-DMAEMA is cytotoxic, but its cytotoxicity and hemolycity are reduced when DMAEMA is copolymerized with non-charged comonomers [15] (e.g. MMA). This strategy has been exploited to produce non-hemolytic bactericidal materials [16, 17].

S. Piotto et al. (Eds.): BIONAM 2019, LNBE, pp. 3–15, 2020.
https://doi.org/10.1007/978-3-030-47705-9_1

Indeed, linear and branched copolymers of MMA and DMAEMA (molar fraction = 25%) with mPEG segments (mPEG-*b*-poly(MMA-*ran*-DMAEMA)) gave not cytotoxic vesicles in the range concentration 10^{-7}–10^{-9} M in the presence of HepG2 tumor cells and MRC5 normal cells [18]. The release of antitumor paclitaxel from these polymeric vesicles is controlled by a pH-dependent swelling instead of disaggregation [18, 19]. The pH-sensitive swelling strongly depends on the topology of the copolymer: the DMAEMA units are randomly distributed in the hydrophobic part of the polymersomes enabling the swelling at acidic pH due to the increased electrostatic repulsions.

More in general, physical and chemical properties of polymers in solution strongly depend on their structure, which can be modulated by covalently linked substituents of different solubility [20] or also by copolymerization of comonomers producing a unique back-bone (e.g. MMA and DMAEMA). In addition, if the polymer in solution is a weak polyelectrolyte, properties such as ionization degree can be influenced by different parameters such as polymer conformation and confinement [21], polyelectrolyte concentration, chain rigidity, the formation of intra- and inter-chain charged hydrogen bonds (c-H-bonds) [22] and architecture [23].

Besides, polyelectrolyte adsorption onto charged nanoparticles, and concurrent effects such as spatial partitioning of ions may be influenced by details of the polyelectrolyte structure (linear or star–like) and size [24]. Such an issue can be fundamental in understanding the bactericidal activity of a polymer when in contact with the outer envelope of a bacteria cell [25].

In any case, a control over the architecture and chemical composition of polyelectrolytes is fundamental to produce pH-sensitive polymers with improved performance and reduced cytotoxicity.

To this end, in this work we focused on the synthesis of MMA and DMAEMA copolymers because of their potential applications in different fields such as the interaction to DNA for the formation of polyplexes [13], pH-sensitive systems for drug delivery [18, 19] and formation of inherently bactericidal materials [17, 25]. The synthesis of smart materials, however, is strictly correlated to the control over the chemical composition, so we report the synthesis of MMA/DMAEMA copolymers in different experimental conditions. The composition of the final copolymer, and the reactivity ratios have been estimated to evaluate the relative reactivity of the two comonomers in ARGET ATRP reactions (Activators Regenerated by Electron Transfer for Atom Transfer Radical Polymerization).

It is well known that controlled polymerization processes are required to design copolymers with a well-defined architecture. ATRP (Atom Transfer Radical Polymerization) is one of the most powerful controlled/living radical polymerization techniques: the living nature depends on the onset of a dynamic equilibrium between dormant species and active radical species. ATRP usually requires a transition metal catalyst in its lowest oxidation state, a ligand to complex and solubilize the metal, an alkyl halide as the initiator (R-X) [26, 27]. A variant of ATRP is ARGET ATRP where the metal is introduced in its highest oxidation state and the active species is continuously regenerated from a reducing agent like glucose, ascorbic acid and the FDA-approved tin (II) 2-ethylhexanoate ($Sn(EH)_2$) [28, 29]. In ARGET ATRP, the amount of metal catalyst can be decreased by thousand times compared to conventional ATRPs and the procedure can tolerate a large excess of reducing agent. This new procedure avoids the deoxygenation of reaction mixtures and simplifies the ATRP process [30]. ARGET ATRP has been successfully applied for the synthesis of homopolymers of styrene, methyl methacrylate, butyl acrylate [28, 31] and DMAEMA [30], and for the copolymerization of styrene and acrylonitrile [32], styrene and methyl acrylate [33], or methyl acrylate with olefins [34].

2 Experimental Section

Copper bromide ($CuBr_2$), 2,2'-bipyridine (bpy), 2–bromoisobutyryl bromide (BMPB), ascorbic acid, Al_2O_3, DMF were purchased by Sigma-Aldrich and used without any further purification. Methylmethacrylate (MMA) and 2(dimethylamino) ethyl methacrylate (DMAEMA) (Sigma-Aldrich) were passed through a column filled with basic alumina prior to use to remove the inhibitors. All manipulations involving air-sensitive compounds were carried out under nitrogen atmosphere using Schlenk techniques.

Stock solutions of ascorbic acid, Cu (II) and bpy were prepared separately dissolving respectively 20 mg, 2.2 mg and 17.1 mg in 10 mL of DMF. The concentrations of the prepared solutions are: $[CuBr_2] = 1 \times 10^{-3}$ M, $[bpy] = 1 \times 10^{-2}$ M and $[\text{ascorbic acid}] = 1 \times 10^{-2}$ M. In a typical run, 100 μL of $CuBr_2$ stock solution, 100 μL of bpy stock solution, 0.5–4.5 mL of MMA and DMAEMA (according to the initial feed ratio), 1.2 μL of 2–bromoisobutyryl bromide (BMPB), 100 μL of ascorbic acid stock solution were added in a 50 mL glass flask, under nitrogen atmosphere. After the addition of the last reactant (ascorbic acid), the mixture was thermostated at 60 °C and magnetically stirred. The reaction was stopped with n-hexane. The precipitated copolymers were filtered and dried in vacuum at 40 °C for 12 h.

The number-average molecular weight M_n and polydispersity index ($Đ = M_w/M_n$) were determined by gel permeation chromatography (GPC) using a Waters Breeze GPC system equipped with a refractive index (RI) detector and four Styragel columns (range 1000–1,000,000 Å) in tetrahydrofuran (THF) as eluent at a flow rate of 1.0 mL/min^{-1}. The calibration curve was established with polystyrene standards.

The mole fractions of MMA and DMAEMA in the final copolymer were evaluated through ^{1}H NMR spectroscopy. NMR spectra were recorded on a Bruker Avance 400 MHz spectrometer at 25 °C. The samples were prepared by introducing 15 mg of copolymer in 0.4 mL of $CDCl_3$ into a tube with an outer diameter of 0.5 mm.

The conversions and the logarithmic concentrations of the monomers in the kinetic plots were evaluated from the yield and chemical composition of the copolymers.

3 Results and Discussion

The synthesis of pH-sensitive copolymers composed of MMA and DMAEMA is illustrated in the reaction scheme of Fig. 1. DMF was employed just to solubilize the metal catalyst, its ligand and the reducing agent; no additional solvent was included in the reaction mixture to reduce as far as possible the presence of organic solvent. Since the amount of metal catalyst is below ppm ($CuBr_2 = 1 \times 10^{-4}$ mmol), it is unnecessary the removal of the metal from the final copolymer; the purification procedure is thus simplified.

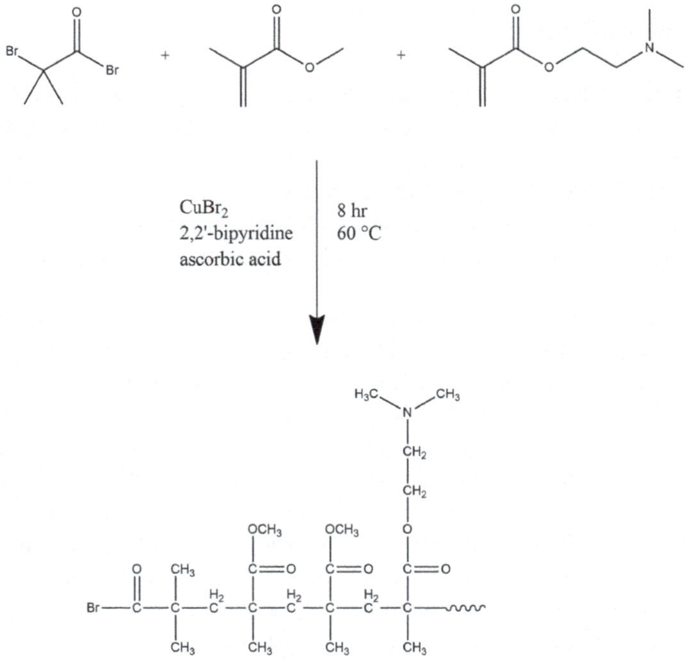

Fig. 1. Reaction scheme for the synthesis of poly(MMA-DMAEMA).

As a first step, the reaction was performed reaching different reaction times to build a kinetic plot. At short polymerization times, the plot of ln[M_0/M] vs polymerization time shows a downward curvature followed by a linear increase, a typical trend of a slow initiation reaction.

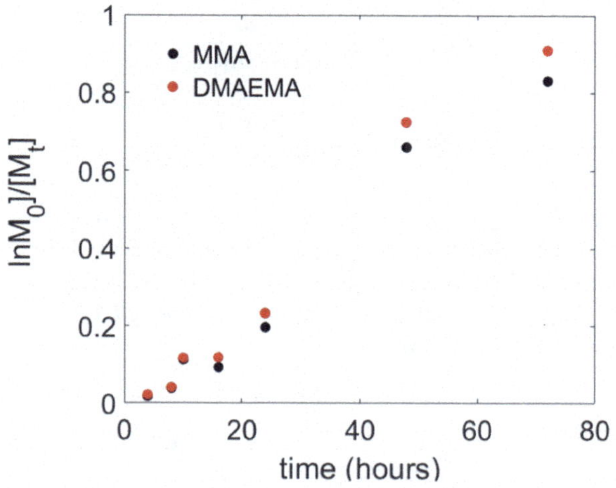

Fig. 2. First order kinetic plot in ARGET ATRP copolymerization of MMA and DMAEMA. Experimental conditions: BMPB initiator, Cu/bpy/BMPB = 1:10:100, $CuBr_2 = 1 \times 10^{-4}$ mmol, $f_{MMA} = 0.61$ and $f_{DMAEMA} = 0.39$. Bulk copolymerization. $V_{tot} \approx 5$ mL, $T_p = 60$ °C.

In order to establish the monomers distribution in the final copolymer, a set of copolymerization runs were carried out changing systematically the molar ratios of MMA and DMAEMA (conditions indicated in Table 1).

Table 1. Copolymerization of MMA and DMAEMA: experimental data used for the evaluation of the final composition[a]

Entry	f^b_{MMA}	F^c_{MMA}	conv. MMA (%)	Yield (g)	M_n (kDa)	$Đ$ (M_w/M_n)
1	0.15	0.13	20	1.0715	37.2	2.7
2	0.29	0.27	22	1.1090	69.2	1.7
3	0.52	0.47	18	0.9243	39.2	1.8
4	0.61	0.60	3.8	0.1834	45.5	1.6
5	0.71	0.67	21	1.0488	26.8	2.3
6	0.86	0.84	2.7	0.1347	20.4	2.0

[a]Experimental conditions: Cu/bpy/BMPB = 1:10:100, $CuBr_2 = 1 \times 10^{-4}$ mmol. Bulk copolymerization. $V_{tot} \approx 5$ mL, $T_p = 60$ °C, $t_p = 8$ h. [b]Mole fraction of MMA in the feed. [c]Mole fraction of MMA in the copolymer.

The copolymer composition was determined from ^{1}H-NMR spectral analysis of the copolymer (Fig. 3) using the Eqs. (1) and (2) [35]:

$$F_{MMA} = \frac{2I_{MMA}}{2I_{MMA} + 3I_{DMAEMA}} \quad (1)$$

$$F_{DMAEMA} = \frac{3I_{DMAEMA}}{2I_{MMA} + 3I_{DMAEMA}} \quad (2)$$

where F_{MMA} and F_{DMAEMA} are respectively the molar fractions of MMA and DMAEMA in the final copolymer, I_{DMAEMA} is the integration of the methylene group of the signal relative to DMAEMA units (-O-CH_2-CH_2-N$(CH_3)_2$), labeled as **f** in the NMR spectrum), I_{MMA} is the integration relative to the -OCH_3 unit of MMA (peak labeled as **c** in the NMR spectrum).

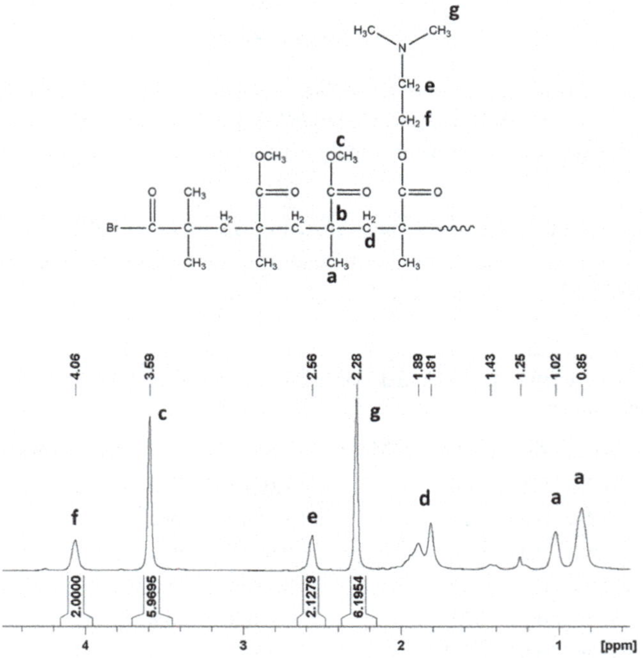

Fig. 3. ^{1}H-NMR spectrum for a MMA/DMAEMA copolymer.

The plot of MMA molar fraction in the final copolymer (F_{MMA}) vs the feed (f_{MMA}) is roughly linear as expected from an ideal copolymerization (Fig. 4).

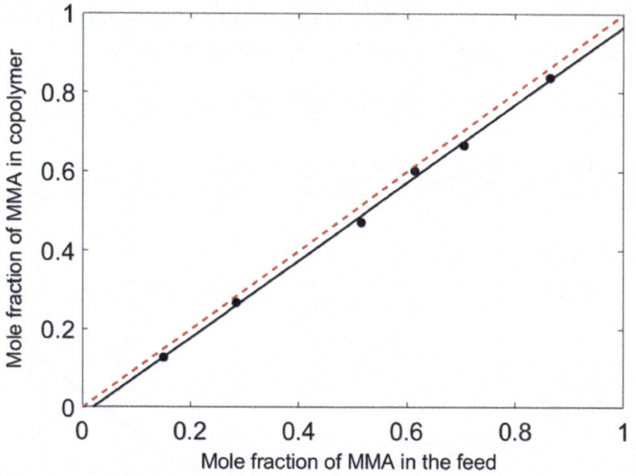

Fig. 4. Plot of MMA mole fraction in the copolymer (F_{MMA}) and in the feed (f_{MMA}). The theoretical curve for an ideal random copolymerization is indicated with a red dashed line.

The monomer feed ratios and the final copolymer compositions of Table 1 were used to estimate the reactivity ratios of the two monomers defined as $r_1 = k_{MMA,MMA}/k_{MMA,DMAEMA}$ and $r_2 = k_{DMAEMA,DMAEMA}/k_{MMA,DMAEMA}$ with $k_{i,j}$ the kinetic rate constants for the reaction between two possible radical sites during the propagation step (scheme in Eqs. 3–6 where M_1 is the monomer MMA, M_2 is the monomer DMAEMA, M_1^* and M_2^* are the reactive terminal units of the growing chain).

$$M_1 + M_1^* \xrightarrow{k_{MMA,MMA}} M_1M_1^* \quad (3)$$

$$M_2 + M_1^* \xrightarrow{k_{MMA,DMAEMA}} M_1M_2^* \quad (4)$$

$$M_2 + M_2^* \xrightarrow{k_{DMAEMA,DMAEMA}} M_2M_2^* \quad (5)$$

$$M_1 + M_2^* \xrightarrow{k_{DMAEMA,MMA}} M_2M_1^* \quad (6)$$

The evaluation of reactivity ratios is helpful for the tuning of the copolymer properties to attain the desired applications (drug delivery systems, coatings [36, 37]). In this work, the reactivity ratios r_1 and r_2 were preliminary calculated with the Mayo-Lewis equation [38] (Eq. 7).

$$F_1 = \frac{r_1 f_1^2 + f_1 f_2}{r_1 f_1^2 + 2 f_1 f_2 + r_2 f_2^2} \quad (7)$$

where $f_1 = f_{MMA}$, $f_2 = f_{DMAEMA}$, $F_1 = F_{MMA}$ and $F_2 = F_{DMAEMA}$, r_1 and r_2 are respectively the reactivity ratios of MMA and DMAEMA.

The Mayo-Lewis equation was rearranged into the linear form proposed by Fineman and Ross [39] (FR, Eq. 8), with slope r_1 and intercept r_2.

$$G = r_1 F - r_2 \tag{8}$$

where $G = x[1-(1/y)]$, $F = x^2/y$, $x = f_1/f_2$, $y = F_1/F_2$.

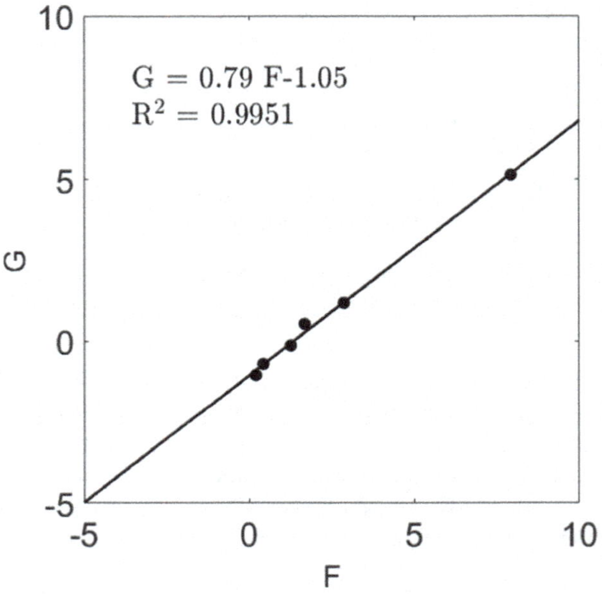

Fig. 5. Fineman-Ross plot for the system poly(MMA-DMAEMA).

The reactivity ratios estimated from the FR plot (Fig. 5) are $r_1 = 0.79$ (±3%) and $r_2 = 1.05$ (±8%). However, since most of the conversions reported in Table 1 are higher than 10% and just entry 4 and entry 6 of Table 1 could be strictly used in the Fineman Ross method, the reactivity ratios were also calculated with Eq. (9) that is the integrated form of the Mayo-Lewis model, the Meyer-Lowry equation [40]:

$$\frac{M}{M_0} = \left(\frac{f_1}{(f_1)_0}\right)^{\alpha}\left(\frac{1-f_1}{1-(f_1)_0}\right)^{\beta}\left(\frac{(f_1)_0-\delta}{f_1-\delta}\right)^{\gamma} \tag{9}$$

where M/M_0 is the total monomer conversion, $\alpha = r_2/(1 - r_2)$; $\beta = r_1/(1 - r_1)$; $\gamma = (1 - r_1 r_2)/((1 - r_1)(1 - r_2))$; and $\delta = (1 - r_2)/(2 - r_1 - r_2)$. The reactivity ratios evaluated from the Meyer-Lowry plot, $r_1 = 0.76$ (±6%) and $r_2 = 1.08$ (±4%) are in agreement with the Fineman-Ross ratios, thus the monomer conversion does not influence the reactivity between the two monomers as expected from a quasi-ideal copolymerization. The value of r_1 estimated indicates the preference for MMA to react with DMAEMA, whereas the value of r_2 referred to DMAEMA, shows that homopropagation is enhanced over cross-propagation. The product $r_1 r_2 \approx 1$ indicates a random distribution. The reactivity ratios also allowed us to predict the mean sequence length and thus the distribution of the two monomers in the final copolymer through the Eqs. (10–11) [41, 42]:

$$\overline{l_1} = r_1 \frac{f_1}{f_2} + 1 \tag{10}$$

$$\overline{l_2} = r_2 \frac{f_2}{f_1} + 1 \tag{11}$$

where f_1 and f_2 are respectively the mole fractions of MMA and DMAEMA in the feed, $\overline{l_1}$ is the mean sequence length of MMA and $\overline{l_2}$ is the mean sequence length for DMAEMA.

The mean distributions for different feed compositions are reported in Table 2. The increase of f_{MMA} (in parallel with the decrease of f_{DMAEMA}) in the feed leads to a gradual increase of the length of MMA units (and decrease of DMAEMA units) in the final copolymer.

Table 2. Mean sequence length in MMA/DMAEMA copolymers.

f_{MMA}	f_{DMAEMA}	$\overline{l_1}$	$\overline{l_2}$	$\overline{l_1}$: $\overline{l_2}$	Distribution[a]
0.15	0.85	1.14	6.93	1:6	M DDDDDDM
0.29	0.71	1.31	3.63	1:3	M DDDM
0.52	0.48	1.84	1.99	1:1	M DM
0.61	0.39	2.25	1.66	1:1	M DM
0.71	0.29	2.88	1.44	2:1	M DMM
0.86	0.14	6.01	1.16	5:1	M DMMMMM

r_1 = 0.79, r_2 = 1.05, M = MMA and D = DMAEMA
[a] Few cases of distribution are reported in the table.

The reactivity ratios evaluated with the two methods are summarized in Table 3 and compared with the literature.

The reactivity ratios determined from ARGET ATRP bulk copolymerizations are similar to the reactivity ratios found for the free radical copolymerization of MMA and DMAEMA in bulk [43], where the reactivity ratios are r_1 = 0.71 (±7%) and r_2 = 1.25 (±12%). However, these results differ from the reactivity ratios calculated from conventional ATRP polymerizations conducted with mPEG-Br linear macroinitiator and toluene as solvent reported by three of the authors in a previous work [35]. The reactivity ratios, determined in the above-cited paper from the Meyer-Lowry equation, were r_1 = 0.36 (±10%) and r_2 = 2.76 (±15%) indicating a gradient like microstructure.

The mechanism put forward to explain the marked reactivity of DMAEMA was the formation of a complex between Cu, bpy and DMAEMA (complexation supported by NMR spectroscopy and DFT calculations). In this work, the higher reactivity of

DMAEMA was also demonstrated, however the different microstructure found can depend on the different initiator employed (it may play a role in the complexation of the metal catalyst) and the absence of solvent (in presence of solvent there may be a partitioning process of the monomer between the solution and the radical end of the polymeric chain, the so-called "bootstrap" effect [44]).

Table 3. Reactivity ratios of MMA and DMAEMA calculated with different methods

Method	Solvent	r_{MMA}	r_{DMAEMA}	Ref.
Kelen-Tudos	bulk	0.71 (±7%)	1.25 (±12%)	[43] [a]
Meyer-Lowry	toluene	0.36 (±10%)	2.76 (±15%)	[35] [b]
Fineman-Ross	bulk	0.79 (±3%)	1.05 (±8%)	This work
Meyer-Lowry	bulk	0.76 (±6%)	1.08 (±4%)	This work

[a] The copolymers were prepared with a free radical polymerization.
[b] The copolymers were prepared with a conventional ATRP using m-PEG-Br macroinitiator.

4 Conclusions

pH-sensitive random copolymers were synthesized with a living radical polymerization, the ARGET ATRP technique. Compared to the traditional ATRP processes reported previously in literature, a low amount (ppm) of metal catalyst was employed and the monomers were not dried and distilled before use reducing the times of preparation. A future investigation may concern the effect of temperature on the polymerization rate: lower temperatures reduce the energetic costs and follow the direction of environment friendly synthetic strategies. The estimation of reactivity ratios allowed us to assign a random distribution of the monomers in the final copolymer. A random distribution of MMA and DMAEMA is desirable for the formation of pH-sensitive polymersomes with mPEG segments: in this way, the amine groups of DMAEMA are distributed in the hydrophobic bilayer allowing the swelling-deswelling behavior pH-dependent. The use of copolymers synthesized from ARGET ATRP for the preparation of pH-sensitive vesicles [45] and the release of pharmacologically active molecules are the next aspects to be explored.

References

1. Meng, F., Zhong, Z., Feijen, J.: Stimuli-responsive polymersomes for programmed drug delivery. Biomacromolecules **10**, 197–209 (2009)
2. Checot, F., Lecommandoux, S., Klok, H.-A., Gnanou, Y.: From supramolecular polymersomes to stimuli-responsive nano-capsules based on poly (diene-b-peptide) diblock copolymers. Eur. Phys. J. E **10**, 25–35 (2003)
3. Qin, S., Geng, Y., Discher, D.E., Yang, S.: Temperature-controlled assembly and release from polymer vesicles of poly (ethylene oxide)-block-poly (N-isopropylacrylamide). Adv. Mater. **18**, 2905–2909 (2006)

4. Cerritelli, S., Velluto, D., Hubbell, J.A.: PEG-SS-PPS: reduction-sensitive disulfide block copolymer vesicles for intracellular drug delivery. Biomacromol **8**, 1966–1972 (2007)
5. Tong, X., Wang, G., Soldera, A., Zhao, Y.: How can azobenzene block copolymer vesicles be dissociated and reformed by light? J. Phys. Chem. B **109**, 20281–20287 (2005)
6. Roy, D., Cambre, J.N., Sumerlin, B.S.: Future perspectives and recent advances in stimuli-responsive materials. Prog. Polym. Sci. **35**, 278–301 (2010). https://doi.org/10.1016/j.progpolymsci.2009.10.008
7. Intiso, A., Miele, Y., Marchettini, N., Proto, A., Sánchez-Domínguez, M., Rossi, F.: Enhanced solubility of trichloroethylene (TCE) by a poly-oxyethylene alcohol as green surfactant. Environ. Technol. Innov. **12**, 72–79 (2018). https://doi.org/10.1016/j.eti.2018.08.001
8. Garza-Arévalo, J.I., Intiso, A., Proto, A., Rossi, F., Sanchez-Dominguez, M.: Trichloroethylene solubilization using a series of commercial biodegradable Ethoxylated fatty alcohol surfactants. J. Chem. Technol. Biotechnol. **94**, 3523–3529 (2019). https://doi.org/10.1002/jctb.5965
9. Miele, Y., Medveczky, Z., Hollo, G., Tegze, B., Derenyi, I., Horvolgyi, Z., Altamura, E., Lagzi, I., Rossi, F.: Self-division of giant vesicles driven by an internal enzymatic reaction. Chem. Sci. **11**(12), 3228–3235 (2020)
10. Urban, M.W.: Stimuli-Responsive Materials: From Molecules to Nature Mimicking Materials Design. Royal Society of Chemistry, London (2019)
11. Yanfeng, C., Min, Y.: Swelling kinetics and stimuli-responsiveness of poly (DMAEMA) hydrogels prepared by UV-irradiation. Radiat. Phys. Chem. **61**, 65–68 (2001)
12. Huang, Y., Yong, P., Chen, Y., Gao, Y., Xu, W., Lv, Y., Yang, L., Reis, R.L., Pirraco, R.P., Chen, J.: Micellization and gelatinization in aqueous media of pH-and thermo-responsive amphiphilic ABC (PMMA 82-b-PDMAEMA 150-b-PNIPAM 65) triblock copolymer synthesized by consecutive RAFT polymerization. RSC Adv. **7**, 28711–28722 (2017)
13. Felgner, P.L., Barenholz, Y., Behr, J.P., Cheng, S.H., Cullis, P., Huang, L., Jessee, J.A., Seymour, L., Szoka, F., Thierry, A.R.: Nomenclature for synthetic gene delivery systems. Hum. Gene Therapy **8**, 511–512 (1997)
14. Duncan, R., Izzo, L.: Dendrimer biocompatibility and toxicity. Adv. Drug Delivery Rev. **57**, 2215–2237 (2005)
15. van de Wetering, P., Moret, E.E., Schuurmans-Nieuwenbroek, N.M., van Steenbergen, M.J., Hennink, W.E.: Structure- activity relationships of water-soluble cationic methacrylate/methacrylamide polymers for nonviral gene delivery. Bioconjug. Chem. **10**, 589–597 (1999)
16. Matrella, S., Vitiello, C., Mella, M., Vigliotta, G., Izzo, L.: The role of charge density and hydrophobicity on the biocidal properties of self-protonable polymeric materials. Macromol. Biosci. **15**, 927–940 (2015)
17. Vigliotta, G., Mella, M., Rega, D., Izzo, L.: Modulating antimicrobial activity by synthesis: dendritic copolymers based on nonquaternized 2-(dimethylamino) ethyl methacrylate by Cu-mediated ATRP. Biomacromol **13**, 833–841 (2012)
18. Villani, S., Adami, R., Reverchon, E., Ferretti, A.M., Ponti, A., Lepretti, M., Caputo, I., Izzo, L.: pH-sensitive polymersomes: controlling swelling via copolymer structure and chemical composition. J. Drug Target. **25**, 899–909 (2017)
19. Barrella, M.C., Di Capua, A., Adami, R., Reverchon, E., Mella, M., Izzo, L.: Impact of intermolecular drug-copolymer interactions on size and drug release kinetics from pH-responsive polymersomes. Supramol. Chem. **29**, 796–807 (2017)

20. Mella, M., Izzo, L.: Structural properties of hydrophilic polymeric chains bearing covalently–linked hydrophobic substituents: exploring the effects of chain length, fractional loading and hydrophobic interaction strength with coarse grained potentials and Monte Carlo simulations. Polymer **51**, 3582–3589 (2010). https://doi.org/10.1016/j.polymer.2010.05.013
21. Mella, M., Izzo, L.: Modulation of ionization and structural properties of weak polyelectrolytes due to 1D, 2D, and 3D confinement. J. Polym. Sci., Part B: Polym. Phys. **55**, 1088–1102 (2017). https://doi.org/10.1002/polb.24351
22. Tagliabue, A., Izzo, L., Mella, M.: Impact of charge correlation, chain rigidity, and chemical specific interactions on the behavior of weak Polyelectrolytes in solution. J. Phys. Chem. B **123**, 8872–8888 (2019). https://doi.org/10.1021/acs.jpcb.9b06017
23. Tagliabue, A., Izzo, L., Mella, M.: Absorbed weak polyelectrolytes: impact of confinement, topology, and chemically specific interactions on ionization, conformation free energy, counterion condensation, and absorption equilibrium. J. Polym. Sci. Part B: Polym. Phys. **57**, 491–510 (2019)
24. Mella, M., Tagliabue, A., Mollica, L., Izzo, L.: Monte Carlo study of the effects of macroion charge distribution on the ionization and adsorption of weak polyelectrolytes and concurrent counterion release. J. Colloid Interface Sci. **560**, 667–680 (2019)
25. Izzo, L., Matrella, S., Mella, M., Benvenuto, G., Vigliotta, G.: Escherichia coli as a model for the description of the antimicrobial mechanism of a cationic polymer surface: cellular target and bacterial contrast response. ACS Appl. Mater. Interfaces **11**, 15332–15343 (2019)
26. Kamigaito, M., Ando, T., Sawamoto, M.: Metal-catalyzed living radical polymerization. Chem. Rev. **101**, 3689–3746 (2001)
27. Wang, J.-S., Matyjaszewski, K.: Controlled/"living" radical polymerization. Atom transfer radical polymerization in the presence of transition-metal complexes. J. Am. Chem. Soc. **117**, 5614–5615 (1995)
28. Jakubowski, W., Min, K., Matyjaszewski, K.: Activators regenerated by electron transfer for atom transfer radical polymerization of styrene. Macromolecules **39**, 39–45 (2006)
29. Min, K., Gao, H., Matyjaszewski, K.: Use of ascorbic acid as reducing agent for synthesis of well-defined polymers by ARGET ATRP. Macromolecules **40**, 1789–1791 (2007)
30. Dong, H., Matyjaszewski, K.: ARGET ATRP of 2-(dimethylamino) ethyl methacrylate as an intrinsic reducing agent. Macromolecules **41**, 6868–6870 (2008)
31. Jakubowski, W., Matyjaszewski, K.: Activators regenerated by electron transfer for atom-transfer radical polymerization of (meth) acrylates and related block copolymers. Angew. Chem. Int. Ed. **45**, 4482–4486 (2006)
32. Wang, G., Lu, M.: ARGET ATRP of copolymerization of styrene and acrylonitrile with environmentally friendly catalyst and ligand. e-Polymers **12**, 1–8 (2012)
33. Plichta, A., Zhong, M., Li, W., Elsen, A.M., Matyjaszewski, K.: Tuning dispersity in diblock copolymers using ARGET ATRP. Macromol. Chem. Phys. **213**, 2659–2668 (2012)
34. Tanaka, K., Matyjaszewski, K.: Copolymerization of (meth) acrylates with olefins using activators regenerated by electron transfer for atom transfer radical polymerization (ARGET ATRP). In: Macromolecular Symposia, pp. 1–9. Wiley Online Library (2008)
35. Mella, M., La Rocca, M.V., Miele, Y., Izzo, L.: On the origin and consequences of high DMAEMA reactivity ratio in ATRP copolymerization with MMA: an experimental and theoretical study#. J. Polym. Sci., Part A: Polym. Chem. **56**, 1366–1382 (2018)
36. Lau, K.K.S., Gleason, K.K.: All-dry synthesis and coating of methacrylic acid copolymers for controlled release. Macromol. Biosci. **7**, 429–434 (2007)
37. Kruft, M.-A.B., Koole, L.H.: A convenient method To measure monomer reactivity ratios. Application to synthesis of polymeric biomaterials featuring intrinsic radiopacity. Macromolecules **29**, 5513–5519 (1996)

38. Mayo, F.R., Lewis, F.M.: Copolymerization. I. a basis for comparing the behavior of monomers in copolymerization; the copolymerization of styrene and methyl methacrylate. J. Am. Chem. Soc. **66**, 1594–1601 (1944)
39. Fineman, M., Ross, S.D.: Linear method for determining monomer reactivity ratios in copolymerization. J. Polym. Sci. **5**, 259–262 (1950)
40. Meyer, V.E., Lowry, G.G.: Integral and differential binary copolymerization equations. J. Polym. Sci. Part A: General Papers **3**, 2843–2851 (1965)
41. Pazhanisamy, P., Reddy, B.S.R.: Copolymers of N-cyclohexylacrylamide and n-butyl acrylate: synthesis, characterization, monomer reactivity ratios and mean sequence length. eXPRESS Polym. Lett. **1**, 391–396 (2007)
42. Igarashi, S.: Representation of composition and blockiness of the copolymer by a triangular coordinate system. J. Polym. Sci., Part C: Polym. Lett. **1**, 359–363 (1963)
43. Camail, M., Essaoudi, H., Margaillan, A., Vernet, J.L.: Copolymérisation radicalaire de méthacrylates de 2-aminoéthyle avec le méthacrylate de méthyle. Eur. Polym. J. **31**, 1119–1125 (1995)
44. Harwood, H.J.: Structures and compositions of copolymers. Macromol. Symp. **10–11**, 331–354 (1987)
45. Miele, Y., Mingotaud, A.-F., Caruso, E., Malacarne, M.C., Izzo, L., Lonetti, B., Rossi, F.: Hybrid giant lipid vesicles incorporating a PMMA-based copolymer. Biochim. Biophys. Acta (BBA) - Gen. Subjects. 129611 (2020). https://doi.org/10.1016/j.bbagen.2020.129611.. In press

Morphological and Properties Characterization of Poly(3-Hydroxybutyrate-Co-3-Hydroxyvalerate)/Halloysite Nanotubes Bionanocomposites

Salima Kennouche[1,2], Nadjet Dehouche[1(✉)], Mustapha Kaci[1], and José-Marie Lopez-Cuesta[2]

[1] Laboratoire des Matériaux Polymères Avancés (LMPA), Faculté de Technologie, Université de Bejaia, 06000 Bejaia, Algeria
salima.kennouche@yahoo.fr, dehouche_nadjet@yahoo.fr, kacimu@yahoo.fr

[2] Centre des Matériaux des Mines d'Alès, IMT-Mines d'Alès, 6 Avenue de Clavières, 30319 Alès, Cedex, France
jose-marie.lopez-cuesta@mines-ales.fr

Abstract. The paper reports some experimental data on the effect of halloysite nanotubes (HNTs) content ratio on morphology and thermal properties of poly(3-hydroxybutyrate-Co-3-hydroxyvalerate) (PHBV) bionanocomposites prepared by melt mixing at 2.5, 5 and 10 wt%. The performance of the bionanocomposite samples was evaluated on the basis of neat PHBV. Scanning electron microscopic (SEM) analysis of the fractured surface of the bionanocomposite samples showed the presence of many filler aggregates at 10 wt%. However, a quite homogeneous dispersion of HNTs in PHBV was observed at 2.5 and 5 wt% with the presence of some individual nanotubes. The results indicated also that adding HNTs to PHBV led to an increase in both the crystallization rate and thermal stability of PHBV/HNTs bionanocomposites, being however more pronounced at 5 wt%, which appeared as the optimized filler content ratio.

1 Introduction

In recent years, an increasing number of biodegradable polymers have been developed, aiming to solve the environmental problem caused by the disposal of large volumes of non-biodegradable materials and also to limit the carbon footprint of polymeric materials [1].

In this regard, polyhydroxyalcanoates (PHAs) are fully renewable and biodegradable thermoplastics [2]. PHAs are aliphatic polyesters naturally synthesized via microbial process on sugar-based medium, where they act as carbon and energy storage material in bacteria [3]. Among the PHAs family, poly(3-hydroxybutyrate-Co-3-hydroxyvalerate) (PHBV) copolymer is one of the most widely studied ones due to its potential use in packaging, automotive, biomedical, and agricultural applications [4]. PHBV offers many advantages including thermoplasticity, natural origin and biocompatibility. However, the development and extension of PHBV to other industrial fields is rather limited due mainly to its slow crystallization rate, high crystallinity, high

S. Piotto et al. (Eds.): BIONAM 2019, LNBE, pp. 16–23, 2020.
https://doi.org/10.1007/978-3-030-47705-9_2

sensitivity towards thermomechanical degradation (decomposition temperature is near to melting point) [5, 6] and high cost [7].

Incorporation of a few amount of mineral clays to polymer matrix often leads to improved mechanical, thermal and barrier properties [8]. In this context, a huge number of papers have been published on successful applications of HNTs in bionanocomposites with various polymer matrices [9]. In this topic, a recent review paper [10] reported a number of examples documenting the significant potential of halloysite to compete with montmorillonite, which is currently the most commercially exploited clays for the preparation of polymer bionanocomposites. The chemical composition of HNTs is an hydrated aluminosilicate clay ($Al_2Si_2O_5$ $(OH)_4$ nH_2O) extracted from natural deposits [1]. The HNTs surface contains siloxane bonds with only few hydroxyl groups, suggesting better dispersion in polymeric matrix than other natural silicates, but also, a potential ability for the formation of hydrogen bonding. The tubular structure of HNTs is believed to be the result of hydrothermal alteration or surface weathering of aluminosilicate. Due to several characteristic features such as nanoscale lumen, relatively high length-to-diameter ratio, and low hydroxyl group density on their surface, HNTs are considered as promising competitors and cheaper alternatives to both carbon nanotubes and organo-modified layered silicates [11]. Furthermore, HNTs exhibit interesting reinforcing characteristics such as thermal stability and rheological properties in many polymeric matrices [12]. In many cases, HNTs are incorporated to the polymer matrix without any chemical modification [13]. Most papers published on the use of HNTs in polymeric matrices as reinforcing materials dealt with commercial ones [14]. However, a few papers are available on the use of Algerian HNTs as reinforcement in bionanocomposite materials for engineering applications. Therefore, the objective of this paper was to investigate the effect of HNTs content ratio on morphology and physical properties of PHBV bionanocomposites prepared by melt mixing at 2.5, 5 and 10 wt%. The performance of the bionanocomposites was evaluated on the basis of improved morphology, thermal stability and crystallinity.

2 Experimental Part

2.1 Materials Used

PHBV containing 5–8 mol. % HV was supplied by NaturePlast under the name of PHI 002. The main physical properties of PHBV are the following: density = 1.25 g/cm^3, T_m = 170 °C, T_g = 5 °C and the onset degradation temperature is 200 °C.

Algerian halloysite (HNTs) was collected from Djebel Debbagh in Guelma located in the Northern-East of Algeria and supplied by SOALKA Company (Algerian Company of Kaolins). The chemical composition and physical properties are detailed in previous paper [8].

2.2 Sample Preparation

Prior to processing, HNTs were ground in a laboratory mill and sieved to select particles having an average diameter of 40 μm. Moreover, both PHBV and HNTs were dried overnight under vacuum at 60 and 80 °C, respectively. HNTs were mixed together with PHBV by using a twin-screw mini-compounder (micro compounder X Plore) operating at 80 rpm and having the following characteristics: chamber volume 15 cm^3, sample weight 12–13 g, maximum temperature 500 °C. The processing temperature was ranged from 170 to 180 °C and the residence time in the extruder was around 5 min. HNTs were incorporated at content ratio of 2.5, 5 and 10 wt%. For comparison, neat PHBV was processed under the same experimental conditions as the bionanocomposite materials.

2.3 Technical Characterization

2.3.1 Scanning Electron Microscopy (SEM)

SEM analysis was carried out on the cryofractured surfaces of the bionanocomposite samples using an FEI Quanta 200 FEG microscope in environmental mode to observe the dispersion state of HNTs in PHBV matrix with the filler content ratio.

2.3.2 Wide Angle X-Ray Diffraction (WAXD)

Crystallinity measurements of neat PHBV and its bionanocomposites were carried out by using an XRD diffractometer of Model BRUKER AXS advance diffractometer. The CuKα (λ = 1,5406) radiation source was operated at 45 kV and with a current of 40 mA. The corresponding data were collected over a 2θ range from 2.5 to 65° at a step size of 0.00743°.

2.3.3 Thermogravimetric Analysis (TGA)

TGA of neat PHBV and PHBV bionanocomposites was performed using a Perkin Elmer (Pyris 1 TGA) apparatus. The mass of samples was approximately 15 mg. TGA thermograms were recorded in the temperature range 30–800 °C at 10 °C/min under nitrogen atmosphere.

2.3.4 Differential Scanning Calorimetry (DSC)

A Perkin Elmer pyris-1 DSC with nitrogen as purge gas was used to investigate the crystallization behavior of PHBV/HNT bionanocomposites. Samples of almost 13 mg were encapsulated in aluminum pans and heated from −30 to 200 °C at heating and cooling rates of 10 °Cmin^{-1}. The thermal history of the samples was erased by a preliminary heating (−30–+200 °C range). The measurements were made in the first cooling scan and in the second heating scan. The degree of crystallinity (Xc) of neat PHBV and bionanocomposite samples was determined from the second melting enthalpy values according to Eq. (1):

$$X_c = \frac{\Delta H_m}{w_p \times \Delta H_m^0} \times 100 \tag{1}$$

where ΔH_m (Jg^{-1}) is the melting enthalpy, w_p is the polymer weight fraction (PHBV) in the sample and ΔH_m^0 is the melting enthalpy of pure crystalline PHBV (ΔH_m^0 = 146 Jg^{-1} [15]).

3 Results and Discussion

3.1 Morphological Analysis of PHBV/ HNT Bionanocomposites

SEM is employed to investigate how HNTs particles are dispersed in PHBV matrix. In this respect, Fig. 1 shows SEM micrographs of the fractured surface of neat PHBV and its bionanocomposites. Neat PHBV shown in Fig. 1a exhibits an irregular fractured surface due to its crystalline structure in agreement with the literature data [16]. Figure 1b, c and d show selected micrographs of the bionanocomposite samples based HNTs at content ratio of 2.5, 5 and 10 wt%, respectively. From SEM micrographs recorded at lower magnification, it is observed a good dispersion of HNTs in PHBV matrix. However at 10 wt%, large HNTs aggregates are clearly observed as a result of strong hydrogen bonding between HNT particles [17]. Similar results were reported by Du et al. [18] and Marius et al. [17]. Moreover, it can also be observed from Fig. 1 b, c and d a homogeneous dispersion of HNTs with individual nanotubes, however more pronounced for the samples loaded at 5 wt%. A preferential orientation of nanotubes following elongation flow in the extension direction is noticed for all bionanocomposites as reported by Lecouvet et al. [19].

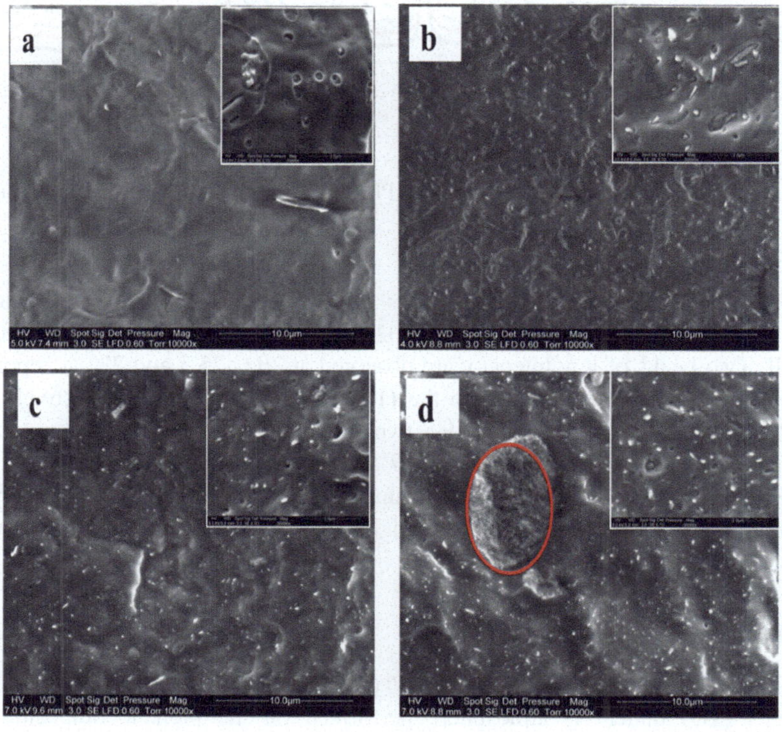

Fig. 1. SEM micrographs of: **a)** neat PHBV **b)** PHBV/HNTs (2.5 wt%), **c)** PHBV/HNTs (5 wt%) and **d)** PHBV/HNTs (10 wt%).

To further illustrate the dispersion state of HNTs and their interactions with PHBV matrix, WAXD patterns of the bionanocomposites samples are shown in Fig. 2 as a function of loading rates.

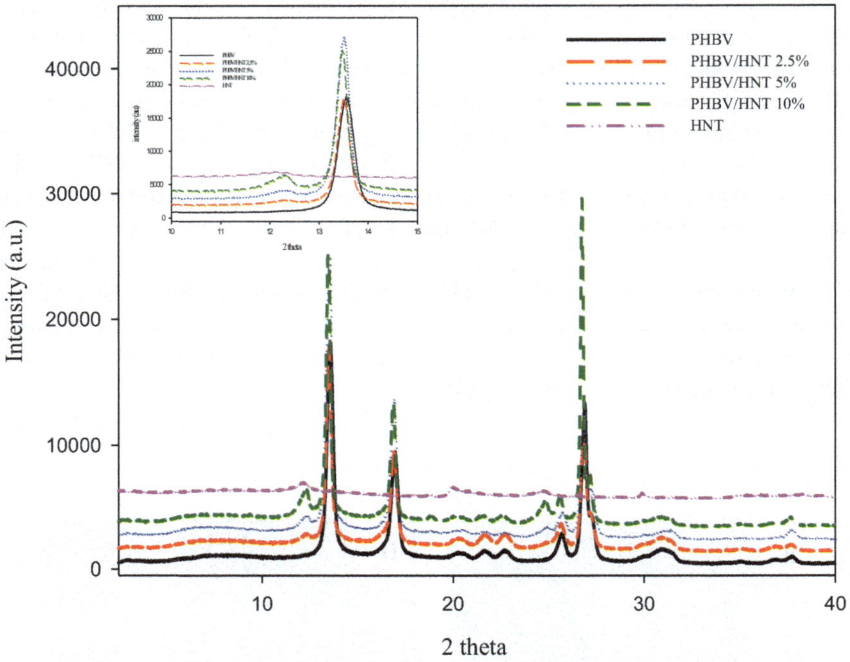

Fig. 2. WAXD patterns of neat PHBV and PHBV/HNTs nanocomposites at various filler content ratios.

WAXD pattern of neat PHBV shows two strong diffraction peaks located at 2θ = 13.5 and 17° corresponding to the (020) and (110) planes of the orthorhombic unit cell. Two peaks of low intensity are observed at 2θ = 25.5 and 27° as reported in literature [16], which are assigned to the (101) and (111) planes, respectively. As a crystalline material, HNTs exhibit also two other peaks at 2θ = 12.18 and 24.74°, which are related to (001) basal spacing of 7.26 nm and (002) of 3.59 nm, respectively. From Fig. 2, it can be noticed that the main PHBV peaks increases in intensity with increasing the HNTs content. This could be explained as a result of interactions between the polymer matrix and the filler.

3.2 Thermal Stability

To highlight the effect of HNTs on PHBV thermal stability, thermogravimetric measurements (TG) under N_2 atmosphere performed on the bionanocomposites were compared to those of neat PHBV. TGA curves are shown in Fig. 3a, while derivative thermogravimetric (DTG) curves are presented in Fig. 3b. Furthermore, the temperature values corresponding to 5 wt% mass loss ($T_{5\%}$), temperature at maximum degradation rate (T_{mdr}), and the residue at 600 °C, are also reported in Table 1.

As shown in Fig. 3a, during thermal degradation of neat PHBV and its bionanocomposites, the samples show a single step degradation in agreement with literature data [20]. According to Table 1, PHBV starts its degradation process around 260 °C ($T_{5\%}$), reaching a maximum degradation temperature (T_{mdr}) around 282 °C. The addition of HNTs to PHBV matrix leads to an increase in both $T_{5\%}$ and T_{mdr} of the bionanocomposite samples. For PHBV/HNTs samples, $T_{5\%}$ increases by almost 2 to 5 °C depending on the loading rate.

Whereas, it can be seen in Fig. 3b that T_{mdr} shifts towards higher temperatures for the bionanocomposite samples, except the one at 10 wt% where no change is observed compared to the neat PHBV. The increase in thermal stability observed in the bionanocomposites could be attributed to HNTs, which act as heat barriers, thus retarding the thermal degradation [21]. Furthermore, Du et al. [18] suggested that the hollow tubular structure of HNTs could enable the entrapment of degradation products inside the lumens, resulting in effective delay in mass transfer and drastically improve thermal stability of the polymer. TGA results reported in Table 1 indicate also that for the bionanocomposite samples, especially for those filled with 10 wt%, no noticeable change in both $T_{5\%}$ and T_{mdr} is observed due probably to filler aggregates formation in the matrix.

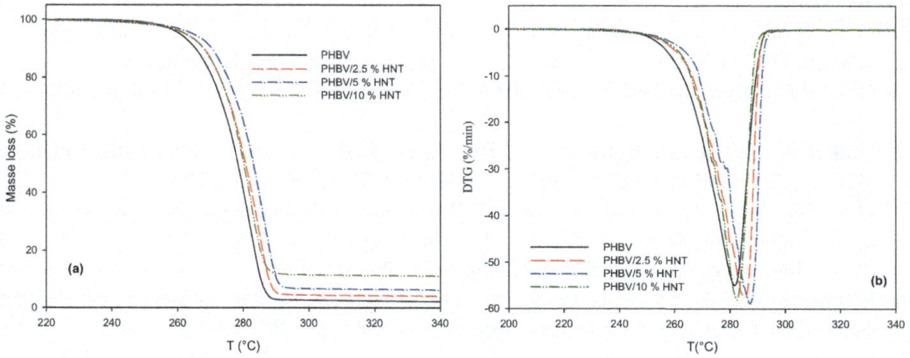

Fig. 3. TGA curves: **(a)** and DTG curves: **(b)** of PHBV/HNTs bionanocomposites.

3.2.1 Crystallinity and Melting Behavior

Table 1 summarizes the main DSC data of neat PHBV and PHBV/HNTs bionanocomposite samples, providing the values of melting temperature (T_m), crystallization temperature (T_c), glass transition temperature (T_g) and crystalline index (X_c).

The DSC thermograms of neat PHBV and PHBV bionanocomposites (not shown) exhibit single melting and crystallization peaks. The incorporation of HNTs has no significant effect on crystallization process of PHBV. However, the values of T_g and T_m decrease slightly with addition of HNTs.

Table 1. DSC and TGA data of PHBV and PHBV/HNTs bionanocomposites.

System	HNTs (wt%)	T_c (°C)	T_m (°C)	T_g (°C)	X_c (%)	ΔH_m (J/g)	$T_{5\%}$ (°C)	T_{mdr} (°C)	Residue (%)
PHBV	0	119	172	9.1	68	99.22	260.6	282	1.34
PHBV/HNTs	2.5	120	171	8	68.7	97.8	262.5	285.3	3.23
	5	120	170	6.9	68.8	95.5	264.7	287.2	5.11
	10	120	170	6	68.3	89.7	262.4	282.5	9.25

4 Conclusion

This work aimed to evaluate the effect of filler content on the morphology and physical properties of PHBV/HNTs filled at 2.5, 5 and 10 wt% without any prior treatment. From the study, the following conclusions can be drawn. Morphological characterization showed that large aggregates of HNTs are formed at 10 wt%, while a homogeneous dispersion of HNTs in PHBV matrix is observed for 2.5 and 5 wt% with the presence of some individual nanotubes. It was also shown that HNTs increase the crystallization rate of PHBV and improve the thermal stability of PHBV/HNTs bionanocomposites, being however more pronounced for the bionanocomposite filled at 5 wt%, which appears as the optimum filler content ratio.

References

1. Schmitt, H., Prashantha, K., Soulestin, J., et al.: Preparation and properties of novel melt-blended halloysite nanotubes/wheat starch nanocomposites. Carbohyd. Polym. **89**, 920–927 (2012)
2. Phukon, P., Saikia, J.P., Konwar, B.K.: Bio-plastic (P-3HB-co-3HV) from bacillus circulans (MTCC 8167) and its biodegradation. Colloid Surf. B **92**, 30–34 (2012)
3. Zhao, H., Cui, Z., Wang, X., et al.: Processing and characterization of solid and microcellular poly(lactic acid)/polyhydroxybutyrate-valerate (PLA/PHBV) blends and PLA/PHBV/ Clay nanocomposites. Comp. B **51**, 79–91 (2013)
4. Rivera-Briso, A.L., Serrano-Aroca, Á.: Poly(3-Hydroxybutyrate-co-3-Hydroxyvalerate): enhancement strategies for advanced applications. polymers **10**, 732–759 (2018)
5. Hassaini, L., Kaci, M., Benhamida, A., et al.: The effects of PHBV-g-MA compatibilizer on morphology and properties of poly(3-hydroxybutyrate-Co-3-hydroxyvalerate)/olive husk flour composites. J. Adhes. Sci. Technol. **30**, 2061–2080 (2016)
6. Yua, H.Y., Qinb, Z.Y., Suna, B., et al.: Reinforcement of transparent poly(3-hydroxybutyrate-co-3-hydroxyvalerate) by incorporation of functionalized carbon nanotubes as a novel bionanocomposite for food packaging. Comp. Sci. Technol. **94**, 96–104 (2014)
7. Zulfiqar, A.R., Sharjeel, A., Ibrahim, M.B.: Polyhydroxyalkanoates: characteristics, production, recent developments and applications. Int. Biodeter. Biodegrad. **126**, 45–56 (2018)
8. Kennouche, S., Le Moigne, N., Kaci, M., et al.: Morphological characterization and thermal properties of compatibilized poly(3-hydroxybutyrate-co-3-hydroxyvalerate) (PHBV)/poly (butylene succinate) (PBS)/halloysite ternary nanocomposites. Eur. Polym. J. **75**, 142–162 (2016)

9. Kumarjyoti, R., Subhas, C.D., Aphiwat, P., et al.: Up-to-date review on the development of high performance rubber composites based on halloysite nanotube. Appl. Clay Sci. **183**, 105300–105314 (2019)
10. Rawtani, D., Agrawal, Y.K.: Multifarious applications of halloysite nanotubes: a review. Rev. Adv. Mater. Sci. **30**, 282–295 (2012)
11. Carli, L.N., Crespo, J.S., Mauler, R.S.: PHBV nanocomposites based on organomodified montmorillonite and halloysite: the effect of clay type on the morphology and thermal and mechanical properties. Comp. A **42**, 1601–1608 (2011)
12. Liu, M., Guo, B., Du, M., et al.: Natural inorganic nanotubes reinforced epoxy resin nanocomposites. J. Polym. Res. **15**, 205–212 (2008)
13. Szpilska, K., Czaja, K., Kud, S.: Halloysite nanotubes as polyolefin fillers. Polimery **60**, 359–371 (2015)
14. Gaaz, T.S., Sulong, A.B., Kadhum, A.A., et al.: The impact of halloysite on the thermo-mechanical properties of polymer composites. Molecules **2200838**, 1–20 (2015)
15. Valentini, F., Dorigato, A., Rigotti, D., et al.: Polyhydroxyalkanoates/fibrillated nanocellulose. J. Polym. Environ. **27**, 1333–1341 (2019)
16. Zembouai, I., Kaci, M., Bruzaud, S., et al.: A study of morphological, thermal, rheological and barrier properties of poly(3-hydroxybutyrate-co-3- hydroxyvalerate)/polylactide blends prepared by melt mixing. Polym. Test **32**, 842–851 (2013)
17. Murariu, M., Dechief, A.L., Paint, Y., et al.: Polylactide (PLA)-halloysite nanocomposites: Production, morphology and key-properties. J. Polym. Environ. **20**, 932–943 (2012)
18. Du, M., Guo, B., Lei, Y., et al.: Carboxylated butadiene–styrene rubber/halloysite nanotube nanocomposites: interfacial interaction and performance. Polymer **49**, 4871–4876 (2008)
19. Lecouvet, B., Sclavons, M., Bourbigot, S., et al.: Water-assisted extrusion as a novel processing route to prepare polypropylene/ halloysite nanotube nanocomposites: structure and properties. Polymer **52**, 4284–4295 (2011)
20. Javadi, A., Srithep, Y., Pilla, S., et al.: Microcellular poly(hydroxybutyrate-co-hydroxyvalerate)-hyperbranched polymer–nanoclay nanocomposites. Polym. Eng. Sci. **51**, 1815–1826 (2011)
21. Liu, M., Zhang, Y., Zhou, C.: Nanocomposites of halloysite and polylactide. Appl. Clay Sci. **75–76**, 52–59 (2013)

Supercritical Assisted Electrospray for the Production of Controlled Size Loaded PVP Microparticles

Lucia Baldino and Stefano Cardea(✉)

Department of Industrial Engineering, University of Salerno, Via Giovanni Paolo II, 132, 84084 Fisciano, SA, Italy
scardea@unisa.it

Abstract. The production of polymeric micro- and nanoparticles with regular morphology and controlled size and distribution is a relevant target for many fields, from pharmaceutical to nutraceutical, from energetics to fine chemical. In this work, an improved micronization process is proposed, called supercritical assisted electrospray, for the production of biocompatible polyvynilpyrrolidone (PVP) particles loaded with quercetin. This supercritical process can largely improve the traditional electrospray performance, thanks to the addition of supercritical CO2 in the liquid polymeric solution at the beginning of the process. Operating in this manner, an expanded liquid is formed, characterized by a reduced viscosity and surface tension. Repeatable PVP microparticle diameters and distributions were obtained, ranging between 0.47 ± 0.24 µm and 6.01 ± 3.60 µm at PVP concentrations from 1% to 7% w/w, and between 1.71 µm ± 1.07 µm and 2.08 µm ± 1.18 µm, for 1% and 3% w/w PVP particles loaded at 7% w/w quercetin, respectively, working at 120 bar, 35 °C and 30 kV.

1 Introduction

Various sectors of the process industry (e.g., pharmaceutical, nutraceutical and biomedical), highlighted the necessity to produce micro- and nanoparticles of controlled size and distribution for different applications. However, the main conventional micronization techniques, such as: spray-drying, spray freeze-drying, spray freeze into liquid, freeze-drying, precipitation from liquid antisolvents, solvent evaporation from emulsions, jet milling, coacervation and centrifugal extrusion, suffer of numerous disadvantages. In particular, particles shape and size are not perfectly regular, with a wide size distribution; they can degradate due to thermal and mechanical stresses, and can be also contaminated by organic solvent residues or other toxic substances after processing [1].

An innovative technique to produce micro- and nanoparticles is the electrodynamic atomization, also called electrospray (ESPR). In the electrodynamic atomization, a liquid jet break up, typically consisting of a polymeric solution, is obtained with a

S. Piotto et al. (Eds.): BIONAM 2019, LNBE, pp. 24–31, 2020.
https://doi.org/10.1007/978-3-030-47705-9_3

droplet formation through a high electric field potential applied to the tip of the injector and to a grounded metal collector. After that, the corresponding micro- and nanoparticles are obtained by solvent evaporation. Unlike traditional techniques, high temperatures are not required in ESPR, and mechanical forces do not stress particles. However, an incomplete solvent evaporation can be obtained and the operating window is limited due to viscosity and surface tension of the starting polymeric solution [1–4].

The application of supercritical fluids (particularly supercritical CO_2, SC-CO_2) as an alternative to traditional processes represents a field of research in continuous evolution. The properties of supercritical fluids are intermediate between those of a liquid and those of a gas, since the density and the solvent power are similar to a liquid, while the viscosity, the surface tension, and the diffusivity are comparable to those of a gas. These properties can be modulated through operative pressure and temperature variations. Moreover, since this kind of processes can be performed at a temperature close to ambient, the degradation of thermolabile compounds is avoided [1, 4].

SC-CO_2 was also applied in electrospinning with the aim of improving solvent removal from the spinned polymeric fibers. Okamoto and Wahyudiono [2] performed the electrospinning of a PVP:PLLA solution (80:20) in dichloromethane, trying to reduce its viscosity using high-pressure CO_2. Operating in this way, fibers with diameters ranging between 0.60 and 0.79 μm were produced. Wahyudiono et al. [3] produced PVP hollow fibers by electrospinning of a solution in pressurized CO_2. Different PVP morphologies were obtained changing the operative pressure: solid fiber at 30 bar, hollow core fibers at 50 bar, and balloon-like structures at 80 bar. However, these techniques needed of complex process arrangement with respect to the traditional one since a high pressure precipitator was required. In order to overcome these problems, Baldino et al. [1] proposed a new technique, called electrospray assisted by SC-CO_2. In this case, SC-CO_2 was added to the polymeric liquid solution before precipitation, largely reducing its viscosity and surface tension (i.e., cohesive forces) thanks to the formation of an expanded liquid [4–6]. Moreover, a high pressure vessel precipitator was no more required and materials collection can be performed at room conditions. Using this technique, it was possible to process high molecular weight polymers, selecting larger liquid flow rates than in traditional processes, and different polymer concentrations, obtaining, at the same time, smaller particles and faster solvent evaporation thanks to the contribution of SC-CO_2. In particular, PVP at different concentrations by weight in ethanol were prepared, and nano- and microparticles with an average diameter between 0.55 and 2.25 μm were obtained, with good repeatability and narrow distribution, especially operating at 120 bar. In a subsequent work [4], the same authors studied the main parameters effect on the supercritical assisted electrospray and found that an increase in the electric potential difference did not significantly modify the PVP particle size distributions; however, the distributions became sharper as the electric potential difference increased. The solution viscosity increased with PVP concentration: therefore, first large and very large particles were formed; then, the process was no more able to produce jet break-up, microfibers were obtained and the process was converted to electrospinning, producing microfibers with diameters down to about 1.4 μm. Therefore, pressure and PVP percentage played opposite roles in generating the different morphologies (i.e., particles or fibers) and, for this reason, electrospray and electrospinning assisted by SC-CO_2 can be considered like the two faces of the same medal.

Quercetin is a bioflavonoid generally found in fruit and vegetable, characterized by several healthy properties, such as free-radical scavenging activity, and potential use for the prevention and the treatment of pathologies like atherosclerosis, chronic inflammation, and others [7–11]. However, quercetin can suffer numerous chemical changes during pharmaceutical processing and storage, due to the effects of oxygen, temperature, pH, etc. Besides, this flavonoid shows a poor water solubility and, thus, reduced bioavailability [11].

Taking into account these considerations, the aim of this work is to produce controlled size PVP microparticles loaded with quercetin, by SC-CO_2 assisted electrospray, as carrier for pharmaceutical formulations thanks to the biocompatibility and the hydrosolubility of this polymer, enhancing drug bioavailability.

2 Materials and Methods

Polyvinylpyrrolidone (PVP, Mw 1,300,000), quercetin ($\geq$95%) and ethanol (purity > 99.9%) were bought from Sigma Aldrich. CO_2 (purity 99.9%) was supplied by Morlando Group s.r.l. (Sant'Antimo, NA, Italy).

PVP powder was dissolved in ethanol at different concentrations by weight (ranging from 1 to 7% w/w), at room temperature and using a magnetic stirrer at 100 rpm. After the obtainment of homogeneous solutions, quercetin at 7% w/w with respect to PVP was added. The solutions were mixed for other 30 min at 100 rpm and at room temperature, until a complete quercetin dissolution was observed.

SC-CO_2 assisted electrospray plant consisted of a stainless-steel high-pressure vessel with an internal volume of about 70 mL, in which a solution to be treated was loaded. The vessel was closed and CO_2 was pumped from the bottom using a high-pressure pump (Gilson, mod. 305, Middleton, WI, USA). When the desired operative pressure was reached, the system was left 10 min for equilibration, to favor the mixing between SC-CO_2 and the liquid polymeric solution to obtain an expanded liquid. In order to allow a constant pressure operation, after this time, nitrogen was added from the top of the vessel at the same pressure selected for processing. A Teflon disk located on the top of the liquid solution avoided the contact and the diffusion of N_2 in the liquid solution. When an ON/OFF valve (Swagelok ON/OFF, Nordival s.r.l., Rovato (BS), Italy) was opened, the polymeric solution started to be delivered from the vessel to a 100 μm internal diameter injector, responsible of the atomization process. The area near the injection system was heated using cable heaters (Watlow, 240 V, 275 W) to counter balance the Joule-Thomson effect, related to a fast CO_2 depressurization. Temperature was measured by thermocouples and regulated using PID controllers (mod. 305, Watlow, Corsico (MI), Italy); whereas pressure in the feeding vessel was measured by a test gauge (mod. MP1, OMET, Lecco, Italy). The voltage generator was a FUG Elektronik (mod. HCP 35-3500, Schechen, Germany). The collector was composed by two adjacent stainless-steel blocks, covered by a thin aluminum foil. An experiment ended when the whole content of the feeding vessel was discharged; each experiment was repeated twice.

PVP-based particles were sputter coated with gold (Agar Auto Sputter Coater mod. 108 A, Stansted, UK) at 30 mA for 80 s and analyzed using a field emission scanning electron microscope (FE-SEM, mod. LEO 1525, Carl Zeiss SMT AG, Oberkochen, Germany) to study their morphology.

Sigma Scan Pro 5.0 (Jandel scientific, San Rafael, QC, Canada) and Origin 9.1 (Microcal, Northampton, MA, USA) were used to measure the average diameter of the particles and to calculate their size distributions. Approximately 300 particle diameters were measured for each calculation.

3 Results and Discussion

The first part of this study was focused on the production of empty PVP particles, in order to confirm the results obtained in a previous work of the same research group [4]. In particular, a polymer concentration of 1, 3 and 7% w/w in ethanol was selected, maintaining constant these operative parameters previously optimized [4]: 120 bar pressure, 35 °C temperature, 30 kV voltage, 100 μm injector diameter, 25 cm injector-collector distance.

SEM analyses on the collected material in each experiment demonstrated that regular spherical particles were produced in all cases, as reported in Fig. 1. These particles were characterized by a mean diameter of: 0.47 ± 0.24 μm, 1.41 ± 0.88 μm and 6.01 ± 3.60 μm for 1, 3 and 7% w/w PVP starting solution, respectively.

Once verified the process reproducibility, in the second part of the experimentation, quercetin at 7% w/w with respect to PVP was dissolved in the ethanolic solutions. This quercetin loading was selected taking into account the solution viscosity increase when a second component was added in the starting solution: i.e., a too viscous solution, particularly in the case of 7% w/w PVP system, can produce fibers instead of particles due to the Taylor cone formation, favored by a cohesive force increase.

(c)

Fig. 1. SEM pictures of microparticles produced using ethanolic solutions at: (a) 1% w/w, (b) 3% w/w and (c) 7% w/w PVP.

Morphological analyses on these loaded samples demonstrated that spherical microparticles were obtained when quercetin was added to the 1% and 3% w/w PVP solutions, as shown in Fig. 2a–b; a beads-on-string morphology was instead observed for the electrosprayed solution of quercetin loaded 7% w/w PVP (see Fig. 2c).

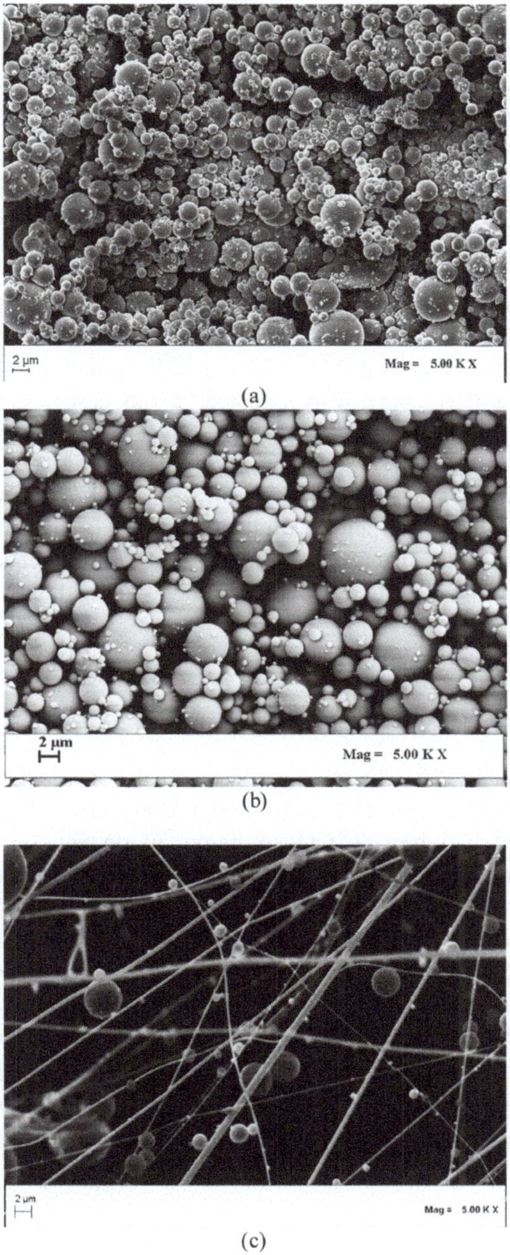

Fig. 2. SEM pictures of 7% w/w quercetin loaded samples produced using ethanolic solutions at: (a) 1% w/w, (b) 3% w/w and (c) 7% w/w PVP.

Moreover, the loaded particles mean diameter was larger than the one measured for the empty PVP particles. In particular, it ranged from 1.71 μm ± 1.07 μm to 2.08 μm ± 1.18 μm, for 1% and 3% w/w PVP loaded particles, respectively. The mean diameter increase measured for the loaded PVP microparticles was due to the quercetin addition to the polymeric solution that produced a consequent viscosity increase and, therefore, larger microparticles. The effect of viscosity cohesive force was particularly observed for the 7% w/w PVP samples loaded with quercetin. SEM picture shown in Fig. 2c demonstrates that the transition from particles to fibers was obtained; i.e., the final solution viscosity of the PVP+quercetin system favored the Taylor cone formation instead of jet break-up, producing this hybrid morphology. These results are also an indirect demonstration of the effective quercetin encapsulation in the PVP microstructures.

In the last part of the work, the effect of the operative pressure on the particles size was investigated. With this purpose, ethanolic solutions at 3% w/w PVP loaded with 7% w/w quercetin were processed at 80, 120 and 160 bar, and maintaining constant the other operative parameters previously adopted. In all cases, spherical micrometric particles were collected; however, the pressure effect was noted on the particle size distribution. In particular, a reduction of particles size was observed from 2.21 μm ± 1.27 μm to 1.37 μm ± 0.72 μm, working from 80 to 160 bar, respectively, accompanied by a sharpening of the particles size distribution. This result confirms the disruptive role of pressure: i.e., the pressure difference between the high pressure vessel and room conditions, favored jet break-up and, thus, the obtainment of smaller particles.

4 Conclusions and Perspectives

Electrospray assisted by SC-CO_2 was proposed in this work for the production of quercetin loaded PVP microparticles, to be used as drug carrier. The experimental tests demonstrated as the addition of SC-CO_2 to the polymeric solution at the beginning of the experiment permitted the use of a high molecular weight polymer (M_w 1,300,000) thanks to the formation of an expanded liquid characterized by a reduced viscosity and surface tension. However, the solution viscosity was confirmed a relevant parameter to be controlled in order to obtain a desired sample morphology and particle size. An opposite role was instead played by pressure that favored the production of smaller microparticles using the same PVP weight concentration in the starting solution.

In perspective, other parameters, such as temperature and the applied voltage, can be investigated, to further optimize samples morphology, and drug release tests can be performed to determine the release kinetics useful for in vivo treatment.

References

1. Baldino, L., Cardea, S., Reverchon, E.: Supercritical assisted electrospray: an improved micronization process. Polymers **11**, 244–255 (2019)
2. Okamoto, K., Wahyudiono, S.: Fabrication of micro-hollow fiber by electrospinning process in near-critical carbon dioxide. In: AIP Conference Proceedings, vol. 43, p. 1586 (2014)

3. Wahyudiono, S., Machmudah, H., Kanda, S., Okubayashi, M., Goto, M.: Formation of PVP hollow fibers by electrospinning in one-stepprocess at sub and supercritical CO_2. Chem. Eng. Process. **77**, 1–6 (2014)
4. Baldino, L., Cardea, S., Reverchon, E.: A supercritical CO_2 assisted electrohydrodynamic process used to produce microparticles and microfibers of a model polymer. J. CO_2 Utiliz. **33**, 532–540 (2019)
5. Falco, N., Kiran, E.: Volumetric properties of ethyl acetate + carbon dioxide binary fluid mixtures at high pressures. J. Supercrit. Fluids **61**, 9–24 (2012)
6. Kiran, E., De Benedetti, P.G., Peters, C.J.: Supercritical fluids: fundamentals and applications (2012)
7. Cunico, L.P., Turner, C.: Supercritical fluids and gas-expanded liquids. In: The Application of Green Solvents in Separation Processes, vol. 7, pp. 155–177 (2017)
8. Fischer, C., Speth, V., Fleig-Eberenz, S., Neuhaus, G.: Induction of zygotic polyembryos in wheat: influence of auxin polar transport. Plant Cell **9**, 1767–1780 (1997)
9. Alinezhad, H., Azimi, R., Zare, M., Ebrahimzadeh, M.A., Eslami, S., Nabavi, S.F., Nabavi, S.M.: Antioxidant and antihemolytic activities of ethanolic extract of flowers, leaves, and stems of Hyssopus officinalis L. Var. angustifolius. Int. J. Food Prop. **16**, 1169–1178 (2013)
10. Harnly, J.M., Doherty, R.F., Beecher, G.R., Holden, J.M., Haytowitz, D.B., Bhagwat, S., Gebhardt, S.: Flavonoid content of US fruits, vegetables, and nuts. J. Agric. Food Chem. **54**, 9966–9977 (2006)
11. Wang, W., Sun, C., Mao, L., Ma, P., Liu, F., Yang, J., Gao, Y.: The biological activities, chemical stability, metabolism and delivery systems of quercetin: a review. Trends Food Sci. Technol. **56**, 21–38 (2016)

Superhydrophobic Coatings and Artificial Neural Networks: Design, Development and Optimization

Francesco Marrafino[1(✉)], Pio Iannelli[1], Miriam Di Martino[1], Simona Concilio[2], and Stefano Piotto[1]

[1] Department of Pharmacy, University of Salerno, via Giovanni Paolo II, 132, 84084 Fisciano, SA, Italy
fmarrafino@unisa.it
[2] Department of Industrial Engineering, University of Salerno, via Giovanni Paolo II 132, Fisciano, SA, Italy

Abstract. Recently, considerable attention has been devoted to the development of superhydrophobic surfaces due to their advantageous antifouling and antimicrobial capacity. While significant effort has been devoted to fabricating such surfaces, very few polymeric superhydrophobic surfaces can be considered durable against various externally imposed stresses. Pyrogenic hydrophobic silica nanoparticles were used to confer superhydrophobic properties to the coatings. 450 samples were prepared using a layer-by-layer approach, deposing an epoxy resin or PDMS layer as adhesive on a substrate (PC/ABS), followed by one or more layers of silica nanoparticles, or silica-resin mixed layers. The best coating obtained shows a contact angle of 157° and a tape peeling grade resistance. The applied method involves the spray deposition of a multilayer coating composed of: silica layer/epoxy resin layer/silica layer, followed by partial curing of the coating (15 min, 70 °C); another silica layer is then sprayed on the surface and is cured for 10 min. Given the high number of parameters involved, process optimization is quite tricky. Artificial Neural Networks are the best tool to deal with multivariate analysis problems and for this reason, data from all the prepared samples were collected into a matrix and was used to train a neural network capable of predicting the degree of hydrophobicity of a silica nanoparticles-based coating.

Keywords: Neural network · Superhydrophobic surface · Antifouling

1 Introduction

The bacteria adhesion and proliferation on abiotic surfaces and the subsequent biofilm formation pose challenges in both healthcare and industrial applications, such as medical implants, petroleum pipelines, aquatic flow systems, textiles and solar panels [1–3]. There are two approaches to the development of a bactericidal or bacteriostatic surface: the introduction of biocidal substances on the polymer surface or the development of a material unsuitable for the growth of microorganisms. The materials of the first type are the most widespread [3–8] and require extreme caution to prevent the

S. Piotto et al. (Eds.): BIONAM 2019, LNBE, pp. 32–40, 2020.
https://doi.org/10.1007/978-3-030-47705-9_4

release of biocides into the environment. Superhydrophobic surfaces (SH) showed excellent self-cleaning [9], anti-icing, anti-biofilm and anti-fouling properties [10–12], preventing protein adsorption, eukaryotic cellular adhesion and proliferation, platelet adhesion/activation and blood coagulation [13, 14], showing a high potential for application in many industrial fields and biomedical devices.

Many functional biological surfaces in nature, e.g., lotus leaves, rose petals, butterfly wings and water striders, have unique wetting properties. The surface roughness caused by unique micro/nanostructures and surface chemistry is two significant factors that affect surface wetting properties [15]. To be defined as superhydrophobic, a surface must have a water contact angle (CA) higher than 150° [16]. Superhydrophobicity is based on two principles [17, 18]: low surface energy of the solid surface and increased surface roughness.

To date, superhydrophobic surfaces have been prepared using a vast array of techniques, such as lithographic processes, sol-gel methods, casting, chemical vapor deposition (CVD), electrospinning, chemical etching, dip-coating, and templating. These approaches, though, show many limitations originating from the complicated and time-consuming preparation processes and expensive equipment, that prevent the large-scale application of such methods.

Thus, low-cost, facile methods for extensive applications are needed. Brush-coating, dip-coating, and spray-coating meet the requirements. However, the mechanical softness of the microstructures on the surface limits the applications of superhydrophobic coatings [19]. The main challenge is to develop a simple, scalable process leading to mechanically robust and chemically inert layers, and an even more significant challenge is to design mechanically and thermodynamically durable, defect-tolerant SH layers [16, 20]. In this work, we show the preparation of a durable SH coating with antifouling capacity.

Fluorinated reagents with low surface free energy are often used in the preparation of superhydrophobic surfaces to improve the hydrophobicity. However, fluorinated compounds are expensive and toxic. For this reason, economic and environmentally friendly fluorine-free compounds, such as alkyl-silanes and long-chain organic compounds, have been employed in the preparation of water-resistant surfaces. Here we present a fluorine-free coating for the production of a durable, self-cleaning, superhydrophobic coating [19].

Various multilayer/mixed layers coatings were prepared, using DGEBA or PDMS as adhesives, providing the surfaces with microscale roughness and robustness, and Aerosil R202 or R504 silica nanoparticles (NPs) to enhance the nanoscale roughness and to confer SH properties. The coatings were prepared via spray deposition. A multilayer epoxy resin/silica NPs superhydrophobic coating with a CA > 150° and a tape peeling grade resistance was obtained. Collected data was used to train an Artificial Neural Network (ANN), in order to optimize the coating and the preparation process. Preliminary results are encouraging because the coatings have great potential for practical application in various fields and allow them to proceed to the realization of prototype systems for biomedical devices.

2 Materials and Methods

Epoxy resin: Araldite GY2600 (Huntsman, Bisphenol A diglycidyl ether). Cross-linking agents: Jeffamine D400 (Huntsman); Jeffamine D230 (Huntsman); Jeffamine EDR148 (Huntsman); Accelerator 400 (Huntsman); Hexamethylenediamine. PDMS: Sylgard 184 (Dowsil), part A (PDMS) + part B (cross-linking agent), mix ratio 10:1. Pyrogenic silica nanoparticles: Aerosil R504 (Evonik), amine-modified surface, nanoparticle diameter approx. 9 nm; Aerosil R202 (Evonik), nanoparticles diameter approx. 11 nm. Coating substrates: PC/ABS, 20 × 30 × 4 mm ca. samples; PLA, 20x30x4 mm ca. samples; polystyrene, 20 × 30 × 4 mm ca. samples. Airbrush: nozzle diameter 0.4 mm; operative pressure 4 bar. Coatings were prepared via multi-step deposition (Fig. 1), a layer-by-layer process. The first layer is the adhesive (epoxy resin), followed by one or more layers with superhydrophobic or protective properties. The layers were deposited via spray coating, keeping the sample at a 15 cm distance. Curing was carried out in the oven at 70 °C to avoid substrate degradation (e.g., PC/ABS). Contact angle measurements were carried out by dropping a 10.0 μl water droplet from a 2 cm height on the samples, properly placed on a flat surface [6]. Abrasion resistance tests were carried out with the tape peeling method.

Fig. 1. Multistep deposition scheme

Neural Network was trained using the GMDH Shell DS 3.8.9 software application, selecting a regression forecasting model. The used core learning algorithm was the GMDH-type Neural Network (Group Method of Data Handling) [21]. Such neural networks, also known as polynomial neural networks, use the combinatorial algorithm to improve neuronal connection. The selected internal function for neurons was the quadratic polynomial function, and the upper limit for the number of network layers created by the algorithm was set at 33. GMDH-type networks generate many layers, and an upper limit of 33 layers is set to reduce the quickly growing computer memory and time consumption, while maintaining good predictive accuracy. The reordering of rows, used to achieve uniform statistical characteristics of training and testing samples and to make them equally informative, was performed with the *odd/even* method, which places all even instances after odd instances. The selected validation strategy was the *k-fold validation* (2 folds): the dataset is split k parts, and then a model is trained for k times, using k-1 parts. The performance of the model is measured at each training using a new remaining part. The selected validation criterion for both the core algorithm and variables ranking was the RMSE (Root-Mean-Squared Error). The variables were ranked by correlation, and only the 20 top ranked variables were considered for model evaluation; the reduction of variables was necessary for quicker processing of the high-dimensional dataset [22].

3 Results and Discussion

The first coatings were prepared according to the scheme: 1) adhesive layer; 2) curing time t; 3) silica or mixed layer.

Adhesive layer is GY2600 + Jeffamine D400 (1:1 ratio); t is 1, 2, 4, 6, 8, 10, 16, or 18 h; silica layer is THF + Aerosil R504 (2.5% wt); mixed layer is THF + Aerosil R504 (2.5% wt) + GY2600 + D400 (resin and crosslinker mixture 0.25% wt).

Aerosil R504, which has amino groups on the surface of the nanoparticles, was used to test the possibility of having a covalent bond between the nanoparticles and the underlying adhesive, to improve coating resistance. Epoxy resin curing was performed in an oven at 70°C. PC/ABS was used as the substrate. Some tests were carried out on polystyrene; however, such samples underwent deformation and degradation with high temperatures, making the material unsuitable for coating deposition. In total, around 80 samples were prepared. The samples did not have superhydrophobic properties, with contact angles ranging from 62° to 96°, and several problems have been observed, including non-uniform curing of the adhesive layer and a recurring phenomenon of coating surface cracking. R504 and R202 silica NPs were dissolved in various solvents (THF, DCM, DMAC, $CHCl_3$, n-hexane, and cyclohexane), then sprayed on PC/ABS samples without adhesive. Cyclohexane and n-hexane resulted in uniform and superhydrophobic coatings, while other solvents showed surface cracking and non-uniform distribution on the surface; n-hexane was then chosen as silica solvent for future tests. Various GY2600 crosslinking agents were tested (Jeffamine D230, Jeffamine EDR148, Accelerator 400, hexamethylenediamine). Dissolution and spray coating tests were then carried out on GY2600 mixed with crosslinking agents. DCM showed relatively fast evaporation in the oven without surface cracking; such a problem was still evident when using other solvents. EDR148 showed the most uniform curing and the shortest curing times among the tested agents, fully curing epoxy resin in 20 min in the oven at 70 °C. Meanwhile, according to the previous scheme, around 80 more samples were prepared, testing the following mixtures:

R504 in cyclohexane (2.5% wt);
R504 in n-hexane (2.5% wt);
GY2600 (5% wt) + D230 (1:3 ratio) in THF/DCM/DMAC/$CHCl_3$;
GY2600 (5% wt) + EDR148 (1:5 ratio) in THF/DCM/DMAC/$CHCl_3$;
GY2600 (5% wt) + hexamethylenediamine (1:6 ratio) in THF/DCM/DMAC;
GY2600 (5% wt) + D230 + Accelerator 400 in DCM;
GY2600 (5% wt) + EDR148 + Accelerator 400 in DCM.

PC/ABS was used as a substrate; CAs ranged between 61° and 140° (nearly SH). The highest CAs were observed on samples prepared by deposing an adhesive layer of GY2600 + EDR148, followed by a layer of R202/R504 in n-hexane. Epoxy resin was partially cured in an oven for 15 min at 70 °C; partial curing is necessary to allow the silica resin layer to sink into the underlying layer, to improve coating resistance. Resistance, however, was mediocre and still needed improvement. Various samples were prepared by spin-coating and dip-coating to test other preparation methods. Surface cracking and non-uniformity of the layers were evident. Moreover, PLA was tested as a substrate; the surface was partially dissolved by solvents. No SH coating could be obtained on PLA samples, with very low contact angles.

A new surface was then designed, according to the following scheme: 1) silica layer; 2) epoxy resin layer; 3) silica layer; 4) curing time t; 5) silica layer 6) curing time t'.

Here, the silica layer is R202 or R504 in n-hexane; the epoxy layer is GY2600 (5% wt) + EDR148 (1:1) in DCM; curing times t, t' is 5, 10, 15, 20 min; curing temperature is 70 °C. Of the 60 prepared samples, CAs and resistance were measured: the best coatings were those with curing time t = 15 min and t' = 10 min, and no noticeable difference was observed between the two types of silica. In Table 1 two examples are reported; GIU05 was prepared with R202 silica, while OTT03 was prepared with R504 silica. CAs were measured before and after 1, 2, 3 and 5 cycles of tape peeling. The amino group of R504 seems to have little to no influence on resistance (i.e., there should be no covalent bonding); resistance is attributed to silica sinking into the epoxy layer. Silica excess allows keeping exposed nanoparticles, avoiding epoxy resin to completely cover the surface, thus depriving it of SH properties while maintaining a reasonable resistance level.

Table 1. Contact angle before and after tape peeling (TP) cycles for samples GIU05 (R202) and OTT03 (R504).

	GIU05 (R202)	*OTT03 (R504)*
CA	157°	151°
TP x1 CA	157°	147°
TP x2 CA	152°	146°
TP x3 CA	151°	145°
TP x5 CA	148°	141°

Additional schemes were designed in order to include more data in the matrix. Several samples were prepared using PDMS as adhesive. Samples were spray-coated according to the scheme: 1) R202 layer; 2) PDMS layer; 3) R202 layer; 4) curing 15 min; 5) R202 layer; 6) curing 10 m.

Samples are superhydrophobic but not resistant. The procedure was replicated with R504, giving similar results. CAs range from 147° to 152°. A thin protective layer of 5%wt PDMS was deposed on R202 and R504 samples to improve resistance. Such a layer, however, lead to a loss of superhydrophobicity, without appreciably modifying resistance.

Samples were spray-coated according to the scheme: 1) PDMS 5%wt or 20%wt layers; 2) silica 2.5%wt layer. Coatings are not superhydrophobic.

More samples were prepared according to the scheme: 1) 1–4 layers of PDMS 5% wt or 10%wt; 2) silica 5%wt mixture, using spray coating and dip coating techniques. Samples are superhydrophobic but not resistant; silica NPs probably prevented PDMS proper curing.

3.1 Artificial Neural Network Training

Coating and preparation process optimizations have many parameters to consider. It is an example of multivariate analysis, not treatable with conventional methods; it was, therefore, essential to use tools, such as artificial neural networks (ANN) and deep learning methods [23]. Data collected from the prepared samples (e.g., reagents, concentrations, number of layers, curing time, resistance, etc.) have been used to build a matrix. This matrix was used to train an ANN, aimed at optimizing the coating itself, using a supervised learning method. The matrix was prepared to make it "readable" by the ANN. For the correct training of a neural network, in fact, it is necessary that data is formatted and labelled appropriately. Artificial neural networks are calculation systems that "learn" to execute commands without being programmed. An ANN is based on a set of connected units or nodes called artificial neurons (Fig. 2). Each connection can transmit a signal to other neurons. An artificial neuron that receives signal processes it and can send a signal to the neurons connected to it. In this case, the initial inputs are the operating parameters, the targets are the contact angle and resistance, while the final output is an equation for the optimization of the used parameters, which provides the best composition to prepare a coating that is both resistant and superhydrophobic.

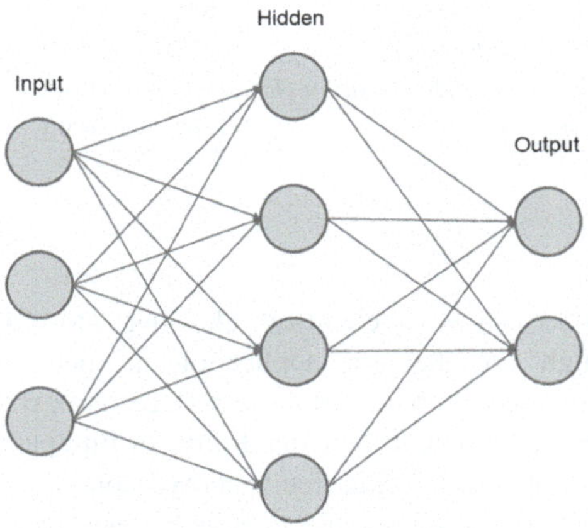

Fig. 2. Example of ANN nodes and layers

The used algorithm is based on supervised learning, i.e., learning that provides the network with a set of labelled inputs to which correspond known outputs [24]. By analysing the data provided, the network learns the link that unites them and, in this way, learns to calculate correct associations between the input data provided. As the network processes outputs, it also proceeds to correct them to improve them, increasing the weight of the parameters that determine the correct outputs and decreasing those that generate invalid values. The supervised learning mechanism, therefore, employs error backpropagation [25].

The training of the network has been carried out, taking into consideration all the operational parameters involved in the preparation of the samples. The matrix initially used is made up of a set of 250 samples, of which 80% was selected, in a random way, as a training set, while the remaining 20% was used as a validation set. Among the various input parameters that make up the matrix we have: the substrate used, the type of silica nanoparticles (R202 or R504), the epoxy resin, the crosslinker, as well as parameters such as the number of layers, the type of mixture, the temperature and curing time and many others. The output parameters on which attention has been focused are two: contact angle and coating resistance.

The first attempts to train the network were found to be poorly precise, due to some gaps in the initial matrix. It was, therefore, necessary to make up for these missing data by preparing new samples and thus obtaining a more complete matrix. The dataset with which the working algorithm was obtained is therefore made up of a total of 450 samples, divided into 80% training set and 20% validation set. This algorithm will be used for the preparation of superhydrophobic coatings with improved resistance. Table 2 shows the most critical data, which summarizes the degree of accuracy and efficiency obtained by the trained neural network.

Table 2. Trained ANN prediction accuracy

	Model fit	*Predictions*
Number of observations	360	90
Mean absolute error	10.86	9.70
Coefficient of determination (R^2)	0.75	0.85
Correlation	0.87	0.92

4 Conclusions

Superhydrophobic surfaces show excellent self-cleaning, anti-biofilm, and anti-fouling properties, with a high potential for application in many industrial fields and biomedical devices. The main problems of these surfaces are expensive and not widely applicable preparation methods and their durability. In this paper, we have shown a preparation method to obtain a fluorine-free superhydrophobic surface with a contact angle higher than 157° and a reasonable degree of resistance. Moreover, the data obtained with the various preparation schemes have been collected and used to train an artificial neural network, which efficiently allows optimizing the preparation of a silica NPs based SH coating. The practical algorithm will be used for the preparation of superhydrophobic coatings with improved resistance and is currently being evaluated for possible patentability.

Bibliography

1. Zhang, X., Wang, L., Levanen, E.: Superhydrophobic surfaces for the reduction of bacterial adhesion. RSC Adv. **3**(30), 12003–12020 (2013)
2. Piotto, S., Concilio, S., Sessa, L., Iannelli, P., Porta, A., Calabrese, E.C., et al.: Novel antimicrobial polymer films active against bacteria and fungi. Polym. Compos. **34**(9), 1489–1492 (2013)

3. Piotto, S., Concilio, S., Sessa, L., Diana, R., Torrens, G., Juan, C., et al.: Synthesis and antimicrobial studies of new antibacterial azo-compounds active against staphylococcus aureus and listeria monocytogenes. Molecules **22**(8), 1372 (2017)
4. Piotto, S., Di Biasi, L., Sessa, L., Concilio, S.: Transmembrane peptides as sensors of the membrane physical state. Front. Phys. **6**, 48 (2018)
5. Concilio, S., Sessa, L., Petrone, A.M., Porta, A., Diana, R., Iannelli, P., et al.: Structure modification of an active azo-compound as a route to new antimicrobial compounds. Molecules **22**(6), 875 (2017)
6. Concilio, S., Iannelli, P., Sessa, L., Olivieri, R., Porta, A., De Santis, F., et al.: Biodegradable antimicrobial films based on poly(lactic acid) matrices and active azo compounds. J. Appl. Polym. Sci. **132**(33) (2015)
7. Clauss-Lendzian, E., Vaishampayan, A., de Jong, A., Landau, U., Meyer, C., Kok, J., et al.: Stress response of a clinical Enterococcus faecalis isolate subjected to a novel antimicrobial surface coating. Microbiol. Res. **207**, 53–64 (2018)
8. Gwisai, T., Hollingsworth, N.R., Cowles, S., Tharmalingam, N., Mylonakis, E., Fuchs, B.B., et al.: Repurposing niclosamide as a versatile antimicrobial surface coating against device-associated, hospital-acquired bacterial infections. Biomed. Mater. **12**(4), 045010 (2017)
9. Zhang, X.-F., Zhao, J.-P., Hu, J.-M.: Abrasion-resistant, hot water-repellent and self-cleaning superhydrophobic surfaces fabricated by electrophoresis of nanoparticles in electrodeposited sol-gel films. Adv. Mater. Interfaces **4**(13), 1700177 (2017)
10. Genzer, J., Efimenko, K.: Recent developments in superhydrophobic surfaces and their relevance to marine fouling: a review. Biofouling **22**(5), 339–360 (2006)
11. Crick, C.R., Ismail, S., Pratten, J., Parkin, I.P.: An investigation into bacterial attachment to an elastomeric superhydrophobic surface prepared via aerosol assisted deposition. Thin Solid Films **519**(11), 3722–3727 (2011)
12. Privett, B.J., Youn, J., Hong, S.A., Lee, J., Han, J., Shin, J.H., et al.: Antibacterial fluorinated silica colloid superhydrophobic surfaces. Langmuir: ACS J. Surf. Colloids **27**(15), 9597–9601 (2011)
13. Lima, A.C., Mano, J.F.: Micro-/nano-structured superhydrophobic surfaces in the biomedical field: part I: basic concepts and biomimetic approaches. Nanomedicine **10**(1), 103–119 (2015)
14. Lima, A.C., Mano, J.F.: Micro/nano-structured superhydrophobic surfaces in the biomedical field: part II: applications overview. Nanomedicine **10**(2), 271–297 (2015)
15. Guo, Z., Liu, W., Su, B.-L.: Superhydrophobic surfaces: from natural to biomimetic to functional. J. Colloid Interface Sci. **353**(2), 335–355 (2011)
16. Ellinas, K., Tserepi, A., Gogolides, E.: Durable superhydrophobic and superamphiphobic polymeric surfaces and their applications: a review. Adv. Colloid Interface Sci. **250**, 132–157 (2017)
17. Dettre, R.H., Johnson, R.E.: Contact angle hysteresis contact angle, wettability, and adhesion. Adv. Chem.: Am. Chem. Soc. **43**, 136–144 (1964)
18. Barthlott, W., Neinhuis, C.: Purity of the sacred lotus, or escape from contamination in biological surfaces. Planta **202**(1), 1–8 (1997)
19. Zhang, Z.-H., Wang, H.-J., Liang, Y.-H., Li, X.-J., Ren, L.-Q., Cui, Z.-Q., et al.: One-step fabrication of robust superhydrophobic and superoleophilic surfaces with self-cleaning and oil/water separation function. Sci. Rep. **8**(1), 3869 (2018)
20. Scarratt, L.R.J., Steiner, U., Neto, C.: A review on the mechanical and thermodynamic robustness of superhydrophobic surfaces. Adv. Colloid Interface Sci. **246**, 133–152 (2017)
21. Anastasakis, L., Mort, N.: The development of self-organization techniques in modelling: a review of the group method of data handling (GMDH). Res. Rep.-Univ. Sheffield Depart. Autom. Control Syst. Eng. (2001)

22. Copyright 2009–2020 © GMDH, LLC. https://gmdhsoftware.com/
23. Chen, Y.-Y., Lin, Y.-H., Kung, C.-C., Chung, M.-H., Yen, I.H.: Design and implementation of cloud analytics-assisted smart power meters considering advanced artificial intelligence as edge analytics in demand-side management for smart homes. Sens. (Basel). **19**(9), 2047 (2019)
24. Russell, S.J.: Artificial intelligence: a modern approach. Pearson, Harlow (2016)
25. Rumelhart, D.E., Hinton, G.E., Williams, R.J.: Learning representations by back-propagating errors. Nature **323**(6088), 533–536 (1986)

Nucleation and Growth Rate of a Poly(Lactic Acid) in Quiescent Conditions

Valentina Volpe(✉), Fabiana Foglia, Valentina Iozzino, and Roberto Pantani

Department of Industrial Engineering, University of Salerno, via Giovanni Paolo II, 132, 84084 Fisciano, SA, Italy
vavolpe@unisa.it

Abstract. Poly(lactic acid) (PLA) is a biodegradable thermoplastic polyester produced from renewable sources that can be easily processed to obtain articles to be used in many applications, from commodities to electronics. The large influence of the crystalline structures and morphology on the biodegradation rate, as well as on the other material properties, led to an ever-increasing interest in the study of the crystallization behaviour of PLA. In this work, experiments on the crystallization kinetics of PLA were carried out by means of optical observations considering the nucleation and growth phenomena of the PLA crystals under quiescent conditions.

Keywords: Crystallization · Poly(Lactic Acid) · Nucleation and growth rate

1 Introduction

In the last decades, biobased polymers have gained increasing importance, with applications ranging from the biomedical to the agricultural and packaging. Among the bio-based polymers, the most promising alternatives to petrochemical based polymers is the poly(lactic acid) (PLA), which is produced from corn or other removable resources and exhibits special properties including compostability or biocompatibility [1–3]. Nowadays, thanks to the large variety of grades provided by suppliers, it is adopted in various processes, such as extrusion or injection molding. However, one of the drawbacks still exhibited by PLA is its slow crystallization kinetics, which determinates an almost amorphous final part with consequent low mechanical performances [4]. Therefore, the degree of crystallinity is of fundamental importance as it influences the final properties of the product and the degradation process [5]. Several investigations on the crystallization kinetics of PLA under quiescent condition were carried out. In 2001 Di Lorenzo et al. [6] presented data of spherulite growth rate obtained in quiescent conditions with combined isothermal and non-isothermal methods. In 2012 Saeidlou et al. [7] presented an overview of the fundamentals of PLA in quiescent conditions in order to provide comprehensive relations between crystallization kinetics and the main molecular structures characteristics of PLA.

In this work, the nucleation and growth rate of PLA crystals was examined under quiescent conditions by means of optical observations in order to investigate the influence of temperature and time on the crystallization rate.

S. Piotto et al. (Eds.): BIONAM 2019, LNBE, pp. 41–47, 2020.
https://doi.org/10.1007/978-3-030-47705-9_5

2 Materials and Methods

The material adopted in this work was a commercial grade PLA produced by NatureWorks with the trade name of 3251D with a D-isomer content of 1.4% [3]. Before any experiment, the material was dried under vacuum at a temperature of 80 °C for 4 h.

Rheological Tests
A rotational rheometer Haake Mars II (Thermo scientific) allowed to analyze the rheological behavior of the PLA 3251D. The experiments were performed with parallel plates configuration in oscillatory dynamic mode under a nitrogen atmosphere. The experiments were executed at 120 °C, 140 °C, 160 °C and 180 °C with a gap of 1 mm.

Linkam Hot Stage
The crystallization behavior of PLA was investigated by a Linkam CSS450 hot stage with a plate-plate geometry in combination with a polarized optical microscope Olympus BX51. The adopted thermal protocol is shown in Fig. 1.

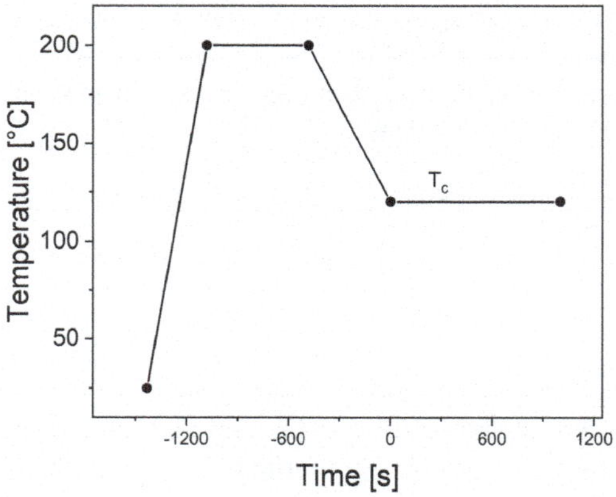

Fig. 1. Thermal protocol adopted for the crystallization experiments.

The sample thickness was set at 350 μm. All samples were melted at 200 °C and kept at this temperature for 10 min in order to cancel their thermo-mechanical history. Subsequently, temperature was lowered at 10 °C/min until the established crystallization temperature. A nitrogen flow was used in order to avoid degradation during the experiments.

For the nucleation density the number of nuclei in the observation window and for the growth rate the radius of the spherulites during time were considered.

3 Experimental Results

Rheology
For rheological characterization, the frequency sweep tests were performed with a stress of 250 Pa in a range of frequency between 0.1 and 100 rad/s. The results of the analysis at different temperatures are shown in Fig. 2, which represents the master curves of the PLA3251D at 120 °C.

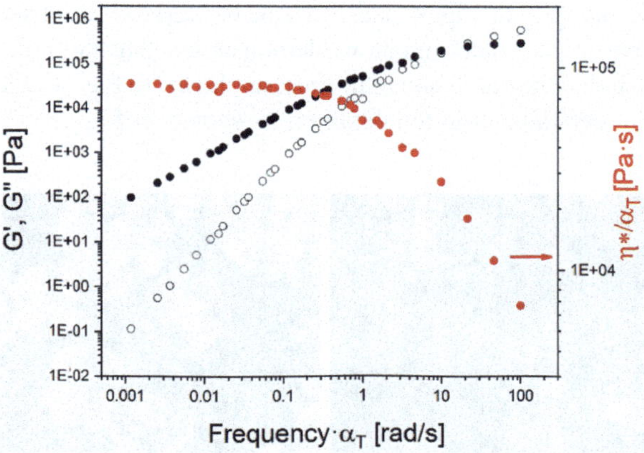

Fig. 2. Master curves of the PLA325D at 120 °C. Black empty circles represent G', black full circles represent G" and full red circles represent the viscosity.

The Williams-Landel- Ferry (WLF) equation describes the dependence of the thermal shift factor from the temperature [8]:

$$\alpha_T = exp\left(\frac{-C_1(T_c - T_0)}{C_2 + T_c - T_0}\right) \tag{1}$$

where T_c is the crystallization temperature,T_0 is the reference temperature and C_1 and C_2 are constant parameters. Figure 3 shows the dependence of the thermal shift factor (α_T) from the temperature evaluated from rheological tests.

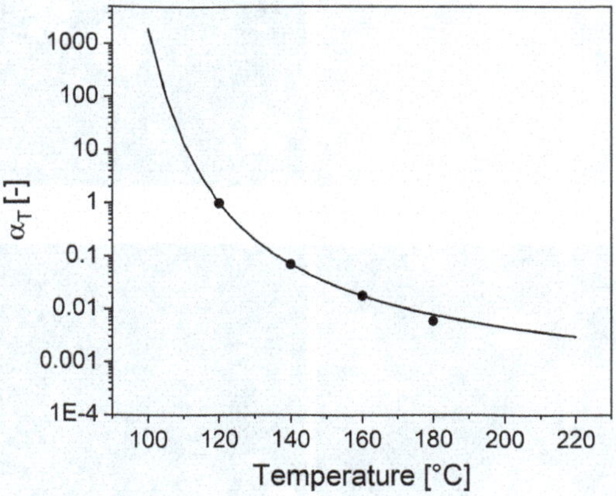

Fig. 3. Thermal shift factors (α_T) evaluated from rheological tests.

Optical Observations

The crystallization under quiescent conditions was analyzed measuring nucleation density, e.g. the number of nuclei in the observed volume, and the growth rate of the nuclei.

From the micrographs in Fig. 4 and 5 it can be possible to observe a spherulitic morphology. Experimental determinations demonstrated that the number of nuclei in the observation window did not change in time, as shown in Fig. 4, while their number decreases with the crystallization temperature, as shown in Fig. 5.

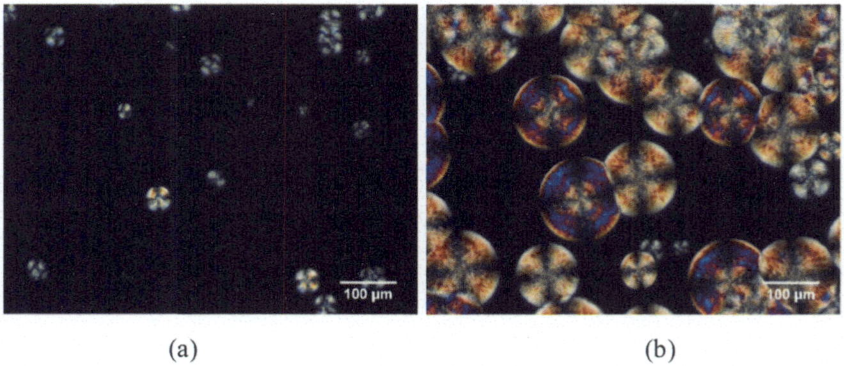

Fig. 4. Optical micrograph of a sample crystallized at 120 °C after 457 s (a) and 1297 s (b) after the reaching of the crystallization temperature.

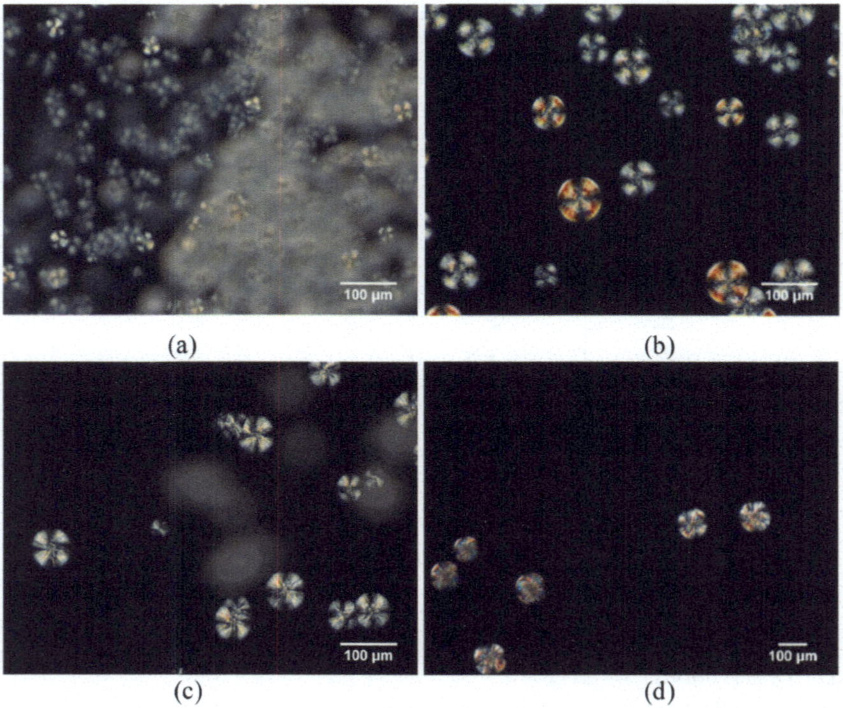

Fig. 5. Optical micrograph of a sample crystallized at different crystallization temperatures: 115 °C (a), 120 °C (b), 125 °C (c) and 130 °C (d).

The observations align with the results of nucleation in quiescent conditions of iPP reported in literature [9]. Figure 6 reports the nucleation density and growth rate in quiescent conditions. The nucleation density (*N*), evaluated as the number of nuclei in the observed volume, decreases with increasing crystallization temperature. The growth rate curve (*G*) is characterized by a double-bell shaped curve as conseguence of the polymorphism of the PLA. The low-temperature behaviour is due to the α' phase and the high-temperature behaviour depends on the presence of the α phase, as reported in different studies [10, 11].

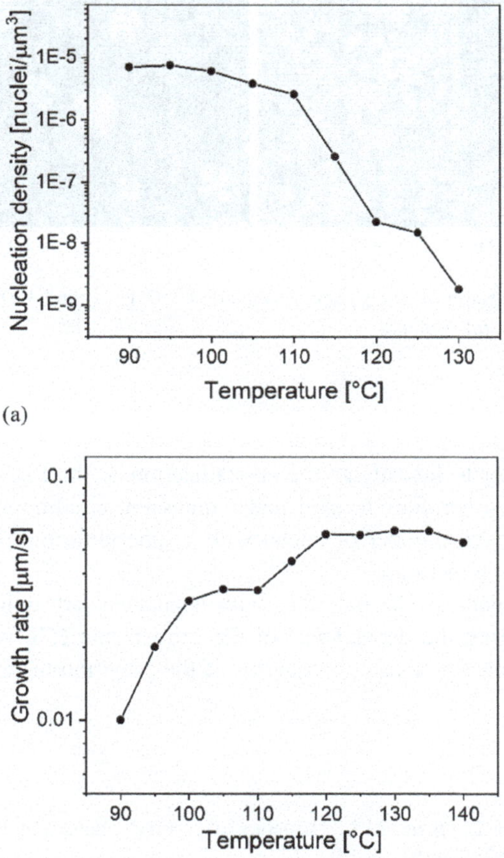

(a)

(b)

Fig. 6. Results of nucleation density (a) and growth rate (b) vs crystallization temperature in quiescent condition.

The morphology of the crystalline structures formed both in case of α and α' phase is the same, as can be observed in Fig. 7, where micrograph collected at 105 °C, temperature at which the α' phase can be found, and 120 °C, at which the α phase can be found, are reported. The difference in number and size is uniquely due to the crystallization temperature. The unit cell in both types of crystalline modifications α and α' is similar, but the segments of the helical chain in α' phase present a conformational disorder. For this reason, the phase α' can be easily transformed into an α form by heating.

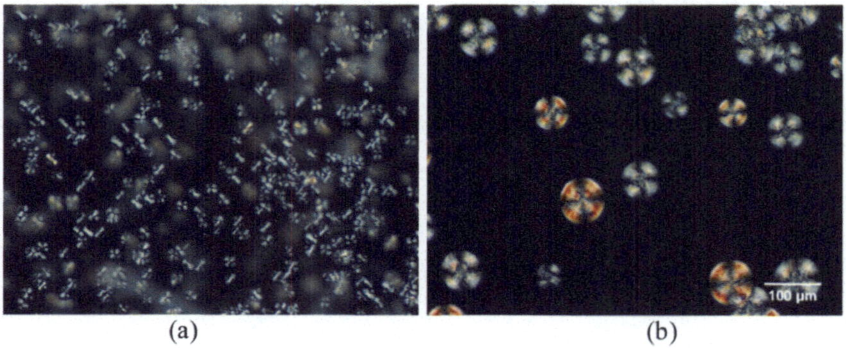

Fig. 7. Optical micrograph of a sample crystallized at 105°C (a) and 120°C (b), corresponding respectively to the α' and α phase

4 Conclusions

In this work, in order to investigate the crystallization of the PLA 3251D, nucleation and growth rate of spherulites formed under quiescent conditions were analyzed. In these conditions a fixed number of nuclei with a spherulitic morphology appeared at each crystallization temperature.

The nucleation density (*N*) increases with decreasing crystallization temperature. The curve representing the dependence of the growth rate (*G*) from the temperature presents a double-bell shape as consequence of the polymorphism of the PLA.

References

1. Androsch, R., Di Lorenzo, M.L.: Kinetics of crystal nucleation of poly(l-lactic acid). Polymer **54**(26), 6882–6885 (2013)
2. Pantani, R., et al.: Foam injection molding of poly(lactic acid) with physical blowing agents. In: Proceedings of Pps-29: The 29th International Conference of the Polymer - Conference Papers, vol. 1593, pp. 397–400 (2014)
3. Volpe, V., et al.: Effect of mold opening on the properties of PLA samples obtained by foam injection molding. Polym. Eng. Sci. **58**(4), 475–484 (2018)
4. Refaa, Z., Boutaous, M., Siginer, D.A.: PLA crystallization kinetics and morphology development. Int. Polym. Process. **33**(3), 336–344 (2018)

5. Gorrasi, G., Pantani, R.: Effect of PLA grades and morphologies on hydrolytic degradation at composting temperature: assessment of structural modification and kinetic parameters. Polym. Degrad. Stab. **98**(5), 1006–1014 (2013)
6. Laura, D.L.M.: Determination of spherulite growth rates of poly(l-lactic acid) using combined isothermal and non-isothermal procedures. Polymer **42**, 9441–9446 (2001)
7. Saeidlou, S., et al.: Poly(lactic acid) crystallization. Prog. Polym. Sci. **37**(12), 1657–1677 (2012)
8. Williams, M.L., Landel, R.F., Ferry, J.D.: The temperature dependence of relaxation mechanisms in amorphous polymers and other glass-forming liquids. J. Am. Chem. Soc. **77**, 3701–3707 (1955)
9. Coccorullo Ivano, P.R.: Titomanlio giuseppe, spherulitic nucleation and growth rates in an iPP under continuous shear flow. Macromolecules **41**(23), 9214–9223 (2008)
10. Lorenzo, M.L.D.: Crystallization behavior of poly(l-lactic acid). Eur. Polym. J. **41**, 569–575 (2005)
11. Androsch, R., Di Lorenzo, M.L.: Effect of molar mass on the α'/α-transition in poly (l-lactic acid). Polymer **114**, 144–148 (2017)

Modeling of Bio and Nanomaterials

Exploring Cation Mediated DNA Interactions Using Computer Simulations

Weiwei He[1,2] and Serdal Kirmizialtin[1(✉)]

[1] Chemistry Program, Science Division, New York University Abu Dhabi, Abu Dhabi, United Arab Emirates
serdal@nyu.edu
[2] Department of Chemistry, New York University, New York, NY 10003, USA

Abstract. Ion mediated interactions between nucleic acid helices are essential for their efficient packaging within tight spaces such as viral capsids, and nucleosomes. Understanding the fundamental rules governing these interactions is the key to design engineering tools in different length scales to achieve supramolecular architectures, leading to novel therapeutics, biosensors, and catalysis. As one of the building blocks for biology and biotechnology, DNA deserves special attention. However, the underlying physical principles governing DNA interactions with itself and its environment are still lacking. In this study, the mechanism of DNA attraction in the presence of divalent ions is investigated. The counter ion distributions and the conformational ensemble of parallel DNA strands are explored by conventional molecular dynamics and metadynamics simulations. Metadynamics simulation allows computing the free energy surface, providing a thermodynamic view to the DNA-DNA interaction. Our analysis reveals a strong correlation between the structure of the counterion atmosphere and inter-DNA interactions. The interaction energy is found to be mediated by the ions that bridge between the two DNA strands. In summary, computer simulations offer unprecedented detail into the dynamics and thermodynamics of DNA interactions that can potentially help to understand the biological DNA and also develop novel DNA based nanotechnologies.

1 Introduction

DNA is the biomolecule that stores and transmits genetic information. In addition to its role in biology, DNA also serves as a building block for nanotechnology [1, 2]. This field has had a remarkable impact on nanoscience and nanotechnology, especially on the molecular self-assembly, biosensor applications, and molecular machines in which DNA plays the key role.

DNA's ability to act both as a structural and functional unit makes a wide range of design strategies [3]. Based on these strategies, DNA nanotechnology can be classified into two groups. The first group is structural DNA nanotechnology (SDN) which makes use of thermal stability reached in equilibrium as the criterion to assemble structural motifs. Connecting units using base pairing and hybridization leads to controlled self-assembled structures [4–8]. The second group is Dynamic DNA nanotechnology (DDN) which operates at non-equilibrium conditions or involves substrate

S. Piotto et al. (Eds.): BIONAM 2019, LNBE, pp. 51–63, 2020.
https://doi.org/10.1007/978-3-030-47705-9_6

binding/unbinding events and conformational changes, just like in biological processes. DDN based nano-devices have revealed promising applications in life sciences [9, 10], and drug delivery [11, 12]. For example, a DNA 'spider' can walk along a predesigned track on a DNA origami [1] where walkers transport cargo along the track. While some DNA motors move by rolling and achieve a speed which is nearly three orders of magnitude faster than conventional DNA motors based on walk [13]. These motors are powered by enzymes. For example, RNase H is used in Ref [13] to hydrolyze the hybridized RNA.

Another direction of the dynamic DNA nanotechnology is biosensors. DNA is a promising material as a sensor because its structure is sensitive to the environment and the binding of cofactors. DNA based sensors have been proposed for sensing temperature, pH, concentration, and crowding [14–18]. The molecular mechanism on how it achieves these roles is hidden in its unique structure, dynamics and its interaction with its environment, making the fundamental research crucial.

Computer simulations and modeling provide great insights into the dynamics and assembly of DNA based technologies. Among the methods available, coarse grained simulations have been widely used [19–21]. These methods allow accessing the experimental timescales and help to understand the inner workings of DNA motors. The limitation comes when coarse graining fails to incorporate the necessary details. The intricate balance between the solvent fluctuations, ion correlations, and sequence effects often make all atom simulations necessary to study DNA motors.

DNA interactions are formulated from a theory of electrostatic interaction between helical molecules that possess different charge patterns [22]. Osmotic stress method has been utilized to investigate the forces between DNA pairs forming arrays [23]. Based on these studies, genomic DNA shows repulsive forces in the presence of monovalent and divalent ions. However, in the presence of poly-valent ions (valence ≥ 3) such as spermine, spermidine, cobalt-hexamine, poly-lysine, and so on, DNA condensation occurs. The governing forces favoring these interactions are likely linked to their ability to better screen the electrostatic repulsions between DNA strands. Interestingly, its cousin, double-stranded RNA helices, resists condensation at the same conditions [24], suggesting that a bare electrostatic argument is insufficient to understand these phenomena. The duplex topology likely plays a role in modulating these interactions. Hydration is also an important key player in intermolecular interactions. Recent studies show evidence that DNA-DNA interactions are sequence specific. Based on computer simulations, a bridging ion mechanism is proposed to account for DNA condensation [25]. The same study also explains sequence specificity by preferential binding of poly-cations to the DNA surface. By designing sequences of repeating bases, the work of Srivastava et. al. [26] show that this preferential binding can also turn into attractive forces even in the presence of divalent cations.

The focus of this work is to study the interactions of a DNA homopolymer made of poly(dA):poly(dT) in atomic detail. To study the DNA and its interactions in atomic detail, we employ computer simulations. In the first part, our focus is on characterizing the DNA counter ion atmosphere formed in $MgCl_2$ solution. In the second part, we discuss the DNA-DNA interactions that lead to DNA condensation in the presence of divalent cations. We investigate the role of counter ions in modulating these interactions. We propose a possible mechanism to explain recent experimental findings.

2 Methods

We built the B-form DNA of $(dA)_{20}$-$(dT)_{20}$ using Nucleic Acid Builder (NAB). DNA refers to the double-stranded $(dA)_{20}$-$(dT)_{20}$ in this study. The DNA was then centered in a simulation box with initial dimensions of 11.8, 11.8, 6.8 nm. The structure was solvated with water molecules, Mg^{2+} and Cl^{-} ions. MD simulations were carried out using GROMACS suite of programs [27]. Amber99sb_parmbsc0 force field [28] represented the DNA interactions. To model water, we used TIP3P [29] and for ions, Yoo and Aksimentiyev parameters were employed [30]. The nonbonded interactions were treated with a cut-off of 1.1 nm with dispersion corrections. The PME method [31] was implemented to calculate electrostatic interactions. The periodic boundary conditions were applied in all directions. The covalent bonds were restrained by LINCS algorithm [32]. Equations of motion was integrated using Leap Frog algorithm [33] with a time step of 2 fs.

After geometry optimization (5000 steps), the system was equilibrated for 2.5 ns in isothermal-isobaric ensemble (NPT) by keeping the temperature constant at around 300 K using velocity scaling [34]. The pressure was kept at 1 bar with Parinello-Rahman barostat [35]. During this time, the volume of the system reached equilibrium. Last frame of the simulation was saved and used as an initial coordinate for the solvent equilibration. Following the volume equilibration, 120 ns long solvent equilibration was employed in the canonical ensemble (NVT). During this process, all heavy atoms of the DNA were harmonically restrained with 1000 kcal/mol/Å^2 force constant, while water and ions were allowed to sample the simulation box freely. The last frame of the simulation was picked as the initial coordinate for the production runs. Employing unrestrained simulations where ion distributions were sampled for 200 ns, we recorded the data for every 2 ps for further analysis.

Later, we studied the interactions of DNA strands aligned in parallel. Duplexes of DNA were placed in a cubic box parallel to the z-axis (Fig. 1A). To avoid end effects, the length along the z-axis was adjusted such that the DNAs were extended to infinite length under periodic boundary conditions. The same settings and equilibration procedure discussed in the single duplex simulations above were employed for the preparation of equilibrium structures of the DNA pairs.

With this setup, we studied the thermodynamics of DNA-DNA interaction. To represent the results in a concise way, we reduced the description of the complex dynamics to a restricted number of variables. The fluctuations along these variables were slow; hence, the sampling was nonergodic on the timescale of the simulations that could be afforded. To overcome the timescale problem, we used metadynamics approach [36] that relies on the introduction of a dynamically changing external bias potential $V(\mathbf{s})$ where $\mathbf{s} \equiv s(\mathbf{R})$, is a set of collective variables (CV) that is a nonlinear function of the atomic coordinates, $\mathbf{R}$. Bold faces represents vectors in this text.

In our choice of metadynamics approach, the fluctuations of the CVs are enhanced by employing the biasing potential

$$V(s) = -\left(1 - \frac{1}{\gamma}\right)F(s) \tag{1}$$

where $F(s)$ is the free energy surface (FES) of the ergodic system which in principle is obtained from unbiased simulations as:

$$F(s) = -k_B T \lim_{t\to\infty} \log(P(s,t)). \tag{2}$$

Here, $P(s,t)$ is the normalized probability distribution function. The ultimate purpose of metadynamics is to estimate Eq. 2 using less amount of computational time. This is achieved by the biasing factor, γ given in Eq. 1. A positive constant γ that is greater than 1 effectively scales down the barrier heights such that the transitions between the metastable states in the FES are enhanced. This can be seen from the following arguments. For the unbiased simulation, the probability distribution is

$$P(s) = \frac{e^{-\frac{F(s)}{k_B T}}}{\int e^{-\frac{F(s)}{k_B T}} ds} \tag{3}$$

while addition of the biasing potential modifies the probability distribution function as:

$$P_V(s) = \frac{e^{-\frac{[F(s)+V(s)]}{k_B T}}}{\int e^{-\frac{[F(s)+V(s)]}{k_B T}} ds} \tag{4}$$

Combining Eqs. 1, 3 and 4, one obtains:

$$P_V(s) = \frac{[P(s)]^{\frac{1}{\gamma}}}{\int [P(s)]^{\frac{1}{\gamma}} ds} \tag{5}$$

which links the biased and unbiased probability distributions. From Eq. 5 one can see that as biasing factor $\gamma \to \infty$, $P_V(s) = \frac{1}{\int ds} = \text{const}$. That is to say, the underlying FES of the biased simulation becomes flat, making all **s** values equally accessible, resulting in the rapid exploration of energy landscapes.

Once the sampling is complete in the biased ensemble, the unbiased FES can be recovered up to a constant term $F(s) = -k_B T \ln P_V(s) - V(s)$. This is achieved by simply reweighting the biased simulation using the relation:

$$P(\mathbf{R}) \propto P_V(\mathbf{R}) e^{\frac{V(s(\mathbf{R}))}{k_B T}} \tag{6}$$

In metadynamics, the biasing kernel is a multidimensional Gaussian function localized in CV space **s**. In the case of Well-tempered metadynamics (WTMD) the kernel iteratively changes during simulation as:

$$V_n(\boldsymbol{s}) = V_{n-1}(\boldsymbol{s}) + We^{-\frac{1}{2}\left\|\frac{\boldsymbol{s}-\boldsymbol{s_n}}{\boldsymbol{\sigma}}\right\|^2} e^{\left[-\frac{1}{\gamma-1}\frac{V_{n-1}(\boldsymbol{s_n})}{k_B T}\right]} \tag{7}$$

where W and $\boldsymbol{\sigma}$ are pre-determined parameters representing the height and the variance of the Gaussian that is going to be deposited at each iteration.

Shown in Fig. 1B is the schematic demonstration of the two collective variables used in this study. Namely, we used the center to center distance of the helices, denoted as *d*. As a second CV, we used the relative azimuthal angle. The first CV controls the distance between the two DNA while the second controls the relative orientation of the major and minor grooves between the two helices, a parameter crucial for making comparisons with Kornyshev-Leikin mechanism [22].

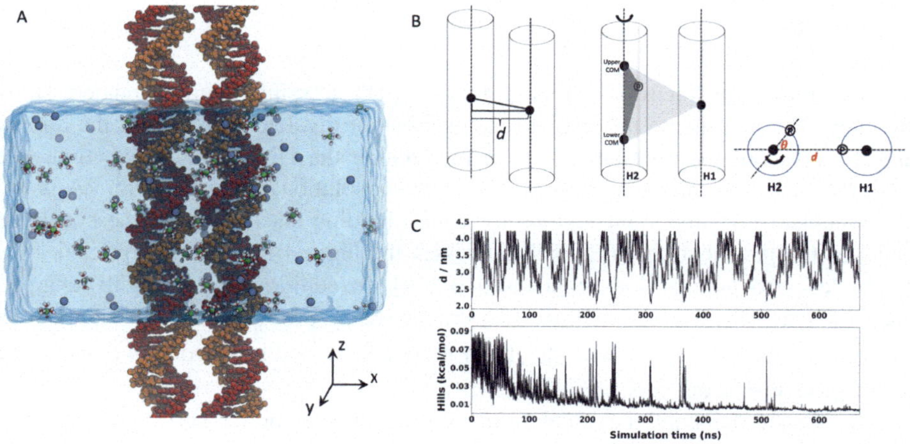

Fig. 1. Two parallel DNA duplexes aligned in the *z* direction. (A) Molecular simulation set up. (B) The order parameters chosen to represent the conformational space. *d* represents the center of mass (COM) in the (*x*, *y*) plane between the two helices (H1 and H2), *θ* represents the rotation of H2 with respect to H1. We restrained H1 orientation in order to compute the azimuthal angle difference easily. The angle *θ* is a dihedral angle from the COM of H1 to the upper and lower COMs of H2, and to an arbitrary phosphate atom on the H2. (C) Time evolution of the collective variable *d* and Gaussian hills monitoring the convergence during Well Tempered Metadynamics simulation.

After the equilibration, as summarized above, the metadynamics method was applied using the Plumed plugin [37]. During the simulation, the translational and rotational motion of duplex, H1, was removed using enforced rotation implementation

[38] in Gromacs. To keep H1 and H2 duplexes parallel, a flat bottom potential implemented in Plumed was introduced. The constraint allowed H2 to lie along the z-direction, a necessary condition to ensure the DNAs to be parallel as in the experimental conditions. The long-range intermolecular interactions are known to decay to zero at long intermolecular distances. To efficiently sample the conformational space, we introduce an upper wall potential to confine the two DNA duplexes between $0 < d < 4.2$ nm. During the well-tempered metadynamics, Gaussians were deposited every 1 ps with a height starting at $W = 0.08$ kcal/mol and we gradually decreased it by a bias factor of 8. The widths of the Gaussians were set to $\sigma_d = 0.05$ nm for the inter-DNA distance and $\sigma_\theta = 0.06$ *rad* for the dihedral angle, respectively. Figure 1C. shows enhanced fluctuations in d during metadynamics simulation. To ensure convergence, we monitored the time evolution of the Gaussian heights deposited during the simulation length of about 700 ns. The metadynamics simulation was considered converged when the heights of the deposited Gaussians in the space of the CVs decayed and reached a threshold value < 0.005 kcal/mol (Fig. 1C below).

3 Results

In this study, our focus is on poly(dA):poly(dT) forming double-stranded DNA structure. We investigate the spatial organizations of divalent ions around the duplex and the result of these organizations in modulating the DNA-DNA interaction. We study DNA cation interactions in $MgCl_2$ aqueous solution. Shown in Fig. 2A is the DNA duplex with important regions for cation binding sites marked. The backbone includes the phosphate group which possesses the highest negative charge and, as a result, leads to counter ions condensation [39]. In addition to the highly charged backbone, the helical structure of the two strands creates grooves. From the simulations we compute the average widths. The major groove has a width of about 14 Å while the minor groove is about 5 Å.

As ions are dynamic, they explore many different sites on the surface of the DNA duplex in addition to sampling positions at the volume occupied primarily by the water. It is reasonable to look into the cation's average behavior. This is achieved by monitoring the position of each cation in time. We show the average Mg^{2+} ion distribution around the DNA duplex in Fig. 2. The distribution from different projections provides a detailed description of the spatial organization of these ions. The concentration profile in a cylindrical coordinate $c(\rho)$ serves as an ideal coordinate to study helices. Here, ρ represents the distance from the center of the helix (Fig. 2B *left*). Details of computing $c(\rho)$ from MD trajectory can be found elsewhere [40]. We focus more on the results of this analysis here.

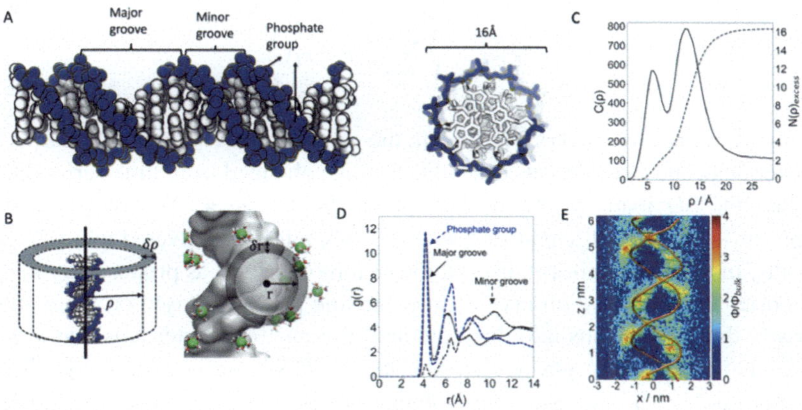

Fig. 2. (A) The structure of double stranded DNA with important regions for ion distributions are marked. (B) The definition of ρ and r. (C) Cylindrical concentration profile along the long axis of dsDNA with dashed line represents the excess number of ions accumulated as the distance from the center of the DNA, r changes. (D) Radial distribution function of magnesium ions from the major group atoms (black), phosphate oxygens (blue dashed) and minor groove atoms (black dashed). (E) Magnesium ion localization projected on to the (x, z) plane and superimposed onto the duplex structure.

The cylindrical concentration profile shows a well-defined Mg^{2+} ion structure around the helix with two distinct peaks, representing two layers of ion atmosphere around the helical duplex. This is similar to the A-form RNA duplex studied [41] with the only difference that the height of the first peak in DNA is smaller, suggesting weaker major groove binding. To further characterize these two peaks, we compute ion distributions in radial coordinates (Fig. 2D). The definition of radial distance from an arbitrary group is sketched (Fig. 2B *right*). Using radial distribution functions (RDF), we quantify the preferential binding of Mg^{2+} into major and minor groves as well as to the phosphate group. RDF analysis indicates that the major groove and the phosphates show strong cation binding. The peak for minor groove located at $r \approx 10$ Å indicates that direct binding to the minor grooves is not possible for magnesium ions. This is likely due to the narrow width of the minor grooves in comparison to the size of a hexahydrated magnesium of diameter ≈ 8 Å. Based on this analysis the first peak located at around $\rho \approx 5$ Å in Fig. 2C corresponds to the major groove binding, while the peak at $\rho \approx 13$ Å corresponds to the Mg^{2+} ions around the phosphate groups. After the second peak, the concentration profile shows a decay reaching to the bulk concentration value of 100 mM at $\rho \approx 20$ Å. From this analysis, we conclude that the apparent diameter of a B-form DNA is about 20 Å in $MgCl_2$ solution.

To visualize the structure of the cation cloud surrounding the duplex, we computed magnesium density profiles in cartesian coordinates. For an arbitrary point in space $\boldsymbol{r} \equiv (x, y, z)$, Mg^{2+} ion number density, $\phi(x, y, z)$, can be computed from the trajectory as:

$$\phi(\boldsymbol{r}) = \left\langle \frac{1}{V} \sum_i \delta(\boldsymbol{r} - \boldsymbol{r}_i) \right\rangle \tag{8}$$

where, the sum goes over every Mg ion i in the simulation box of volume V. $\delta(x)$ is the Kronecker delta and $\langle\,\rangle$ represents ensemble average obtained from time-series data. We zoom into the 3D map by slicing $\phi(\boldsymbol{r})$ along the x-z plane. We choose a surface centering at $-4\,\text{Å} \leq y \leq 4\,\text{Å}$ that divides the DNA into two halves. Figure 2E shows our results. Instead of a uniform distribution of ions densities, as proposed by the mean field theories for DNA, we observe distinct binding patterns. Based on these analyses we classify the counter ions into three different groups: (i) counter ions that discreetly bound to the electrostatic pockets created by the DNA surface topology, (ii) those bound but mobile in the apparent volume of the macroion $8\,\text{Å} < \rho < 20\,\text{Å}$, and (iii) counter ions that move at the outside of the DNA apparent volume ($\rho > 20\,\text{Å}$). Most emphasis has been invested in the last two groups when DNA-DNA interactions are concerned in mean-field approaches [39] as they comprise the majority of the ions in the polyelectrolyte system. The discrete binding, on the other hand, gives rise to sequence specificity in nucleic acid interactions.

In addition to providing atomic details, computer simulations also allow us to test the accuracy of the mean field theories. From the $c(\rho)$, one can compute the excess number of Mg ions condensed onto the DNA. The excess number can simply be obtained as:

$$N_{\text{excess}} = 2\pi l \int_0^{\sigma_{bulk}} \sigma' [c(\sigma') - c_{bulk}] d\sigma' \tag{9}$$

where, σ_{bulk} is the distance from the center of the DNA to where the bulk concentration, c_{bulk} is reached. This number in our case is $\approx 20\,\text{Å}$. l is the length of the DNA. We find

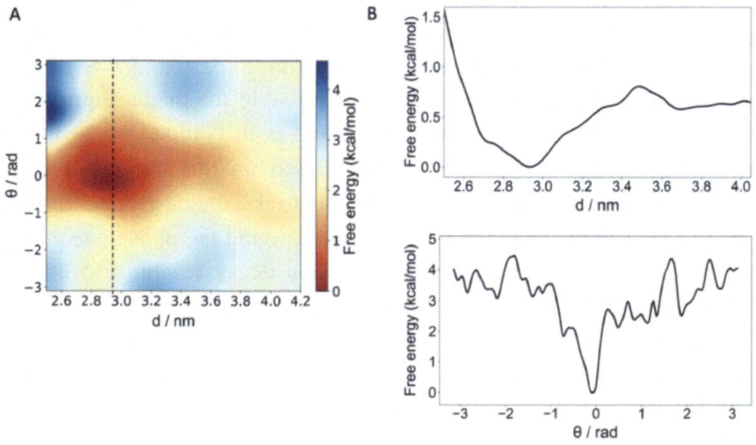

Fig. 3. (A) Free Energy Surface (FES) projected on the collective variables (d, θ). Dashed line is guide to eye showing the slice of the FES taken in angle space. (B) The free energy change projected on d (above) and FES projected on θ (below).

N_{excess} to be $+33e$ in 100 mM $MgCl_2$ that corresponds to a dissociation constant of $33/38 \approx 0.86$, which is estimated to be 0.88 in counter ion condensation theory [39]. The agreement between the mean field theory and atomic simulations is remarkable.

Now, we turn our attention to the mechanism of DNA-DNA interaction in divalent ions. We look into the Mg ion distributions for a pair of dsDNAs aligned in parallel. We compute the free energy of inter-strand interactions modulated by cation distribution. The sequence we choose shows attraction in earlier studies [26]. Therefore, it poses a challenge to the mean field theoretical studies which predict repulsion for the same conditions [22].

We projected the free energy surface explored by Well-tempered Metadynamics (WTMD) onto the two collective variables (d, θ). Figure 3 summarizes our results. The global free energy minimum is observed at (2.87 nm, −0.1 rad) indicating an attraction for the two DNA pairs. To quantify this attraction, we divide the conformational space into two regions. The bound state is defined as $d < 3.4$ nm. The unbound state is when DNAs sample distances $d \geq 3.4$ nm. The binding free energy is the free energy difference between the bound and unbound states as $\Delta F_{B,U} = F_U - F_B$ which we can obtain from the FES by:

$$\Delta F_{B,U} = -k_B T \, ln \frac{\int_B ds e^{-\frac{F(s)}{k_B T}}}{\int_U ds e^{-\frac{F(s)}{k_B T}}}. \tag{10}$$

The Eq. 10 gives rise to $\Delta F_{B,U} \approx -0.4$ kcal/mol. This finding is consistent with recent experiments [42]. Interestingly, FES minima suggest a preferential orientation for the two DNA pairs. At the limit of far DNA distances, the two helices sample angular space equally. At $d < 3.0$ nm, the orientation of the helices is locked to the range $-1 < \theta < 1$ rad. The optimum angle corresponds to an orientation where the minor groove of H1 shows proximity to the major groove of H2 (Fig. 4B).

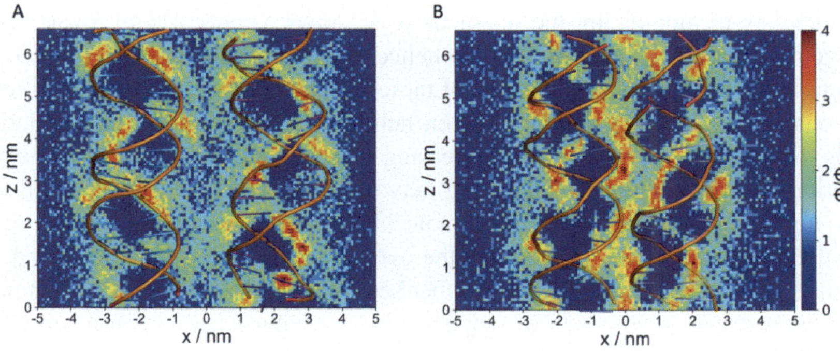

Fig. 4. Magnesium ion density profile when projected onto the (x, z) plane superimposed to the duplex structure. To focus on the counter ions bound to the DNA surface, a thickness of $-4\,\text{Å} \leq y \leq 4\,\text{Å}$ is used. Cation density, (A) when the two helices are at d = 4.2 nm, (B) when helices are at d = 2.9 nm.

To understand the role of cations to the adhesion phenomenon observed, we analyze the spatial distribution of cations around the duplexes. Similar to Fig. 2E, we project the 3D cation density onto (x, z) plane. This time, the two DNA pairs are analyzed for the pairs in different inter-DNA distances (Fig. 4). When the two helices are far apart ($d \sim 4.2$ nm) (Fig. 4A) the cation clouds of parallel helices are observed to be weakly coupled. Ion densities show similar patterns to the isolated DNA (Fig. 2E), suggesting that at this distance the cation cloud is unperturbed. Upon the approach of the two pairs ($d \sim 2.9$ nm), we observe a strong re-organization of the cation distributions at the interface (Fig. 4B). Interestingly, more discrete binding sites are formed. By averaging the number of ions between the interface we found about 2 extra cations that bridge between the two helices. We believe that these ions help to provide extra stability by reducing the electrostatic repulsions between negatively charged DNA surfaces and also by creating charge fluctuations.

4 Conclusions and Discussion

Genomic DNA shows repulsion in the presence of alkali earth metal ions of valence ≤ 2. Homopolymeric double-stranded DNAs show condensation even in the presence of divalent ions. We show that the spatial organization of magnesium ions around a double-stranded DNA can be probed by molecular simulations. Cations show localization around major grooves and phosphates, giving rise to charge patterns on the DNA surface. Our approach allows us to benchmark the counterion condensation theory which we find to be in agreement with our simulation in the case of divalent ions.

For poly(dA)-poly(dT) we observe an adhesion between helices, an observation consistent with the recent study [26]. Besides, we demonstrate that the free energy landscape of inter-DNA conformational space shows distance and orientation dependence. The spacing and helical orientations are coupled at short distances.

Molecular simulations of cation distributions around DNA helped us to uncover the role of cations in modulating the interactions of poly(dA):poly(dT) in divalent ions. The electrostatic field created by the two helices resulted in additional discrete cation binding sites at the interface. The effect of these new sites is three folds (i) they recruit more ions from solution to bridge between helices (ii) they increase the translational entropy of the trapped cations by increasing multiplicity, (iii) they create charge fluctuations due to bridging ions hopping between the sites at the interface. All these effects favor attraction between macromolecules. Further studies are underway to assess the contribution of each factor to the overall free energy change observed. The generality of the conclusions drawn for other sequences and divalent cation conditions will be investigated. The impact of sequence, helix structure and cation fluctuation in stabilizing the biomolecular complexes demands further investigations.

This intriguing physical phenomenon has important implications in biology and biotechnology. For example, these interactions can be engineered to assemble and control the motion of DNA based molecular motors. The molecular details of cation-DNA interactions will help to design sequences with controlled intermolecular interactions and orientations.

Molecular simulations have been providing a fundamental understanding of complex biomolecular interactions. Despite the invaluable insights they offer, the current approach of molecular simulations sometimes fails to capture the accurate intermolecular forces in nucleic acid systems. Unfortunately, many of the results reported in simulations are force-field dependent, precluding their use in the absence of experimental data blindly. Currently, most of the effort in forcefield optimization is limited to the torsional angles. The non-bonded terms of widely used forcefields fail to capture inter-DNA interactions measured experimentally. For that purpose, we used a modified version of Amber forcefield [25] and ion parameters that represent dsDNA-dsDNA interactions accurately. However, the effect of these modifications to other nucleic acid topologies such as junctions, loops has not yet been tested. Further studies are necessary to optimize nucleic acid forcefield parameters for more general use. Another challenge in molecular simulations is the time scale issue, as we highlighted in the methods section. We used Well-tempered metadynamics to overcome this challenge in this study. Defining collective variables a priori is a limitation in our current approach. This can become a problem when the dynamics of the macromolecule is not trivial and/or the chain conformational dynamics are coupled with slow solvent degrees of freedom. Further studies are underway to determine the collective variable on the fly for nucleic acid – ion molecular systems.

Acknowledgments. We thank NYUAD for their generous support for the research. This research was carried out on the High-Performance Computing resources at New York University Abu Dhabi and the AD181 faculty research grant to S.K. Dr. Nawavi Naleem for careful reading of the work and Dr. Amit Srivastava for discussions.

References

1. Seeman, N.C., Sleiman, H.F.: DNA nanotechnology. Nat. Rev. Mater. **3**, 1–23 (2017)
2. Yurke, B., et al.: A DNA-fuelled molecular machine made of DNA. Nature **406**, 605–608 (2000)
3. Wang, W., et al.: Complex wireframe DNA nanostructures from simple building blocks. Nat. Commun. **10**, 1–8 (2019)
4. Winfree, E., et al.: Design and self-assembly of two-dimensional DNA crystals. Nature **394** (6693), 539–544 (1998)
5. Seeman, N.C.: DNA enables nanoscale control of the structure of matter. Q. Rev. Biophys. **38**(4), 363–371 (2005)
6. Seeman, N.C.: From genes to machines: DNA nanomechanical devices. Trends Biochem. Sci. **30**(3), 119–125 (2005)
7. Seeman, N.C.: Nanotechnology and the double Helix. Sci. Am. **290**(6), 64 (2004)
8. Aldaye, F.A., Palmer, A.L., Sleiman, H.F.: Assembling materials with DNA as the guide. Science **321**(5897), 1795–1799 (2008)
9. Sherman, W.B., Seeman, N.C.: A precisely controlled DNA biped walking device. Nano Lett. **4**(7), 1203–1207 (2004)
10. Omabegho, T., Sha, R., Seeman, N.C.: A bipedal DNA Brownian motor with coordinated legs. Science **324**, 67 (2009)

11. Vries, J.D., et al.: DNA nanoparticles for ophthalmic drug delivery. ScienceDirect **157**, 98–106 (2017)
12. Hu, Q., et al.: DNA nanotechnology-enabled drug delivery systems. Chem. Rev. **119**(10), 6459–6506 (2019)
13. Yehl, K., et al.: High-speed DNA-based rolling motors powered by RNase H. Nat. Nanotechnol. **11**(2), 184 (2015)
14. Murade, C., Shubeita, G.T.: A molecular sensor reveals differences in macromolecular crowding between the cytoplasm and nucleoplasm. Biophys. J. **116**(3), 315a (2019)
15. Ohmichi, T., et al.: DNA-based biosensor for monitoring pH in vitro and in living cells. Biochemistry **44**, 7125–7130 (2005)
16. Gareau, D., Desrosiers, A., Valleé-Belisle, A.: Programmable quantitative DNA nanothermometers. Nano Lett. **16**, 3976–3981 (2016)
17. Jonstrup, A.T., Fredsøe, J., Andersen, A.H.: DNA hairpins as temperature switches, thermometers and ionic detectors. Sensors **13**, 5937–5944 (2013)
18. Takahashi, S., et al.: Characterization of intracellular crowding environments with topology-based DNA quadruplex sensors. Anal. Chem. **91**, 2586–2590 (2019)
19. Ouldridge, T.E., Louis, A.A., Doye, J.P.K.: Structural, mechanical, and thermodynamic properties of a coarse-grained DNA model. J. Chem. Phys. **134**(8), 02B627 (2011)
20. Ouldridge, T.E., et al.: Optimizing DNA nanotechnology through coarse-grained modeling: a two-footed DNA walker. ACS Nano **7**(3), 2479–2490 (2013)
21. Ouldridge, T.E.: DNA nanotechnology: understanding and optimisation through simulation. Mol. Phys. **113**(1), 1–15 (2015)
22. Kornyshev, A.A., Leikin, S.: Electrostatic zipper motif for DNA aggregation. Phys. Rev. Lett. **82**(20), 4138–4141 (1999)
23. Rau, D.C., Lee, B., Parsegian, A.: Measurement of the repulsive force between polyelectrolyte molecules in ionic solution: hydration forces between parallel DNA double helices. Proc. Natl. Acad. Sci. U.S.A. **81**, 2621–2625 (1984)
24. Li, L., et al.: Double-stranded RNA resists condensation. Phys. Rev. Lett. **106**(10), 108101 (2011)
25. Yoo, J., et al.: Direct evidence for sequence-dependent attraction between double-stranded DNA controlled by methylation. Nat. Commun. **7**, 11045 (2016)
26. Srivastava, A., et al.: Sequence effects on Mg+2 Ion mediated DNA - DNA interactions. Biophys. J. **116**(3), 360a (2019)
27. Abraham, M.J., Murtola, T., Schulz, R., Páll, S., Smith, J.C., Hess, B., Lindahl, E.: GROMACS: high performance molecular simulations through multi-level parallelism from laptops to supercomputers. SoftwareX **1–2**, 19–25 (2015)
28. Zgarbova, M., et al.: Refinement of the sugar-phosphate backbone torsion beta for AMBER force fields improves the description of Z- and B-DNA. J. Chem. Theory Comput. **11**(12), 5723–5736 (2015)
29. Jorgensen, W.L., et al.: Comparison of simple potential functions for simulating liquid water. J. Chem. Phys. **79**(2), 926–935 (1983)
30. Yoo, J.J., Aksimentiev, A.: Improved parametrization of Li+, Na+, K+, and Mg2+ Ions for all-atom molecular dynamics simulations of nucleic acid systems. J. Phys. Chem. Lett. **3**(1), 45–50 (2012)
31. Essmann, U., et al.: A smooth particle mesh Ewald method. J. Chem. Phys. **103**(19), 8577–8593 (1995)
32. Hess, B.: P-LINCS: a parallel linear constraint solver for molecular simulation. J. Chem. Theory Comput. **4**(1), 116–122 (2008)
33. Van Gunsteren, W.F., Berendsen, H.J.C.: A leap-frog algorithm for stochastic dynamics. Mol. Simul. **1**(3), 173–185 (1988)

34. Bussi, G., Donadio, D., Parrinello, M.: Canonical sampling through velocity rescaling. J. Chem. Phys. **126**(1), 014101 (2007)
35. Parrinello, M., Rahman, A.: Crystal-structure and pair potentials - a molecular-dynamics study. Phys. Rev. Lett. **45**(14), 1196–1199 (1980)
36. Barducci, A., Bussi, G., Parrinello, M.: Well-tempered metadynamics: a smoothly converging and tunable free-energy method. Phys. Rev. Lett. **100**(2), 020603 (2008)
37. Bonomi, M., et al.: PLUMED: a portable plugin for free-energy calculations with molecular dynamics. Comput. Phys. Commun. **180**(10), 1961–1972 (2009)
38. Kutzner, C., Czub, J., Grubmueller, H.: Keep it flexible: driving macromolecular rotary motions in atomistic simulations with gromacs. Biophys. J. **102**(3), 171a (2012)
39. Oosawa, F.: Polyelectrolytes. Dekker, New York (1971)
40. Kirmizialtin, S., et al.: The ionic atmosphere around A-RNA: Poisson-Boltzmann and molecular dynamics simulations. Biophys. J. **102**(4), 829–838 (2012)
41. Kirmizialtin, S., Elber, R.: Computational exploration of mobile ion distributions around RNA duplex. J. Phys. Chem. B **114**(24), 8207–8220 (2010)
42. Qiu, X.Y., et al.: Inter-DNA attraction mediated by divalent counterions. Phys. Rev. Lett. **99**(3), 038104 (2007)

Harvesting Free Energy Landscapes in Biological Systems

Darren Wragg[1,2], Angela Casini[2], and Stefano Leoni[1(✉)]

[1] School of Chemistry, Cardiff University, Park Place, Cardiff CF103AT, UK
leonis@cf.ac.uk

[2] Department of Chemistry, Technical University of Munich, Lichtenbergstr. 4, 85748 Garching, Germany
angela.casini@tum.de

Abstract. A sophisticated mechanistic investigation of biological processes is the starting point for the calculation of accurate free energy figures, which enable comparison with experimental results. Applied to drug discovery, the use of molecular methods can guide the formulation and optimisation of novel drugs, specifically targeting molecular processes as they are learned from simulations. This is illustrated here based on two examples, the study of secondary DNA structures (G4s) as target for small gold-based molecules, and the investigation of the mechanisms of glycerol permeation *via* the membrane channel aquaglyceroporin-3 (AQP3). Both approaches leverage the enhanced sampling efficiency of accelerated molecular dynamics.

1 Introduction

The discovery of novel drugs is nowadays supported by different and complementary computational approaches [1]. In particular, the use of molecular dynamics (MD) techniques, a cornerstone of modern computational chemistry. This technique is instrumental to mechanistic elucidations, which are the starting point for a rational approach to drug design. The accurate predictions of trends in substrate-drug interactions, and the reliable computation of relevant quantities, such as binding energies, critically depend on the accuracy of available force field and on the sampling efficiency of the methods used. For biological systems, the current paradigm is dominated by so-called additive force fields, which guarantee substantial coverage of the chemical space of interest, including proteins, nucleic acids, lipids and carbohydrates. While the capacity of determining correct total energies, and thereupon interatomic forces is a prerequisite for a predictive approach, just propagating a given system in time may often not suffice for the calculation of correct binding energies. While some discrepancies may result from incorrect energy estimations due to the chosen force field, the main problem remains the overall sampling efficiency of molecular dynamics methods. The correct statistical weight of a substrate-drug pose results from the reconnaissance of all possible poses. Their sampling is however hampered by an intrinsic timescale problem. Since different poses are separated by activation energies, starting the

S. Piotto et al. (Eds.): BIONAM 2019, LNBE, pp. 64–77, 2020.
https://doi.org/10.1007/978-3-030-47705-9_7

simulation from one pose using straightforward MD will offer limited chances for the drug to find its way spontaneously to another minimum (pose), even an immediately adjacent one, on the timescale of a typical molecular dynamical simulation. Longer simulation times, albeit increasingly impractical, may allow the observation of a pose change, however, the event of interest will still be infrequent, i.e. rare, on the timescale of the simulation. This rare event scenario will simply result in the permanence of the system in long-lived metastable states, and sudden (if at all) changes of configurations. The calculation of populations and free energies requires, therefore, specific methods in order to fully elucidate mechanisms and extract the quantitative information required to support and direct drug discovery. Here, examples of the application of metadynamics to two biological systems will be discussed, including i) the mechanism of gold(I) N-heterocyclic carbene (NHC) compounds binding to DNA G-quadruplexes [2] and ii) the mechanisms of aquaglyceroporin-3 glycerol permeation [3], both tackled using metadynamics techniques [4].

2 The Mechanism of Gold(I) NHC Compounds Binding to DNA G-Quadruplexes (G4s)

DNA can adopt different structures from the canonical right-handed double helix. In particular, guanine-rich DNA sequences can form secondary structures termed G-quadruplexes (G4s) [5]. Therein, four guanine bases assemble into a pseudo-planar tetrad (G-quartet) which is held together by one or more nucleotide strands and stabilised by metal ions. Recent bioinformatics studies have shown that there are ca. 716,000 DNA sequences in the human genome, which can form G4 structures [6]. These sequences are present in telomeres and promoter regions of oncogenes. These non-canonical DNA structures have been the subject of intense study in the last decade due to their association with a number of biological processes such as telomere maintenance, gene regulation, and replication [7, 8].

Formation of the quadruplex structure in promoter regions of oncogenes can control transcription and hence the expression of the corresponding oncogenes [9]. Furthermore, stabilising G4s in telomeres results in indirectly inhibiting telomerase activity, affecting cell aging and cancer cells mortality [8]. In this context, G4s emerge therefore as promising targets for anticancer drug development, as they can be stabilised by small molecules with associated anticancer effects [10]. One example is the tri-substituted acridine derivative, BRACO-19, a telomeric G4 stabiliser, which displays *in vitro* anticancer activity in prostate cancer [11]. Despite the great advances in the development of G4 stabilisers, important challenges still remain. Understanding highly selective binding of small molecules to specific quadruplex structures over duplex DNA and other G4s is a task of top priority towards identifying the next generation of targeted anticancer drugs with reduced side effects.

Our pioneering work into G4 stabilisers identified small-molecule organometallic Au(I) compounds with NHCs ligands, as potent, selective stabilisers of telomeric G4s

[12, 13]. We recently reported on a bis-NHC gold(I) complex - [Au(9-methylcaffein-8-ylidene)$_2$]$^+$ (AuTMX$_2$, Fig. 1) - containing caffeine as NHC ligand [12]. The compound showed selective anti-proliferative effects against human ovarian cancer cells over non-tumorigenic ones. X-ray diffraction analysis of the adduct formed by AuTMX$_2$ and Tel23, a 23-nucleotide telomere repeat sequence, indicated that the compound binds non-covalently between neighbouring G4 units [13].

In order to rationally optimize the metallodrug selectivity for its biological target, *in silico* methods, including MD approaches, are highly valuable as they can elucidate both mechanisms and energetics underlying the ligand/target recognition process. Therefore, metadynamics was applied to evaluate the binding mode of AuTMX$_2$ with two different G-quadruplex structures, namely the human telomeric sequence *h*Telo (pdb 2HY9 [14]) and the *C-KIT*1 oncogene promoter sequence (pdb 4WO2 [15]). Mechanisms and binding energetics were compared with those calculated for the neutral mono-carbene complex [Au(9-methylcaffein-8-ylidene)I] (AuTMX-I, Fig. 1). To validate the accuracy of our calculations we also evaluated the G4 stabilization properties of the Au(I) compound using FRET (fluorescence resonance energy transfer) DNA melting assays.

Metadynamics can achieve a rapid reconnaissance of free energy minima, by sampling free energy surfaces along selected reaction coordinates, so-called collective variables (CVs), while a history-dependent potential discourages the calculation from revisiting already sampled, more probable regions. In this case, a distance and a pseudo torsion angle, monitoring the orientation of the molecule on the G4, represented the set of CVs.

Trajectories of AuTMX$_2$ on hTelo showed two possible binding sites (state I and II), with the first one (state I) having the lowest energy (ca. −37 kJ/mol). Figure 1 shows a multiple CV plot of the resulting free energy surface. Interestingly, state I shows two minima (a and b), corresponding to the same Au^+-K^+ distance (ca. 0.8 nm) but with two different torsion angles. AuTMX$_2$ is virtually in the same position above the tetrad but, since the gold complex can rotate around its centre, two different torsion angles are selected (see position of the caffeine ligands in Fig. 1).

A)

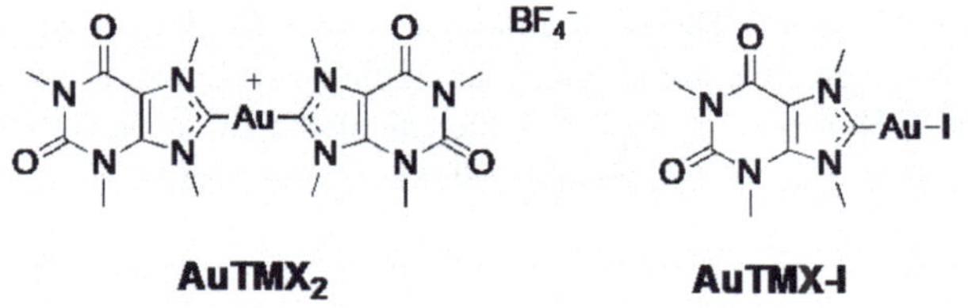

B)

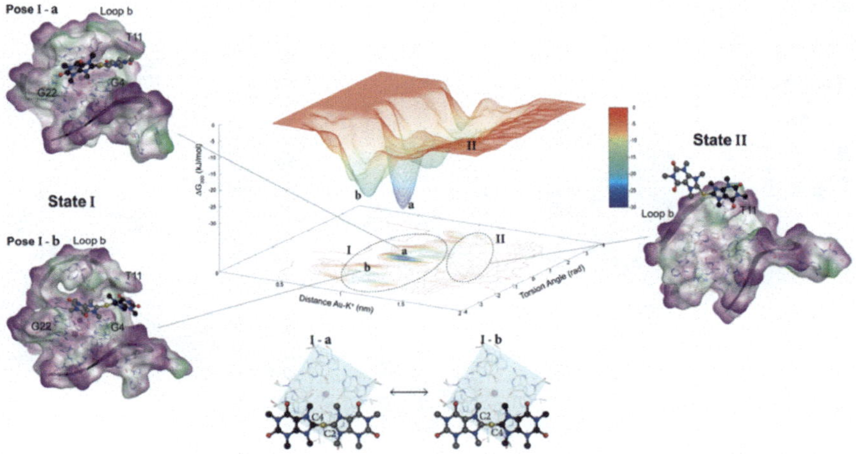

Fig. 1. A) Structures of the Au(I) N-heterocyclic carbene compounds AuTMX$_2$ and AuTMX-I. B) Multiple collective variable (distance in nm, pseudo-torsion angle in rad) plot of free energy surface of AuTMX$_2$ interactions with hTelo (centre). Two states are highlighted (I and II) and two poses for state I are shown as a and b. States I-a, b and II are shown in translucent molecular surface coloured according to lipophilicity (green: lipophilic, pink: hydrophilic). G4 structure is shown as sticks and ribbon, with hidden backbone for clarity. AuTMX$_2$ is shown in ball and stick, with each caffeine ligand coloured differently (black and grey). C2 and C4 highlight the carbon atom positions in AuTMX$_2$ in each of the related poses I-a or I-b.

In state I, AuTMX$_2$ is interacting with both an adenine (A13) in the loop b region, and two of the guanine bases in the upper tetrad (G4 and G22), with strong π stacking interactions between the NHCs of the gold complex and the aromatic rings of G22 (Fig. 1). Instead, the higher energy state II (ca. −14 kJ/mol, Table 1) corresponds to a position, in which the gold complex does not interact with any of the guanine bases in the upper tetrad, but exclusively with the loop thymine (T11) (Fig. 1). In this second state, the loop covers the top of the G4-tetrad, hindering any possible interactions between the gold complex and the G-tetrad.

Interestingly, a similar two-states behaviour was observed for AuTMX$_2$ binding to *C-KIT*1, whose structure and surface are very different to that of *h*Telo. In *C-KIT*1, the gold complex interacts with the top of the tetrad and the rings of flanking bases with no obstruction.

Multi-CV calculations were also run for the neutral mono-NHC complex AuTMX-I on both *h*Telo and *C-KIT*1. The compound was shown to interact *via* π-π and π-alkyl interactions with the guanine tetrad (G22) and the loop (A13) in hTelo, as observed for $AuTMX_2$ (state I). However, as expected, the calculated ΔG_{MD} was lower compared to $AuTMX_2$, due to the lack of the second caffeine ligand (Table 1).

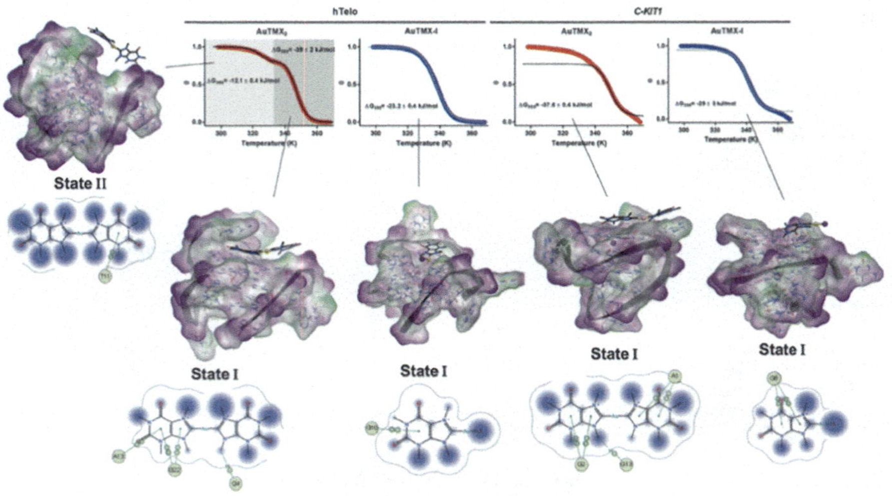

Fig. 2. Folded fraction (θ) of *h*Telo and *C-KIT*1 with either $AuTMX_2$ (red) or AuTMX-I (blue). Corresponding fits are shown as black or grey lines, and the calculated Gibbs-free energy (ΔG_{300}) is shown in each plot. The most representative poses for each combination of G4-gold complex are shown against hydrophobic (green)/ hydrophilic (magenta) surfaces. Below each pose the interactions observed between the complexes and the G4s, corresponding to arene-arene or arene-H interactions are shown.

The observed multiple poses of $AuTMX_2$ prompted further investigation of the stabilisation properties of the two gold-based complexes. This was achieved through the determination of the Gibbs free energy (ΔG_{exp}) from FRET DNA melting profiles. The difference in DNA melting temperature (ΔT_m) of *h*Telo and *C-KIT*1 induced by the binding of the two Au(I) NHC complexes was monitored, which enabled the quantification of the compounds' stabilisation properties of G4-DNA.

$AuTMX_2$ shows a characteristic melting profile for *h*Telo, featuring a two-step melting pattern, where a small increase in fluorescence is initially observed before a steep increase after ca. 65 °C. $AuTMX_2$ with *C-KIT*1 shows, on the contrary, a gradually incrementing curve, rather than an initial steep ramp or two-step curve. In order to calculate binding energies (ΔG_{exp}), experimental data were normalised to folded fraction (θ) of G4-DNA and fitted, while the enthalpy (ΔH) for the process was derived from the resulting fit [2]. The two-step melting profile of $AuTMX_2$ with *h*Telo was separated into two independent melting curves for the fitting. ΔG_{exp} was calculated for both compounds for each G4 structure, including the *h*Telo two-step melting curve (Table 1). The resulting fitted curves are shown in Fig. 2 and the data indicate that the greater stabilisation of the G4 structure by the compound, the lower the energy of binding (when considering the most stable mode, state I for *h*Telo).

Table 1. Gibbs free energy values, experimental (ΔG_{exp}) and calculated (ΔG_{MD}), for $AuTMX_2$ and AuTMX-I interactions with *h*Telo and *C-KIT*1 (units *kJ/mol*, T = 300 K.)

System	*h*Telo		*C-KIT*1	
	$^a\Delta G_{MD}$	ΔG_{exp}	$^a\Delta G_{MD}$	ΔG_{exp}
$AuTMX_2$ (state I)	-37 ± 7	-39 ± 2	-45 ± 3	-37.6 ± 0.4
$AuTMX_2$ (state II)	-14 ± 3	-12.1 ± 0.4	-12 ± 3	-15 ± 1
AuTMX-I	-28 ± 3	-23.2 ± 0.4	-30 ± 5	-29 ± 3

[a]Data are obtained from simulations using a distance CV.

ΔG_{exp} are in perfect accordance with ΔG_{MD} values obtained from metadynamics calculations (Table 1). Analysis of the melting curves and ΔG_{exp} values of $AuTMX_2$ of the two G4 models considered clearly suggests the existence of two distinct binding modes, in agreement with the computational results. In *h*Telo, the first binding mode corresponds to the lower energy state I (Figs. 1 and 2), owing to molecular interaction with the guanines in the tetrad. The second binding mode corresponds to the higher energy (state II) and involves loop/flanking base interactions and/or interactions with part of the tetrad. The flatness of the top tetrad of *C-KIT*1 allows the compound to probe the entire top-surface of the G4 without hindrance. Since the stacking of the complex with the guanines is more favourable (state I), we argue that this is the interaction most likely to occur *in vitro*. Notably, our metadynamics results also point towards the existence of a second binding site for AuTMX-I with *h*Telo, involving the loop C. This interaction may be exploited in the future for optimised compound's selectivity in *h*Telo stabilisation.

3 Unveiling the Mechanisms of Glycerol Permeation *via* Aquaglyceroporin-3 (AQP3) by Metadynamics

Aquaporins (AQPs) [16] are a family of membrane spanning proteins responsible for the control of cellular water flux. AQPs are organized into two subfamilies: the orthodox aquaporins, which conduct exclusively water, and the aquaglyceroporins which conduct glycerol efficiently and generally have lower water conductance [19]. AQPs have become relevant drug targets for the treatment of a number of diseases [17, 18] and new roles continue to emerge as more is known about their structure and selectivity towards certain substrates. Increased understanding of the underlying mechanisms of substrate permeation *via* AQPs by water or other small molecules can contribute greatly to the understanding of the molecular mechanisms of certain diseases, especially those related to AQP over-expression and to the development of selective modulators capable of acting as either chemical probes or as possible therapeutics [17, 18]. Furthermore, this knowledge is crucial to the design of new selective AQPs inhibitors to be used as chemical probes to study the protein function as well as to develop new targeted therapeutic agents.

In humans, 13 AQP isoforms (AQP0-AQP12) have been characterised, among which aquaporin-3 (AQP3) is known to allow the transportation of H_2O_2 [20], glycerol [21], ammonia [22] and urea [21, 22]. AQP3 is present in a number of tissues and is also the most abundant aquaporin isoform in skin cells [23]. Additionally, it has been found to be expressed in various cancer types. Therefore, it has been hypothesised that AQP3 and its substrates (glycerol and H_2O_2) may play a crucial role in tumour growth and proliferation [18]. In fact, AQP3 has been found to be expressed in a number of cancer cell lines including lung, skin and breast cancer. Therefore, it has been postulated that AQP3, and its substrate glycerol, may play a role in tumour growth and metastasis.

Overall, the mechanisms of water and glycerol conductance *via* AQP3 are of great interest for the development of novel anticancer therapeutics, and the elucidation of the mechanisms of solute transport by AQPs by means of advanced computational techniques could enable accurate free energies mapping of pore permeation, as well as for a comparison of water and glycerol conductance mechanisms. Within this framework, a number of valuable computational studies have contributed to the mechanistic understanding of both water and glycerol conduction in AQPs [24–26].

An AQP monomer features an hourglass shaped pore formed by six transmembrane helices, connected by five loops (A through E) and two semi-helices meeting at the centre of the structure. Close to the extracellular side, the aromatic/arginine selectivity filter (ar/R SF), the narrowest point of the pore, is responsible for size-exclusion of molecules [28]. Underneath, the pore centre is defined by two highly conserved asparagine-proline-alanine (NPA) motifs contained in the B and E loops and semi-helices responsible for exclusion of charged solutes through the formation of an electrostatic barrier [29]. Water transport through AQPs has been understood to be a single-file mechanism [24, 30]. MD studies have helped elucidate the role of the selectivity filters and their effects on the permeation of water and other small uncharged solutes [24, 31]. The size and shape of the pore constrain water orientation and determine the internal water-water interactions [32].

Recent studies on carbon nanotubes as bio-mimic channel systems identified an optimal pore size threshold for single-file water transport of 0.8 nm, which significantly affects transport rate and molecular translocation direction [33]. Whilst not as complex as AQPs, carbon nanotubes show the efficiency of single file water transport over bulk water flux upon molecular confinement. Moreover, biophysical studies suggest that unitary water channel conductance (p_f) depends exponentially on the number (N_H) of available hydrogen bond donors and acceptors within the pore [34]. However, different

AQPs, despite having the same N_H, show markedly different p_f, which could be explained by the dehydration penalty that water molecules face upon entering the single-file region of the pore [32]. The single file water permeability p_f and the Gibbs activation energy barrier $G_t^{\ddagger}$ for facilitated water transport *via* AQPs are intrinsically linked, and factors other than hydrogen bonding may play a role, such as a positive surface charges at the channel mouth.

To date, minimal information is available concerning glycerol permeation of human AQPs. An early equilibrium MD study of glycerol-saturated bacterial glycerol facilitator (GlpF) by Schulten and co-workers proposed a mechanism for glycerol conductance [35]. In 2002, the same authors reported on steered-MD simulations of glycerol permeation through the same aquaporin isoform [36], which revealed glycerol channel hydrogen bonding interactions along with the stereoselective nature of the channel. In 2008, Hub and de Groot studied AQP1 and GlpF selectivity for O_2, CO_2, NH_3, glycerol, urea, and water permeation using classical MD [37], focusing on the description of the key role played by the ar/R site, acting as a filter permeated only by small polar solutes.

MD calculations can provide free-energies of glycerol permeation *via* AQP3 by collecting the relevant equilibrium configuration probability distributions. However, they provide no kinetic information of solute conductance, which is required for an unbiased mechanistic analysis over a physiologically significant timeframe. Metadynamics was therefore, used to reconstruct the free-energy profile for the process of interest from multiple independent runs, each allowing for manifold glycerol and water permeation events *via* the AQP3 pores. Simulation lengths of 200 ns, and the inclusion of 20 glycerol molecules in a single simulation run, provided enough time for conformational changes to be observed within the tetramer during water and frequent glycerol permeation through the pores.

Metadynamics accelerates rare event occurrence along collective variables (CV) [38]. For both water and glycerol pore permeation, the CV was defined as the distance between centre-of-mass (COM) of glycerol molecule and a plane defined by four stable atoms inside the channels. A large variation of the CV corresponded to successful translocation events, from which 2D FES could be obtained [38] for each substrate under investigation. The FES highlights the energy expenditure during a permeation event, providing a highly detailed energy profile as the molecule crosses the pore (Fig. 3), allowing the matching of local interactions to energy barriers involved in solute conductance. As there are no directional constraints on any of the solvent molecules and periodic boundary conditions (PBC) are applied in all directions, both uptake and efflux permeation can be observed independently.

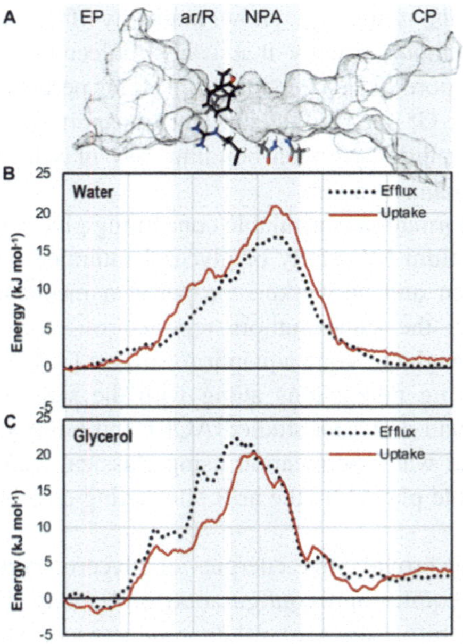

Fig. 3. (A) AQP3 internal pore solvent exclusion surface, indicating the position of the amino acid residues that constitute the ar/R SF and NPA, with amino acids shown in black (ar/R SF) or gray (NPA). Extracellular (EP) and cytoplasmic (CP) pockets are also highlighted. Free energy of water (B) and glycerol (C) uptake (solid line) and efflux (dashed line) based on averaged FES data from multiple successful permeation events.

Free energy profiles for water and glycerol were obtained for both uptake and efflux pathways, through each AQP3 pore. Despite the narrow section size at ar/R, our simulations show that the electrostatic NPA motif provides the highest energy barrier for permeation for both water and glycerol. Whilst the ar/R SF constitutes a region of steric hindrance, introducing both size and orientation constraints for a chance to permeate, it appears to be relatively low in energy demand when compared to the NPA region (Fig. 3). The latter is the area of the pore where the highest ΔG values are reached for both water and glycerol substrates, namely ΔG_{NPA}-water $\approx$ 26 kJ/mol^{-1} and ΔG_{NPA}-glycerol $\approx$ 40 kJ/mol^{-1}, respectively.

The collected trajectories reinforce the established flipping mechanism of water passing though the NPA region of the pore, as observed in other AQPs [30, 39], which also validated the CV choice. In a total of 2.4 μs of combined simulation time, 30 water and 28 glycerol molecules (out of 100 water and 140 glycerol molecules, respectively) were observed to successfully traverse one of the four pores in either direction. Noticeably, in the absence of an osmotic gradient, and without directional bias, the number of permeation events was imbalanced between uptake and efflux, and more significantly so in the case of water: efflux events were 80% more successful for water and 60% for glycerol molecules. In order to understand the molecular pathways, and to identify which amino acid interactions are common to both water and glycerol, the H-bond network for each successful permeation event calculated by metadynamics was analysed (Fig. 4).

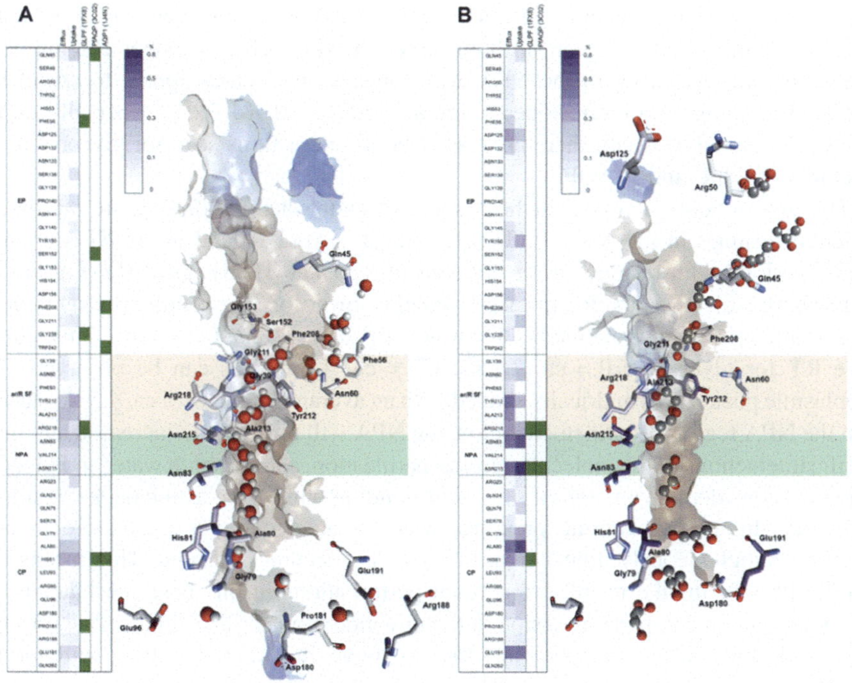

Fig. 4. Water (A) and glycerol (B) permeation routes, H-bond patterns and RT (%) from metadynamics. Average H-bond RT (%) of glycerol during uptake and efflux are shown in a gradient purple colour. Amino acids involved in H-bonds are explicitly shown. Carbons are colour-mapped (white to purple) according to their corresponding RT (short to long). Green boxes represent the positions of water and glycerol molecules within the crystal structures of GlpF (pdb 1FX8) and PfAQP (pdb 3C02). Pore colour based on hydrophobicity, blue = hydrophilic, brown = hydrophobic.

For water permeation residence times (RT) of water molecules (ca. 1.4 ns) are observed inside the ar/R SF, where waters form H-bonds mostly with Tyr212, Ala219 and Arg218. Residence times within the NPA are on the average shorter (ca. 0.9 ns). In this region, water molecules interact with Asn83, Val214 and Asn215. In the extracellular pocket (EP), water appears not to bind to any specific residues and has a residence time of ca. 1.6 ns, whereas, in the cytoplasmic pocket (CP), water hydrogen bonds mainly to three amino acids (Ala80, His81, Glu96) and the residence time can increase up to ca. 2.2 ns (Fig. 4A). In Fig. 4 key H-bonding interactions are highlighted (green) and matched to the corresponding amino acids in hAQP3. GlpF shows a H-bond with the Arg in the ar/R SF (Arg218 in AQP3), while PfAQP and AQP1 both show H-bonds to the highly conserved His located in the cytoplasmic pocket (His81 in hAQP3). Other H-bonding interactions are more specific to each particular isoform, due to the variation in the amino acid composition of the pores.

In the case of glycerol permeation, its size and increased number of available hydrogen bonding groups (3 -OH) directly affects its permeation time. Accordingly, increased H-bonding residence times were observed, compared to water (Fig. 4). The longest RT involving specific residues (ca. 2.3 ns) was observed for Asn83 and Asn215 within the NPA region, in both directions (Fig. 4B). A longer RT implies a higher free energy barrier in the NPA area. Glycerol molecules appear to spend less time in H-bonding with the residues in the ar/R SF (ca. 0.9–1.5 ns), especially during uptake. Compared to water, there are less H-bond interactions seen for glycerol in the structures of GlpF and PfAQP.

The amino acids involved in both glycerol and water H-bonding are similar, in particular residues in the ar/R SF (Tyr212, Ala213, Arg218) and in the NPA (Asn83, Asn215), along with residues in the cytoplasmic pocket (His81, Glu191) (Fig. 4).

For both molecules, the RT in each H-bond is lower for extracellular pocket than for cytoplasmic pocket (cumulative RT for water: EP = ca.1.6 ns, CP = ca. 2.2 ns; cumulative RT for glycerol: EP = ca. 2.9 ns, CP = ca. 4.7 ns), as can be observed. The cytoplasmic pocket is cylindrically shaped with an average diameter of ca. 4.5 Å, starting from the NPA region. Confinement below the NPA will thus lead to increased and more specific interactions of the molecules during permeation, causing both water and glycerol to spend a considerable amount of time in H-bonding with cytoplasmic pocket residues.

In the absence of osmotic pressure, water is able to traverse the pore in both directions simultaneously, rather than in a single direction at a time. This means that overall efflux or uptake results from small perturbations in this base mechanism and does not require a complete inversion in the direction of water flux. Figure 5 shows that water molecules inside the pore are free to move in either direction, without file disruptions (Fig. 5 A, B). To this end, water implements a "hopping mechanism", in which single water molecules are able to switch with adjacent molecules in either direction, along the single file chain. We propose that this process offers, if not an active regulatory mechanism, at least a structure capable of rapidly responding to environmental changes, as both transport directions are active simultaneously.

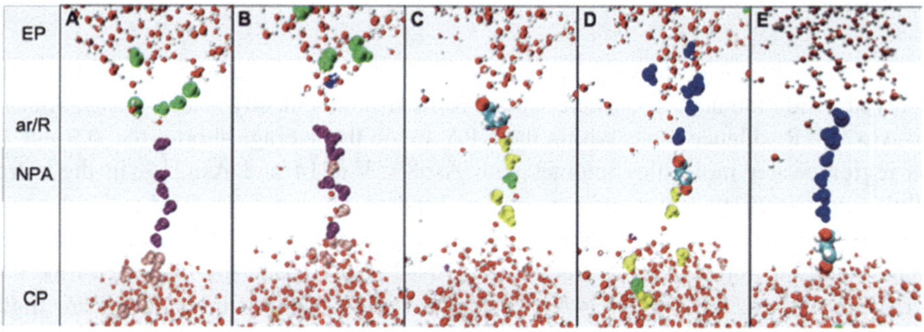

Fig. 5. The passage of glycerol through the AQP3 pore disrupts the "leap-frogging" water chain. (A) Water molecules are coloured based on position: purple - inside the pore, pink – cytoplasmic pocket (CP), green - extracellular pocket (EP). (B) Water molecules start moving from the CP into the channel, replacing the purple molecules. (C) A glycerol molecule enters the ar/R region from the EP, yellow - waters occupying the pore, originating from both EP and IP. (D) Glycerol passes through the NPA region, forcing the waters out of the channel into the CP area, with new water molecules (blue) populating the channel area. (E) The chain of water molecules is re-established as glycerol leaves the pore and the pore is re-populated by single-file water molecules from EP.

This bidirectional water flux is disrupted as a consequence of glycerol solute molecules entering the pore, inserting into the water chain in the direction of permeation (Fig. 5 C, D). Glycerol insertion prevents water crossing in the opposite direction due to its increased size. As soon as glycerol has traversed the pore back into bulk water, the "leap-frog" mechanism within the water chain re-establishes itself by filling the pore as the glycerol passes through (Fig. 5), i.e. without voids or latencies in the basal water transport mechanism.

As previously reported for glycerol conductance of GlpF [36], the small reduction in energy seen in the FES of glycerol at the extracellular and cytoplasmic vestibules (Fig. 3) facilitates glycerol conductance *via* AQP3, increasing the probability of glycerol joining the single file water transport mechanism.

4 Methods

In a representative metadynamics simulation, the collective variable bias was controlled by the Plumed [38] driver, using Gromacs [40] as molecular dynamics (MD) engine. Therein, constant-temperature, constant pressure MD was performed on TIP3P water solvated G4s and AQP3 models, respectively. Short-ranged interactions were evaluated based on the amber99sb-ildn+slipid force filed, while long ranged Coulomb interactions were accounted for with a particle-mesh Ewald approach. Several independent well-tempered metadynamics runs (50 ns, 1 fs integration step) were performed, from which free energy profiles were integrated and averaged. For the G4 system, the distance between $AuTMX_2$ (Au) and the K^+ ion of the top tetrad represented one component of the CV. A torsion angle component was added to account for the mutual orientation of the G4 and drug molecules. For AQP3, the distance between the centre of mass of glycerol or water, and the plane formed by backbone carbons of Asparagine 215 from the four monomers defined the CV. A Gaussian bias was deposited every 2 ps, with a gaussian width of 0.025 Å and $\tau = 50$. The bias factor was set to 12, ΔT was 3600 K.

5 Conclusions

The examples covered above illustrate how a detailed mechanistic investigation of drug/target interactions and membrane transport processes are a prerequisite for an accurate evaluation of free energies of binding and selectivity. Besides reliable quantitative values, alternative poses (G4s) or elusive details of molecular transport (AQP3) could be elucidated. For the investigated DNA G4s, the obtained computational results fully support the experimental data revealing two ligand binding modes of the gold compound $AuTMX_2$ on the two selected G4 structures. The validation of the binding energy of Au(I) complexes with G4s adducts calculated by FRET DNA melting assays was also achieved by metadynamics methods. This investigational approach can be extended to other types of molecules as G4 stabilizers, including organic compounds, highlighting selectivity features essential to orient the drug design.

Furthermore, the mechanisms of water and glycerol conductance *via* human AQP3 was investigated using metadynamics, accelerating pore permeation statistics. A full mechanistic picture and the underlying free energy could be reliably collected for both water and glycerol. Single-file water permeation by AQP3 appears always to be bidirectional at equilibrium conditions, whilst still maintaining a continuous transport flux when compared to bulk water. Glycerol permeation appears to critically depend on this single file water flux, with transport resulting from bond switches within a dynamic hydrogen bond scaffold created by the interplay of glycerol, water and pore amino acid residues. This reveals a novel scenario, in which solute molecules exploit an existing bidirectional water conduction mechanism in AQP3. Free energy results also suggest a binding affinity between key residues and solute molecules on the exterior pore surface facilitating solute transport.

In the future, complementing metadynamics with quantum mechanics/molecular mechanics (QM/MM) calculations will provide knowledge of the system at an atomistic level and the identification of key interactions with protein residues which may be specifically targeted by newly designed AQP3 inhibitors.

Overall, the reported examples further demonstrate the invaluable role of advanced multiscale modeling and simulation methods, and of its close integration with experiments, for applications in life, health and innovative technologies. We expect that in the coming years metadynamics methods are likely to be central in drug discovery and development programmes.

Acknowledgements. This work has made use of resources of the "Cambridge Service for Data Driven Discovery" (CSD3, http://csd3.cam.ac.uk) system operated by the University of Cambridge Research Computing Service (http://www.hpc.cam.ac.uk) funded by EPSRC Tier-2 capital grant EP/P020259/1. We gratefully acknowledge NVIDIA Corporation for the donation of a Quadro P5000 GPU. We thank ARCCA, Cardiff for additional computational support.

References

1. Amaro, et al.: Nat. Rev. Chem. **2**, 0148 (2018)
2. Wragg, D., De Almeida, A., Bonsignore, R., Kühn, F.E., Leoni, S., Casini, A.: Angew. Chemie **57**, 14524–14528 (2018)
3. Wragg, D., de Almeida, A., Casini, A., Leoni, S.: Chem. – Eur. J. **25**, 8713–8718 (2019)
4. Laio, A., Parrinello, M.: Proc. Natl. Acad. Sci. U.S.A. **99**, 12562 (2002)
5. Monchaud, D., Teulade-Fichou, M.P.: Org. Biomol. Chem. **6**, 627–636 (2008)
6. Chambers, V.S., Marsico, G., Boutell, J.M., Di Antonio, M., Smith, G.P., Balasubramanian, S.: Nat. Biotechnol. **33**, 877–881 (2015)
7. Rhodes, D., Lipps, H.J.: Nucleic Acids Res. **43**, 8627–8637 (2015)
8. Bernal, A., Tusell, L.: Int. J. Mol. Sci. **19**, 1–21 (2018)
9. Neidle, S.: Therapeutic Applications of Quadruplex Nucleic Acids. Academic Press, Boston (2012)
10. Balasubramanian, S., Hurley, L.H., Neidle, S.: Nat. Rev. Drug Discov. **10**, 261–275 (2011)
11. Incles, C., Schultes, C., Kempski, H., Koehler, H., Kelland, L.R., Neidle, S.: Mol. Cancer Ther. **3**, 1201–1206 (2004)

12. Bertrand, B., Stefan, L., Pirrotta, M., Monchaud, D., Bodio, E., Richard, P., Le Gendre, P., Warmerdam, E., De Jager, M.H., Groothuis, G.M.M., et al.: Inorg. Chem. **53**, 2296–2303 (2014)
13. Bazzicalupi, C., Ferraroni, M., Papi, F., Massai, L., Bertrand, B., Messori, L., Gratteri, P., Casini, A.: Angew. Chemie - Int. Ed. **55**, 4256–4259 (2016)
14. Dai, J., Punchihewa, C., Ambrus, A., Chen, D., Jones, R.A., Yang, D.: Nucleic Acids Res. **35**, 2440–2450 (2007)
15. Wei, D., Husby, J., Neidle, S.: Nucleic Acids Res. **43**, 629–644 (2015)
16. Preston, G.M., Carroll, T.P., Guggino, W.B., Agre, P.: Science **256**, 385–387 (1992)
17. Verkman, A.S., Anderson, M.O., Papadopoulos, M.C.: Nat. Rev. Drug Discov. **13**, 259–277 (2014)
18. Aikman, B., De Almeida, A., Meier-Menches, S.M., Casini, A.: Metallomics **10**, 696–712 (2018)
19. Soveral, G., Nielsen, S., Casini, A.: Aquaporins in Health and Disease: New Molecular Targets for Drug Discovery. CRC Press, Taylor & Francis Group (2017)
20. Bienert, G.P., Chaumont, F.: Biochim. Biophys. Acta - Gen. Subj. **2014**, 1596–1604 (1840)
21. Ishibashi, K., Sasaki, S., Fushimi, K., Uchida, S., Kuwahara, M., Saito, H., Furukawa, T., Nakajima, K., Yamaguchi, Y., Gojobori, T.: Proc. Natl. Acad. Sci. U.S.A. **91**, 6269–6273 (1994)
22. Litman, T., Sogaard, R., Zeuthen, T.: Ammonia and urea permeability of mammalian aquaporins. In: Handbook of Experimental Pharmacology, pp. 327–358 (2009)
23. Boury-Jamot, M., Daraspe, J., Bonté, F., Perrier, E., Schnebert, S., Dumas, M., Verbavatz, J. M., Bonte, F., Perrier, E., Schnebert, S., et al.: Aquaporins, Germany, pp. 205–217 (2009)
24. de Groot, B.L.: Science **294**(80), 2353–2357 (2001)
25. Tajkhorshid, E., Nollert, P., Jensen, M.Ø., Miercke, L.J., O'Connell, J., Stroud, R.M., Schulten, K.: Science **296**(80), 525–530 (2002)
26. De Groot, B.L., Grubmüller, H.: Curr. Opin. Struct. Biol. **15**, 176–183 (2005)
27. Jin Sup Jung, G.M., Preston, B.L., Smith, W.B., Guggino, P., Agre, J.: Biol. Chem. **269**, 14648–14654 (1994)
28. Krenc, D., Song, J., Almasalmeh, A., Wu, B., Beitz, E.: Mol. Membr. Biol. **31**, 228–238 (2014)
29. Chakrabarti, N., Tajkhorshid, E., Roux, B., Pomès, R.: Structure **12**, 65–74 (2004)
30. Murata, K., Mitsuoka, K., Hiral, T., Walz, T., Agre, P., Heymann, J.B., Engel, A., Fujiyoshi, Y.: Nature **407**, 599–605 (2000)
31. Janosi, L., Ceccarelli, M.: PLoS One **8** (2013) https://doi.org/10.1371/journal.pone.0059897
32. Horner, A., Zocher, F., Preiner, J., Ollinger, N., Siligan, C., Akimov, S.A., Pohl, P.: Sci. Adv. **1**, e1400083–e1400083 (2015)
33. Tunuguntla, R.H., Zhang, Y., Henley, R.Y., Yao, Y.C., Pham, T.A., Wanunu, M., Noy, A.: Science **359**, 792–796 (2018)
34. Horner, A., Pohl, P.: Faraday Discuss. **209**, 9–33 (2018)
35. Jensen, M., Tajkhorshid, E., Schulten, K.: Structure **9**, 1083–1093 (2001)
36. Jensen, M.O., Park, S., Tajkhorshid, E., Schulten, K.: Proc. Natl. Acad. Sci. **99**, 6731–6736 (2002)
37. Hub, J.S., de Groot, B.L.: Proc. Natl. Acad. Sci. **105**, 1198–1203 (2008)
38. Tribello, G.A., Bonomi, M., Branduardi, D., Camilloni, C., Bussi, G.: Comput. Phys. Commun. **185**, 604–613 (2014)
39. Spinello, A., De Almeida, A., Casini, A., Barone, G.: J. Inorg. Biochem. **160**, 78–84 (2016)
40. Abraham, M.J., et al.: SoftwareX **1–2**, 19–25 (2015)

Using Computer Simulations and Virtual Reality to Understand, Design and Optimize Artificial Water Channels

Xavier Martinez[1,2](✉), Arthur Hardiagon[1,2](✉), Hubert Santuz[1,2], Samuel Murail[3], Mihail Barboiu[4], Fabio Sterpone[1,2], and Marc Baaden[1,2](✉)

[1] CNRS, Université de Paris, UPR 9080, Laboratoire de Biochimie Théorique, 13 rue Pierre et Marie Curie, 75005 Paris, France
baaden@smplinux.de

[2] Institut de Biologie Physico-Chimique-Fondation Edmond de Rotschild, PSL Research University, Paris, France

[3] Université de Paris, CNRS, INSERM, Biologie Fonctionnelle et Adaptative UMR 8251, Computational Modeling of Protein Ligand Interactions U1133, 75205 Paris, France

[4] Institut Europeen des Membranes, Adaptive Supramolecular Nanosystems Group, Université de Montpellier, ENSCM, CNRS, Place Eugene Bataillon CC047, 34095 Montpellier, France

Abstract. In biology, metabolite transport across cell membranes occurs through natural channels and pores. Artificial ion-channel architectures represent potential mimics of natural ionic conduction. Many such systems were produced leading to a remarkable set of alternative artificial ion-channels. Far less advances were achieved in the area of synthetic biomimetic water channels, even though they could improve our understanding of the natural function of protein channels and may provide new strategies to generate highly selective, advanced water purification systems. Most realizations have used the selectivity components of natural protein channels embedded in artificial systems. Such biomolecules provide building blocks to constitute highly selective membrane-spanning water transport architectures. The simplification of such compounds, while preserving the high conduction activity of natural macromolecules, lead to fully synthetic artificial biomimetic channels. These simplified systems offer a particular chance to understand mechanistic and structural behaviors, providing rationales to engineer better artificial water-channels. Here we focus on computer simulations as a tool to complement experiment in understanding the properties of such systems with the aim to rationalize important concepts, design and optimize better compounds. Molecular dynamics simulations combined with advanced visual scrutiny thereof are central to such an approach. Novel technologies such as virtual reality headsets and stereoscopic large-scale display walls offer immersive collaborative insight into the complex mechanisms underlying artificial water channel function.

S. Piotto et al. (Eds.): BIONAM 2019, LNBE, pp. 78–99, 2020.
https://doi.org/10.1007/978-3-030-47705-9_8

1 Introduction

1.1 Origins and Brief Overview

Water is fundamental to life. It fulfills a variety of functions related to its complex dynamic properties at the supramolecular level [1]. Most physiological processes require selective traffic between a cell and its environment, where water is essential to enable molecular transport events [2]. Artificial ion-channels have been extensively studied with the expectation to mimic the ionic conduction along protein ion-channels through biological membranes [3–6]. Amazingly, until recently there has been almost no systematic progress in the area of water transport by Artificial Water Channels (AWCs) [7]. Before 2011, these artificial water pore systems, selectively transporting water against ions had not been described. We initiated this unexplored and highly innovative field with a series of studies described in [7–11]. Figure 1 provides three examples of AWCs.

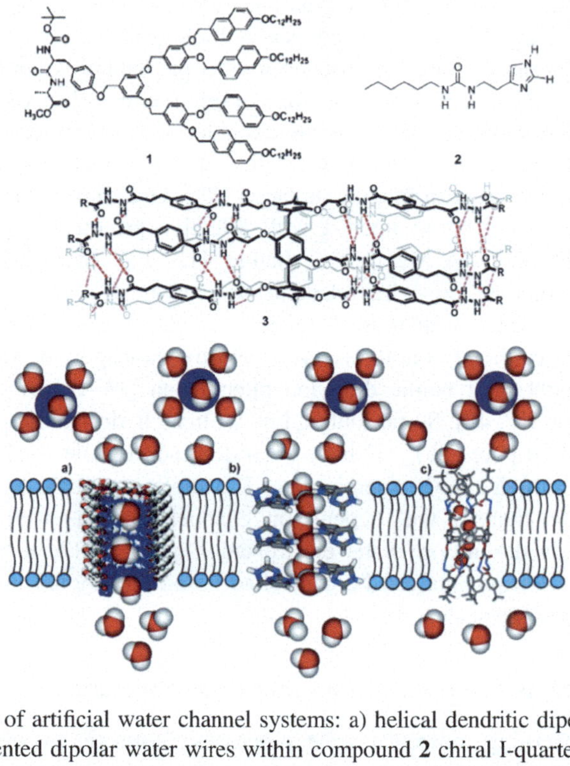

Fig. 1. Selection of artificial water channel systems: a) helical dendritic dipeptides, compound **1**; [12, 13] b) oriented dipolar water wires within compound **2** chiral I-quartet assemblies [7–9] that selectively transport water and protons against ions; c) hydrazide-pilar [5] arene compound **3**, [14–17] as single channels. Neutral and protonated water molecules are depicted as red and white CPK models. Blue spheres symbolize hydrated cations, unable to penetrate the bilayer membrane. Figure reproduced with permission from [8].

Since then, many other groups joined this adventure and many works have been published including a feature paper highlighting the developments of the field during the last 5 years [18]. Important studies of AWCs including molecular dynamics (MD) simulations have been pioneered by several groups around the world, notably in the USA, China, Singapore, Hong-Kong, etc.

Molecular modeling approaches - in particular molecular dynamics in explicit solvent and membrane environments - constitute a pivotal tool to extend the methodology, providing insight at a molecular scale as described in [19–21]. Moving up in scales, the development of multi-scale approaches to describe hydrodynamics in molecular systems has the potential to provide new stimulus for water channel modeling [22, 23].

1.2 Artificial Water Channels Are Conceptually Important

In contrast to isotropic spherical ions, dipolar water molecules shed light on multiple oriented interactions with biological environments. The overall orientation of assembled water dipoles is crucial for the regulation of the selective recognition, exchange and transport of charged biocomponents across biological barriers. For example, the electrochemical potential along the Aquaporin-Z Aqp-channel is governed by encapsulated single-file water wires of opposite dipolar orientation, selectively allowing rapid water diffusion, while protons and ions are excluded [24–28]. In comparison to classical polymeric systems, the water-permeability of AqpZ incorporated in polymeric vesicles is at least one order of magnitude larger [26]. One-dimensional water wires in particular have attracted broad interest with selective proton gating being the key function of the Influenza A M2 proton channel [12, 13]. Entrapped water clusters within complex structures are a key element intervening in a variety of fundamental processes such as water structuration related to water-water or water-host-matrix interactions under confined conditions [47], selective transport of small metabolites [52], and in context of dynamic diffusion phenomena [14–17] of water molecules across the channels [17, 48]. So far, only a few artificial hydrophobic [29], hydrophilic [7–9] or amphiphilic pores [30, 31] have been designed to selectively and efficiently transport water through bilayer membranes. The molecular-scale hydrodynamics of water through the channel will depend on channel-water as well as water-water interactions, and on the in-pore water electrostatic dipolar profile within the channel. These molecular determinants can be controlled based on rational insight into AWC properties at the atomic scale.

1.3 Major Insights Are Gained from Molecular Simulations

In particular concerning molecular-scale processes, atomistic as well as simplified coarse-grained or meso-scale simulations have been pivotal to our understanding of key concepts of water transport in biological and synthetic systems. For instance, seminal work based on molecular dynamics simulations of Aquaporin channels has substantially contributed to understand molecular mechanisms responsible for their efficiency and selectivity [25, 28, 32]. General features of biological pores were studied by Beckstein and Sansom [33, 34], as well as others, in model systems of decorated simplified ion channel pores, providing rationales for designing improved channels. Another model system of great importance concerns the study of water transport in carbon nanotubes [35–37]. These few examples among many show that molecular-level insights are key to our understanding and can be achieved through molecular dynamics type simulations acting as a computational microscope. A few studies of artificial water channels including MD simulations have appeared recently and a Faraday Discussion documents the current status in the field [38–42]. The present discussion will firmly build on such simulation approaches to guide and optimize experimental design.

1.4 Now and Tomorrow in This Newborn Field

Despite the tremendous imagination of synthetic chemists to produce sophisticated architectures confining water clusters, most strategies to transport water selectively have been inspired by the natural aquaporin scaffold as selectivity components, embedded in diverse arrays of bio-assisted artificial systems. The highly selective transport of water in natural systems is occurring along Aquaporin (AQP) channels known for their fast transport rates ($\sim 10^8$–10^9 water molecules/s/channel), and the perfect rejection of ions and protons. Aquaporin water channels are composed of an hourglass structure with a narrowest constriction of 2.8 Å. Each water molecule in AQP water channels forms one hydrogen-bond with the inner wall of the protein and one with an adjacent water molecule. It is of particular interest that the dipolar alignment of water molecules imposed by the water-pore and water-water interactions can control the water *vs.* proton transport selectivity of AQPs. These discoveries have inspired the incorporation of AQPs into artificial membranes for desalination and water-purification applications. These results on the combination of natural proteins with artificial lipid matrices, have demonstrated that natural biomolecules can be used as building blocks for the construction of membranes for water transport. However, their large scale application is not yet possible, because of the high costs of AQP production, their low stability and practical synthetic constraints of membrane fabrication processes.

A possible way forward is to replace AQPs with synthetic channels, a major motivation for our work on AWCs. Recent advancements have made it possible to synthesize artificial water channels featuring efficient, yet selective transport of water based on size exclusion and molecular interactions, such as hydrogen-bonding. Given their completely artificial nature, these AWCs exhibit the following potential advantages:

- Synthetic AWCs are easily tuneable and may be chemically optimized for desirable properties;
- AWCs can be prepared on a large scale at low cost;
- Engineering processing without bio-related steps becomes more simple and reproducible, and they can be easily immobilized in membrane-like supports in a scalable manner;
- AWCs should also be more robust and stable with a longer lifetime after incorporation into membranes for water separation and purification.

However, currently, the research on AWCs is focused mainly on the synthesis and characterization in terms of its water transport property and solute rejection. More focus should be turned on integrating the synthetic channels with a membrane for water separation and purification. Given the importance of developing a highly efficient membrane desalination technology, the potential of synthetic water channels and the knowledge gap for integration of synthetic water/ion channels into membranes, it is envisaged that development of synthetic channel-based biomimetic membranes with enhanced performance compared to current membranes is a challenging goal. It will require to provide answers to complicated biological scenarios related to water pores and to develop bio-inspired desalination processes that could use the non-existing all-made artificial synthetic materials that mimic the mechanisms of natural desalinators. Computational approaches can guide these endeavors.

2 From Aquaporin to I-Quartet Channels

2.1 Learning from Biological Water-Channels as Sources of Inspiration

Potassium channel KcsA K^+ [43], Aquaporin-AQP, [24–28], Influenza A M2-M2IA, [12, 13] or Gramicidin-GA, [44] are well known, non-exclusive examples of proteins in which ions, water molecules and protons are envisioned to diffuse along water filled pores. It is worth to note that most protein channels share some recurrent structural aspects. For instance many are self-assembled from multiple subunits within the trans-membrane domains. These protein subunits are packed around a water filled pore, a crucial element in the functional conformation of these channels. The narrowest region of the pore usually determinesd the selectivity of these molecular channels, showing gating behaviors generated by the structural motion of the external subunits in response to voltage, ligand, pH or other external stimuli. The interactions between the constitutive components of the pore with water (most often a one-dimensional water wire) or with ionic species will impose a specific net dipolar moment controlling the translocation within the narrow region of the pore. Despite a wealth of experimental data on structural details of protein channels, major issues still need to be resolved and understood at the atomic level.

For example, in Aquaporins, once passing the center of the pore occupied by a water molecule, the water molecules undergo a flip-like opposite orientation in the AQP channel. This inversion causes the inversion of the net dipolar moment and hinders proton transport. Therefore, the structure and alignment of the water wires within the synthetic pores may be important in the design of artificial pores creating (pumping) or collapsing (gating) ionic gradients throughout the membrane.

However there are many challenges to use the AQP scaffold as selective transporters related to their:

- Long term stability, which is unclear
- Unconventional processing requirements - use of detergent and specialized aqueous self-assembly
- Production - membrane proteins are challenging to mass produce

Other well-known natural ion-channels jointly transporting water and cations are of primordial inspiration for the design of the artificial channels. For example the dipolar structure of Gramicidin A-GA, helps to overcome the high energy barrier of water and ion translocations across the cell membrane, both being synergistically transported sharing one pathway [44]. Particularly interesting, the dipolar alignment of water molecules imposed by the pore structure can control the ionic conduction, including the translocation of protons via the inner-pore water wires. This oriented single-file water wire - as for the aquaporin pore - imposes a net dipole moment which is particularly influenced by the presence of the ions inside the channel. Depending on the position of the ion within the channel, two linear chains of water molecules of opposite orientation are determining the net dipole potential which is zero on average when the ion is in the middle of the channel. Similarly, the water molecules permeate through the KcsA K^+ channel, together with the K^+ ions in a concerted way [43]. Interestingly, a K^+ depleted KcsA K^+ channel is changing its initial conformation to allow water transport 20 times faster than one dimensional bulk diffusion of water [45]. On the same principle, the proton selectivity and low-pH gating properties are key functions of the M2 Influenza A - M2IA proton channel [12, 13]. Although there is some variability to set off the mechanisms, many structural features are related to the $(His^{37})_4$ selectivity filter [46].

The encapsulation of water clusters in artificial constructs enables the exploration of the collective behaviors of water in conditions very close to biological systems where the confined molecules may be at the limit between solid and liquid states [14–17, 47–51]. One-dimensional water-wires in particular have generated much interest. Their structural features similar to water-clusters present inside the protein channels were closely examined. Self-assembled tubular supramolecular architectures, mutually stabilized by strong non-covalent bonds and encapsulating inner water wires have provided excellent reasons to be considered as valuable models for biomimetic water/proton-channel systems [14–16, 49]. However, most of these structures are highly polar and it remains complicated to achieve their inclusion in hydrophobic bilayer membranes. All these non-exhaustive examples are relating specific features of the self-assembly behaviors of water clusters under confined conditions, toward systems of increasing functional complexity.

Understanding the molecular-scale dynamics of water clusters confined in such structurally simple artificial channels is of fundamental and crucial relevance for the next developments of functional AWCs. The mechanisms of water transport through aquapores are related to structural behaviors of water clusters entering the channels which lose or change part of the H-bonds connecting the water molecules, depending on the hydrophobic or hydrophilic nature of the channel [53]. Spontaneous and continuous diffusion of a one-dimensional ordered water chain in interaction or not with the surrounding aquapore is related to molecular interactions, the inner pore surface profile, the pore diameter and dipolar water orientations.

With all these features in mind we can observe that natural water channels show a variety of sizes and different single-file water net dipole orientations that are directly connected to their specific function. They are selective toward other permeants (ions, protons, molecules) depending strongly on the channel structure and dimensionality, channel-water affinity, the average pore occupancy and the dynamics of the permeating water.

2.2 Biomimetic Artificial Water Channels

Parallel to the investigations on natural water-channels the adventure of AWCs is just starting. Like in natural channels, the water transport will depend on key interactions, such as channel-water, water-water and electrostatics guided by the water in-pore electrostatic dipolar profile within the channel. Reproducing the complex interaction superstructures of proteins is an essential goal. A valid approach is to mimic key parts of active filters, as ingredients to provide chemical selectivity, in order to gain insight into the dynamics of orientating water dipoles. Particularly their role for ions/protons pumping along the channel has to be understood. Going from complex natural to simpler artificial water-channel constructs, we would have the chance to better comprehend mechanistic and structural behaviors of such interactions. Such knowledge may unlock the door to the novel interactive design of water-channels, paralleling that of biomolecular systems. Different synthetic building blocks have been reported until now to generate such systems: water was shown to transport through hydrophilic, hydrophobic or hybrid hydrophobic/hydrophilic channels as briefly summarized in Table 1.

The water diffusion and facilitated transport of protons excluding cationic and anionic transport through bilayer membranes incorporating all-artificial water channels have been reported for the first time by Percec *et al.* [29]. The dendritic dipeptides self-assemble via enhanced peripheral π-stacking to form stable cylindrical helical pores (14.5 Å in diameter), of reasonable thermal stability. Moreover they selectively transport water molecules against ions through self-assembled hydrophobic nanotubes stable in phospholipid membranes. The ion-exclusion phenomena are based on hydrophobic effects which appear to be very important. These pores are envisioned to be the first artificial "primitive aquaporins" which transport water but do not reject protons. Next, the significant contribution by Hu *et al.* [30, 31] shows that polydrazide-pillar [5] arenes tubular structures can be used as single-molecular water channels in bilayer membranes. The transport mechanism is strongly dependent on the length of the former components: the shortest pillar [5] arenes-tetraester, induce the formation of water wires within the stacked molecular cylinders and can successfully be used to translocate protons *via* bilayer membranes. The longer poly-hydrazide-pillar [5] arene compounds differently to shorter ones, present alternating hydrophobic/hydrophilic structural domains along their cylindrical structure which disrupt the formation of water wires within the inner channel core and thus block the proton flux along the discontinuous water phases. Differently to AQP, in which the control of the water/proton translocation is based on reverse-flip net-dipolar profiles within the channel [24], other channels may relate their selectivity to the structural disruption of water-wires.

Table 1. Compounds used as molecular building blocks to create artificial water channels [38]

Compound	Nature of the channel	Net Permeability / Selectivity / Single channel permeability
I-quartet Channels, **2** Alkylureidoimidazoles	Hydrophilic H-bonded imidazole channel (2.6 Å) and hydrophobic peripheral alkyl chains	1.0–4.0 μm/s (shrinking mode) / high selectivity for water reject all ions except protons / 1.5×10^6 water molecules/s/channel
Aquaporins – AQPs	Natural hourglass structure of the pore, with a narrowest constriction of 2.8 Å	High water selectivity. Perfect rejection of ions and protons / 4×10^9 water molecules/s/channel
Aquafoldamers	Hydrophilic helical channel (2.8 Å) via π-π stacking of aromatic units	No permeability reported / high selectivity for water, reject all ions except protons
Bola-amphiphile bis-triazole (TCT), T-channels	Self-assembled helical pores (~2.5–4 Å); double helical channels double net-dipolar orientation	No permeability reported for water / enhanced conduction states for alkali cations and for protons
Tubular pores formed via self-assembly of arylene-ethynylene macrocycle	Hydropobic tubular channel (6.4 Å) via π-π stacking of aromatic units and H-bonding of peptide arms	51 μm/s/ no selectivity for water, high conduction for K^+ and protons / 4.9×10^7 water molecules/s/channel
Peptide appended Pillar [5] arenes, PAPs	Hybrid hydrophobic/hydrophilic cylindrical (~5 Å) unimolecular systems.	30 μm/s (swelling mode, good conduction for alkali cations $3.5\ 10^8$ (swelling mode) and 3.7×10^6 water/s/channel (shrinking)
Hydrazide appended Pillar [5] arenes, PAHs **3**	Hybrid hydrophobic/hydrophilic cylindrical (6.5 Å) unimolecular systems	8.6×10^{-6} μm/s/ no water selectivity, good alkali cation conduction, but not for protons / 40 water molecules/s/channel
Dendritic dipeptide, (4-3, 4-3, 5)-12G2-CH_2-Boc-L-Tyr-L-Ala-OMe, **1**	Hydrophobic H-bonded central pore (14.5 Å); stacked dendritic periphery	No permeability reported / water against ions except protons
Carbon Nanotubes- CTNs	The estimated diameter of CTNs (12, 12) is 16 Å	No ion rejection: stable and high permeability / 9×10^8 water molecules/s/channel

Compounds are approximately ordered according to the channel diameter

2.3 Imidazole I-Quartet Channels

We reported [7–9] that imidazole I-quartet channels can be mutually stabilized by inner water-wires, reminiscent of the oriented water single-file columns observed for AQP and ion-charged GA natural pores. In addition, the I-quartet may be considered as the representative water-quartet superstructure, reminiscent of that of the most known

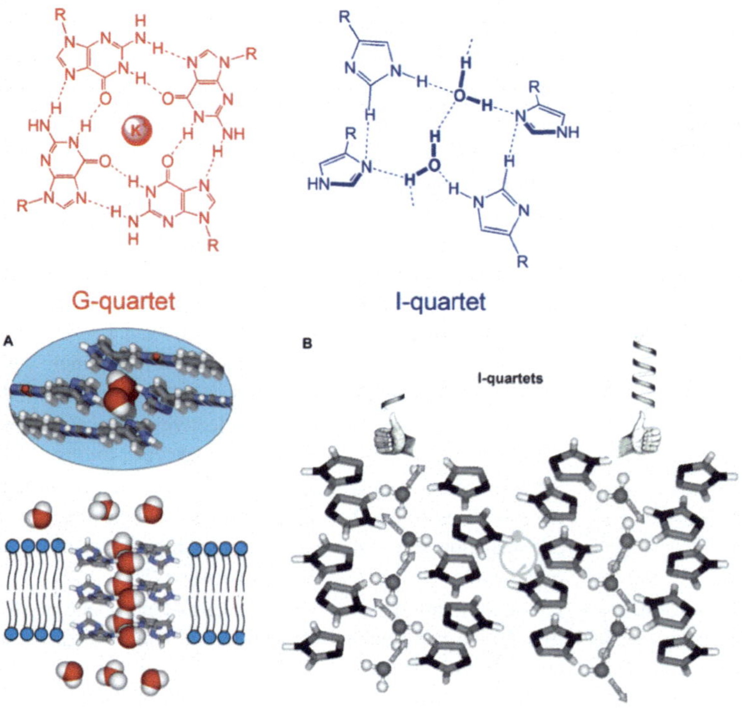

Fig. 2. Cross section and top views of Imidazole I–quartets generating water-channels in which the water molecules present a unique dipolar orientation. The I-quartets presenting supramolecular chirality, can accommodate dipolar water–wires along the length of the channel. Water molecules are shown either in van der Waals or in ball and stick representation.

Guanosine-G-quartet for cations (see Fig. 2 for a comparison). The I-quartet superstructures are stable in solid state and within bilayers leading to functional water-channels. These systems show total ion rejection and the ion-exclusion phenomena are based on dimensional steric effects whereas hydrophobic and hydrodynamic effects appear to be less important.

As observed in aquaporin channels, the confined water wires form one hydrogen bond with the inner wall of the assembled I-quartet scaffold and another one with an adjacent water molecule. Furthermore, the water molecules embrace a unique dipolar orientation thereby preserving the overall electrochemical dipolar potential along the channel. These findings strongly support the idea that water molecules and protons can permeate across the lipid bilayer through pores formed by I-quartet aggregates. These I-quartet constructs provide good reasons to consider that their supramolecular chirality and water induced polarization (unique dipolar orientation) within the channels may be strongly associated [20]. Chiral surfaces are determinant to a more extended scale in asymmetric tissue morphogenesis.

Water is an electrically dipolar molecule and an exceptional (bio)lubricant; it can simply read the spatial information of the asymmetric chiral superstructures to generate asymmetric dipolar-wires of dynamic behaviors. Within this context, chiral synthetic pores of 2.8 Å diameter (like the AQP natural ones!) orienting the dipolar water wires within the channels, give rise to novel strategies to mimic the protein-channel machinery. I-quartet channels offer an exciting opportunity to encapsulate water-clusters in confined chiral space and offer a chance to explore water behaviors very close to pore-confined biological water, exhibiting properties at the limits between solid and liquid states. Moreover all these unique features might deserve to confirm a number of structural descriptors explaining the structure-directed functions of natural channel proteins. The water-free I-quartet-"off form" superstructure described by our group is reminiscent of a closed proton gate conformation of the M2 Influenza A M2IA protein [12, 13]. Slight conformational changes lead to the water assisted opening of the I-quartets through which protons can diffuse along dipolar oriented water-wires in the open state pore-gate region. Our opinion is that the I-quartet system represents a fundamentally important biomimetic superstructure for water/proton transport that can be extensively developed based on new structural motifs.

First results may illustrate these ideas. Previous MD simulation studies of dynamic behaviors of the artificial I-quartet water channels and water molecules under confined conditions in a lipid bilayer environment were reported in [19–21]. An important aspect to bear in mind from these works is the quantitative correlation between the dipolar orientation of water-wires and their effects on their stability once inserted in the bilayer membranes. Water and proton translocation properties through AWCs are interlinked with the water-wire orientation as well. Two extreme cases are illustrated in Fig. 3, where water is either permeating in a random pathway or through highly ordered channels, depending on small modifications on the I-quarted building blocks. These studies provide detailed experimental data correlated with extensive theoretical simulations on water permeability with efficiency similar to that of natural biological systems. The theoretical stability and dynamics of the assembly have been tested first, probing different stoichiometries and box sizes. As next steps, the energetics of water transport will be assessed [54]. This knowledge will enable us to suggest modifications in the channel scaffold to modulate and improve water transport that can subsequently be correlated experimentally. Once a full characterization at an all-atom level has been achieved, we will be able to devise a simplified coarse-grained model in order to study larger assemblies and longer timescales.

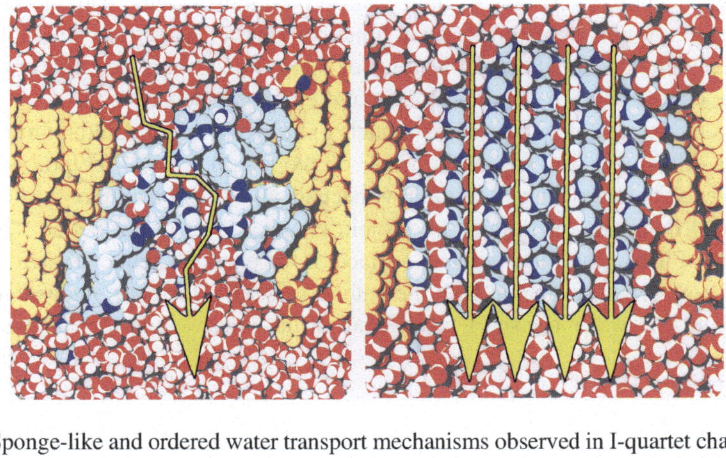

Fig. 3. Sponge-like and ordered water transport mechanisms observed in I-quartet channels [21].

3 Computational Design

3.1 A Roadmap for Improving Our Current Understanding

Here we sketch a computationnally-focused approach exploring unpaved ways to design artificial systems for water transport in bilayer membranes closer to the natural ones with the goal to further improve AWC performance. Their resultant functionality can be dependent on the nature of the scaffold and added functionalities, as their multivalent interactions have been shown to be influenced by such factors as nature of the confining matrix, the stability and the orientation of the channel, etc. In this context, computational studies, rational design and engineering of water in biological systems fulfils the following aims:

- Improve the understanding of the adaptive functional self-organization of molecular components into functional channel platforms by using a self-adaptive strategy and by exploring the synergistic influence of structural self-assembly/platform function and of the evolutive behaviours on the addition of different external stimuli.
- Accurately control functional responses through bilayer membrane platforms developed *via* tuned incorporation of components into channel superstructures.
- Investigation of application avenues towards larger impact of developed bio- and environmental responsive materials for reverse osmosis (RO) and forward osmosis (FO) membranes.

Such an approach is challenging and ambitious but highly feasible since, although based on new supramolecular/constitutional/MD simulation concepts, it relies on substantial preliminary data, expertise, experience and facilities related in particular to the I-quartet channel family of compounds illustrated in Figs. 1b and 2. Molecular dynamics simulations strongly contribute to understand and improve the dynamic behaviors of water molecules under confined conditions. The key is to fine-tune the balance between attractive and repulsive elements. The formation of single-file columns of water confined along the internal pore surfaces is a critical element. It may impose a net-dipole alignment of water molecules. Such an alignment influences the conduction of fluids, envisioned to diffuse along the dipolar hydrophilic pathways. Synthetic architectures generating stable water/proton translocation pathways in bilayer membranes to mimick the natural protein functions can be characterized at high spatial- and time resolutions. The dynamic features of these objects confined in a lipid bilayer create new opportunities to further modify and engineer such systems.

Ideally, the supramolecular water-channels to be synthesized would require a nearly frictionless diffusional pathway of the water on the length scale of the pore trajectory. What is needed is an enhanced liquid flow mechanism at the location of artificial pores - one critical hallmark for mimicking protein channels - or doing better. The strong interactions of the water with the inner surface of the channels reduce the efficiency of water transport, when compared with the natural systems. However, understanding the fine details of the water transport mechanism at the microscopic level is essential to decipher how hydrogen-bond switch events can assist efficient in-pore diffusivity. Straightforward synthetic access to superstructures - artificial Aquaporins - allowing water wires give rise to novel strategies to constitutionally build up very selective water transport devices.

Despite the few advances that have been made in this very young field, one may say that the discovery of novel synthetic systems able to form water-channels is most of the time empirical. But even if the discovery is empirical, the natural proteins may serve us to tailor the functions of the synthetic systems and engineer optimized molecular constructs. This interdependence makes our next challenge still more ambitious: understanding the natural systems transport mechanism is often one way to raise artificial systems to the level of natural functions or even beyond. Then, if we can imagine how the system is designed in our hands (minds) we could ask a number of questions. For this purpose it is particularly important to form structural models in our minds. Advanced visualization is a particularly powerful approach already extensively used to understand biological membrane objects [55]. We will get back to the strengths of novel technological approaches such as using virtual reality gear or high-resolution and large-scale display walls a little later in this discussion [56–60]. The aim of such a visual analysis is to bring novel concepts to the engineering of nanostructured functional systems, together with the understanding of the mechanisms underlying a rational use of water transport functions employed as the nanostructuring tools. Such an understanding would contribute both to fundamental and applicative research. For the former, leading to a better understanding of the natural function of protein channels and solving important biological scenarios of oriented water diffusion along channels. For the latter, learning form natural water purifying systems we might shed light on new strategies to generate highly selective materials/systems for advanced water purification.

The proposed approach, fundamental in nature, should permit to develop not only new knowledge but also to gain a greater insight on Biomimetic Artificial Water Systems, which can serve to understand complicated biological scenarios or to imagine novel membrane systems for water purification.

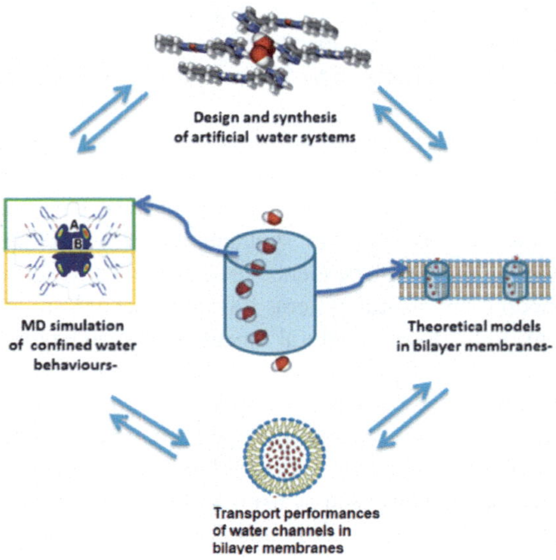

Fig. 4. illustrates the interplay of synthesis, experimental characterization and computational modeling. Water channel systems are synthetically engineered to enable platforms for the selective transport of water with active molecular control and motion. MD simulations and theoretical models of the channels embedded in realistic membrane systems guide the design of active systems through a mechanistic understanding of both the self-assembly process and the transport characterization offering experimental verification of the modeling predictions. Such an undertaking requires theoretical molecular models in realistic lipid bilayers and polymers, the design, synthesis and facilitated transport of water channel systems, and the structural characterization, stability studies of AWCs through bilayer membranes.

3.2 Theoretical Molecular Models in Realistic Lipid Bilayers and Polymers

A computational pipeline permits to explore the molecular organization of the candidate molecules in the membrane, and elucidate the mechanism of water transport. The approach we envision is based on coarse-grained (CG) and atomistic simulations supported by enhanced sampling. For a given compound, MD simulations based on CG models will be carried out so to explore and to individuate via enhanced sampling the possible free energy minima of the aggregates in the membrane. Several initial conditions should be tested (crystal like arrangement, spontaneous early steps aggregation, human design) so to enlarge the exploration. The individuated most stable aggregated states are filtered according to different criteria (energetics, packing, percolation) and will be back-mapped to the atomistic resolution. Further MD will be performed to test their capability for water transport and ion rejection. For the first study-cases the enhanced sampling will be performed at atomistic resolution to control the possible biases of the CG preliminary screening. The workflow will be performed for several stoichiometries. For each compound the combined information from the theoretically deduced structural/functional response of the molecules and from experiments, will suggest molecular modifications to be submitted to the pipeline depicted in Fig. 4. The preliminary CG approach has the advantage to reduce the computational cost for the most time-consuming part of the pipeline, the sampling of self-assembled structures. Such an approach should answer the following main questions:

- what is the structure and stability of AWCs in a membrane (or polymer) environment?
- how do water transport and ion rejection happen at the molecular level?
- how can transport (and AWCs properties in general) be tuned and optimized?

To further understand the structure and stability in a membrane environment, we need to investigate how the channels (self)assemble. The self-aggregation process raises a number of questions. What is the size of the aggregates? Are all channels identical/homogeneous? How does chirality affect the process?

To address water-channel functionality in a saline condition, we need to describe how precisely transport happens. MD simulations may be used to systematically examine the transport through AWCs of novel materials and specific AWCs to elucidate the role of hydration and detailed molecular interaction during nanofiltration. What is the geometry, mechanism and energetics of the process? How are ions rejected, and what happens to protons? These are essential characteristics that will enable to proceed to rational optimization of AWCs.

As briefly mentioned above, we performed MD simulations to understand the structure/activity relationships of the artificial water channels formed by I-quartets. We embedded small (ca. 3 nm wide) I-quartet channel patches based on the X-ray structures in a lipid bilayer environment with the same composition as that used in the experiments to begin each simulation. The MD results showed that I-quartets preferentially located within the bilayer membrane region and stabilized water channels in all simulations, although the degrees of water-occupancy, transport and structuring varied. We observed that the structural dynamics of I-quartet channels was dependent on the lateral pressure applied to the membrane, with higher pressure favoring the structuring of the water-wires and stabilizing the structure of the I-quartets. Very interestingly, different water permeation mechanisms were observed ranging from disordered cross-membrane pathways to strongly regular channel structures as depicted in Fig. 3.

Water transport and ion rejection characterization can be addressed even more quantitatively through Potential-of-Mean-Force type approaches or use of adapted collective variables to enhance the sampling. Osmotic gradients may be used to mimic experimental conditions. These techniques enable to determine quantitative energetics for transport of water, proton and ionic species. Statistical analysis can then provide the key features (geometry, mechanism) of transport. Large-(time&length) scale assessments may be required for selected systems, incorporating effects such as hydrodynamics within a simplified model.

Tuning of AWC properties to optimize transport and stability may be treated in an iterative fashion. First one may investigate a number of modifications such as decorating the transport channel walls. Such approaches were successfully used for understanding ion channel properties (work by Beckstein and Sansom, for instance [33, 34, 37, 61]) as well as e.g. water transport through carbon nanotubes (work by Hummer [35, 36]). We can draw on our own experience with ion channels and the fine characterization of water in biological systems [54]. Each modification should then be analyzed in terms of transport efficiency, ion rejection capability and stability.

3.3 Design, Synthesis and Facilitated Transport of Water Channel Systems

The key to the design of water channels is having stable pores wider than the critical diameter for selective binding to allow free flow of the water molecules. In general two main water translocation mechanisms were identified in selective filter channels working on a size restriction. They are related to:

- hydrophobic pores in which no H-bonding are occurring between the water and the channels and
- hydrophilic pores that specifically bind water through hydrogen-bonds, to the wall of the channel.

In the first case the design might be fairly straightforward; the size of the hydrophobic pore is needed to be <3 Å to block the passage of hydrated ions. This difficulty still lies in the synthetic routes for obtaining hydrophobic channels without H-bonding moieties that might induce the selective transport of water against the ions. Whilst the size restriction is necessary, water specific binding is needed for selectivity. We already find important structural details toward this goal, because of the recent discovery by our group of the artificial I-quartet channels introduced previously, having 100% selectivity against salts and very interesting water permeability of $\sim 1.5 \times 10^6$ water molecules/second/channel, approaching those of natural systems [7, 26]. To the best of our knowledge, the I-quartets are the only channels nowadays presenting total ion-exclusion, based on dimensional steric reasons, whereas water clustering effects appear to be also important [26]. The selective binding of water can be achieved through the synergetic combination of donor-acceptor H-bonding moieties. In both cases binding/transport of cations or anions are prohibited, as such processes would induce an energy penalty from the dehydration of the ions due to the simultaneous donor-acceptor interactions with the hydrogen-bonding channels, mainly leading to the selectivity of the water in competition with other ions.

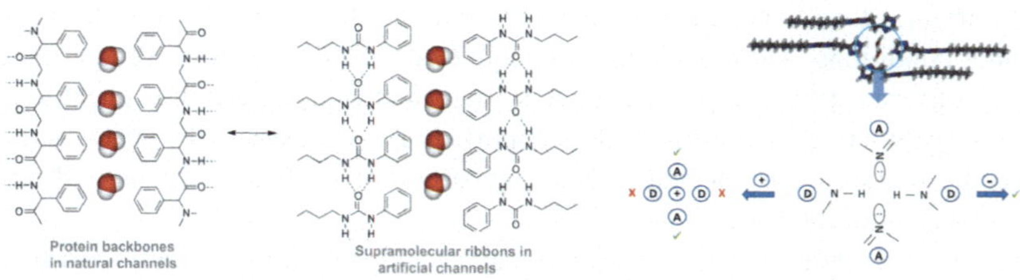

Fig. 5. Biomimetic artificial water channels (left) Natural protein versus artificial supramolecular ribbon self-assembly for water channel generation (right). Donor-acceptor-donor-acceptor H-bonding moieties in donor-acceptor based channels for water recognition incompatible with the binding of cations and anions.

We now intend to reach the next goal of making more competitive channels with controlled transport properties. This objective may require novel synthons combining a particularly broad range of features, decorating the functional polar head with various functionalities in order to enhance self-assembly and to control water binding properties of the polar head. The directional supramolecular guiding H-bonding interaction is reminiscent of the amide moiety's H-bonding in proteins and can be varied in order to modulate the dynamics of the ribbon's self-assembly. The hydrophobic tail provides another design opportunity.

4 Advanced Visual Exploration

4.1 Spatial Complexities in Visually Understanding AWCs

Analyzing water channels through membrane-traversing pores on a 2D standard desktop is a complex task due to the lack of a true perspective view of the spatial structure. Such channels may form intricate three-dimensional ramifications within the membrane. The difficulty resides in identifying precisely and effectively which part of the I-quartet molecules interacts with the water and constitute transport channels. In the models we investigated, several close-by yet independent channels are present, hence one may easily misinterpret which water molecule belongs to which channel in a typical crowded 2D projection. More advanced and intrinsically three-dimensional display hardware solutions provide powerful alternatives with respect to 2D screens to address these issues. We have carried out first experiments with three different setups, comparing a 2D desktop environment, virtual reality headsets for truly immersive in-depth exploration of artificial water channels, as well as a large-scale high-resolution display wall. The VR approach is very powerful to immerse an individual into the molecular world, whereas the display wall is more naturally prone to collaborative exchanges around a complex 3D scene. We use our open-source UnityMol software [62–64] for all experiments described here. The software is available for download and operates in desktop (2D), display wall (3D) and virtual reality headset (immersive 3D with head-tracking) modes.

4.2 Navigating Through AWCs in Three Dimensions

First experiments using 3D glasses on the display wall immediately showed the gain of adding a 3rd dimension to the visual exploration. With respect to the water channel visualization, the 3rd dimension adds depth of field enabling a scientist to more easily identify each channel and to highlight the water molecules present in the interior of a given pore. The large size of the display makes it easy to unite several collaborators with complementary expertise such as a domain scientist, an expert in visualization, a molecular modeling professional etc., such as to optimize the process and jointly achieve the best analysis and visual inspection of the system. In particular when using a surface representation of the channels and pores the 3D visualization provides a strong gain through improved perception of bumps and crevices. By (arbitrarily and geometrically) dividing the channels in two halves and displaying them as surfaces with different attributes, for instance in terms of color or transparency, the visualization is improved and the identification of differences in stabilization and permeation on each "side" is facilitated. Using a cutting plane to expose the channel interior is another approach that works well. When using the virtual reality headset instead, such a visual system preparation becomes less important because one can easily move inside the molecule, finely control and change the point of view to adjust to a given channel of interest instead of moving the whole molecular system into focus. The visual experience is clearly enhanced and requires less preparation to be efficient when moving from a desktop to 3D display wall to VR headset context. In all three cases it has proven difficult to intuitively select interesting parts of the system without an advanced geometrical selection capability, as the pores do not follow a unique and easily identifiable direction in the coordinate system. A prior selection, for example based on relevant residue ids allows to work around this current limitation.

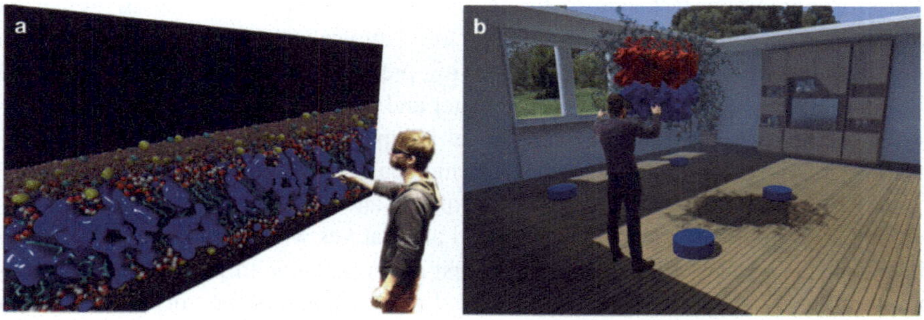

Fig. 6. Visual experiments exploring the properties of the I-quartet channel water conduction pathways on a 3D stereoscopic display wall (a) and in a fully immersive virtual reality environment using a headset (b). Both panels represent illustrative photomontages of such experiments.

4.3 First User Feedback

Although the 3D display hardware capacities are clearly benefical to the exploration, they are by no means sufficient to obtain insight into these complex molecular systems. It is as necessary to finetune the visual representations as it is to add the third dimension. We have collected the first user feedbacks on avenues for improving the visual repertoire for these systems.

The lipid bilayer is a useful visual cue, however it easily saturates the visual channel. An abstract and simplified representation, such as for instance depicting the phosphorus atoms representing the lipid headgroups as a cue for the bilayer provides a better experience. Similarly, representing parts of the system through transparency is required to depict the context without hiding the water channels and not overloading the scene. Highlighting hydrogen bonds in the system further helps to identify stabilizing elements holding the water channels together. By depicting the surface blobs of the water channels throughout trajectory playback, it is easily possible to follow leakage or possibly breakdown of the channels. Such observations naturally generate ideas about key parameters in the system and how to possibly quantify them. In the present case, for instance measuring the pore diameter during trajectory playback to quantify "openness". Some of these features still need to be implemented or refined in the UnityMol software before the next round of experiments can be carried out. Finetuning immersive visual analytic tools is a lengthy and iterative process of which we only have scratched the surface so far.

5 Conclusion

The field of artificial water-channels is still a very young field. High conduction activity obtained with natural compounds can be bio-mimicked using simpler compounds displaying constitutional functions like the natural ones in order to create all-made artificial biomimetic channels by using synthetic approaches. Improved synthetic biomimetic water-channels and pores bear great promise to enhance our insight into the natural function of macromolecular channels. Their development may lead to new strategies for the design of highly selective, advanced water purification systems. The membrane-assembled architectures, mimicking the natural protein functions and generating water/proton translocation pathways in bilayer membranes can be theoretically scrutinized to unravel underlying mechanisms. Because of the spatial complexities, advanced visualization approaches are particularly precious to achieve efficient visual analytics of these membrane-traversing objects.

Acknowledgments. This work was conducted within the ANR-15-CE29-0009 DYNAFUN (Mihail Barboiu) and ANR-18-CE06-0004 WATERCHANNELS (Fabio Sterpone) projects. The study was further supported by the "Initiative d'Excellence" program from the French State (Grant "DYNAMO", ANR-11-LABX-0011 and grant "CACSICE", ANR-11-EQPX-0008). X. M., A.H., H.S. and M.B. thank Sesame Ile-de-France for co-funding the display wall used in this study. Computational work was performed using HPC resources from GENCI-CINES (grant number 2016-072292) to Marc Baaden. X.M. and M.B. thank UCB Biopharma for support. This publication received support from the European Union's Horizon 2020 research and innovation programme under Grant No. 667387 WIDESPREAD 2-2014 SupraChem Lab.

References

1. Ball, P.: Water as an active constituent in cell biology. Chem. Rev. **108**, 74–108 (2008). https://doi.org/10.1021/cr068037a
2. Bellissent-Funel, M.-C., Hassanali, A., Havenith, M., Henchman, R., Pohl, P., Sterpone, F., van der Spoel, D., Xu, Y., Garcia, A.E.: Water determines the structure and dynamics of proteins. Chem. Rev. **116**, 7673–7697 (2016). https://doi.org/10.1021/acs.chemrev.5b00664
3. Gokel, G.W., Mukhopadhyay, A.: Synthetic models of cation-conducting channels. Chem. Soc. Rev. **30**, 274–286 (2001). https://doi.org/10.1039/b008667n
4. Sakai, N., Mareda, J., Matile, S.: Rigid-rod molecules in biomembrane models: from hydrogen-bonded chains to synthetic multifunctional pores. Acc. Chem. Res. **38**, 79–87 (2005). https://doi.org/10.1021/ar0400802

5. Cazacu, A., Tong, C., van der Lee, A., Fyles, T.M., Barboiu, M.: Columnar self-assembled ureido crown ethers: an example of ion-channel organization in lipid bilayers. J. Am. Chem. Soc. **128**, 9541–9548 (2006). https://doi.org/10.1021/ja061861w
6. Fyles, T.M.: Synthetic ion channels in bilayer membranes. Chem. Soc. Rev. **36**, 335–347 (2007). https://doi.org/10.1039/B603256G
7. Le Duc, Y., Michau, M., Gilles, A., Gence, V., Legrand, Y.-M., van der Lee, A., Tingry, S., Barboiu, M.: Imidazole-quartet water and proton dipolar channels. Angew. Chem. Int. Ed. **50**, 11366–11372 (2011). https://doi.org/10.1002/anie.201103312
8. Barboiu, M.: Artificial water channels. Angew. Chem. Int. Ed. **51**, 11674–11676 (2012). https://doi.org/10.1002/anie.201205819
9. Barboiu, M., Gilles, A.: From natural to bioassisted and biomimetic artificial water channel systems. Acc. Chem. Res. **46**, 2814–2823 (2013). https://doi.org/10.1021/ar400025e
10. Barboiu, M., Le Duc, Y., Gilles, A., Cazade, P.-A., Michau, M., Marie Legrand, Y., van der Lee, A., Coasne, B., Parvizi, P., Post, J., Fyles, T.: An artificial primitive mimic of the Gramicidin-A channel. Nat. Commun. **5**, 4142 (2014). https://doi.org/10.1038/ncomms5142
11. Barboiu, M., Cazade, P.-A., Le Duc, Y., Legrand, Y.-M., van der Lee, A., Coasne, B.: Polarized water wires under confinement in chiral channels. J. Phys. Chem. B **119**, 8707–8717 (2015). https://doi.org/10.1021/acs.jpcb.5b03322
12. Schnell, J.R., Chou, J.J.: Structure and mechanism of the M2 proton channel of influenza A virus. Nature **451**, 591–595 (2008). https://doi.org/10.1038/nature06531
13. Hu, F., Luo, W., Hong, M.: mechanisms of proton conduction and gating in influenza M2 proton channels from solid-state NMR. Science **330**, 505–508 (2010). https://doi.org/10.1126/science.1191714
14. Cheruzel, L.E., Pometun, M.S., Cecil, M.R., Mashuta, M.S., Wittebort, R.J., Buchanan, R. M.: Structures and solid-state dynamics of one-dimensional water chains stabilized by imidazole channels. Angew. Chem. Int. Ed. **42**, 5452–5455 (2003). https://doi.org/10.1002/anie.200352157
15. Fei, Z., Zhao, D., Geldbach, T.J., Scopelliti, R., Dyson, P.J., Antonijevic, S., Bodenhausen, G.: A synthetic zwitterionic water channel: characterization in the solid state by x-ray crystallography and NMR spectroscopy. Angew. Chem. Int. Ed. **44**, 5720–5725 (2005). https://doi.org/10.1002/anie.200500207
16. Legrand, Y.-M., Michau, M., van der Lee, A., Barboiu, M.: Homomeric and heteromeric self-assembly of hybrid ureido–imidazole compounds. Cryst. Eng. Comm. **10**, 490 (2008). https://doi.org/10.1039/b717015g
17. Legrand, Y.-M., van der Lee, A., Masquelez, N., Rabu, P., Barboiu, M.: Temperature induced single-crystal-to-single-crystal transformations and structure directed effects on magnetic properties. Inorg. Chem. **46**, 9083–9089 (2007). https://doi.org/10.1021/ic700867b
18. Barboiu, M.: Artificial water channels – incipient innovative developments. Chem. Commun. **52**, 5657–5665 (2016). https://doi.org/10.1039/C6CC01724J
19. Licsandru, E., Kocsis, I., Shen, Y., Murail, S., Legrand, Y.-M., van der Lee, A., Tsai, D., Baaden, M., Kumar, M., Barboiu, M.: Salt-excluding artificial water channels exhibiting enhanced dipolar water and proton translocation. J. Am. Chem. Soc. **138**, 5403–5409 (2016). https://doi.org/10.1021/jacs.6b01811
20. Kocsis, I., Sorci, M., Vanselous, H., Murail, S., Sanders, S.E., Licsandru, E., Legrand, Y.-M., van der Lee, A., Baaden, M., Petersen, P.B., Belfort, G., Barboiu, M.: Oriented chiral water wires in artificial transmembrane channels. Sci. Adv. **4**, eaao5603 (2018). https://doi.org/10.1126/sciadv.aao5603
21. Murail, S., Vasiliu, T., Neamtu, A., Barboiu, M., Sterpone, F., Baaden, M.: Water permeation across artificial I-quartet membrane channels: from structure to disorder. Faraday Discuss. **209**, 125–148 (2018). https://doi.org/10.1039/C8FD00046H

22. Sterpone, F., Derreumaux, P., Melchionna, S.: Protein simulations in fluids: coupling the OPEP coarse-grained force field with hydrodynamics. J. Chem. Theor. Comput. **11**, 1843–1853 (2015). https://doi.org/10.1021/ct501015h
23. Brandner, A.F., Timr, S., Melchionna, S., Derreumaux, P., Baaden, M, Sterpone, F.: Modelling lipid systems in fluid with Lattice Boltzmann Molecular Dynamics simulations and hydrodynamics. Sci. Rep. **9**, 16450 (2019). https://doi.org/10.1038/s41598-019-52760-y
24. Agre, P.: Aquaporin water channels (nobel lecture). Angew. Chem. Int. Ed. **43**, 4278–4290 (2004). https://doi.org/10.1002/anie.200460804
25. Tajkhorshid, E.: Control of the selectivity of the aquaporin water channel family by global orientational tuning. Science **296**, 525–530 (2002). https://doi.org/10.1126/science.1067778
26. Kumar, M., Grzelakowski, M., Zilles, J., Clark, M., Meier, W.: Highly permeable polymeric membranes based on the incorporation of the functional water channel protein Aquaporin Z. Proc. Natl. Acad. Sci. **104**, 20719–20724 (2007). https://doi.org/10.1073/pnas.0708762104
27. Hinds, B.J.: Aligned multiwalled carbon nanotube membranes. Science **303**, 62–65 (2004). https://doi.org/10.1126/science.1092048
28. de Groot, B.L., Grubmüller, H.: The dynamics and energetics of water permeation and proton exclusion in aquaporins. Curr. Opin. Struct. Biol. **15**, 176–183 (2005). https://doi.org/10.1016/j.sbi.2005.02.003
29. Kaucher, M.S., Peterca, M., Dulcey, A.E., Kim, A.J., Vinogradov, S.A., Hammer, D.A., Heiney, P.A., Percec, V.: Selective transport of water mediated by porous dendritic dipeptides. J. Am. Chem. Soc. **129**, 11698–11699 (2007). https://doi.org/10.1021/ja076066c
30. Hu, X.-B., Chen, Z., Tang, G., Hou, J.-L., Li, Z.-T.: Single-molecular artificial transmembrane water channels. J. Am. Chem. Soc. **134**, 8384–8387 (2012). https://doi.org/10.1021/ja302292c
31. Si, W., Chen, L., Hu, X.-B., Tang, G., Chen, Z., Hou, J.-L., Li, Z.-T.: Selective artificial transmembrane channels for protons by formation of water wires. Angew. Chem. Int. Ed. **50**, 12564–12568 (2011). https://doi.org/10.1002/anie.201106857
32. de Groot, B.L., Grubmüller, H.: Water permeation across biological membranes: mechanism and dynamics of aquaporin-1 and GlpF. Science **294**, 2353–2357 (2001). https://doi.org/10.1126/science.1062459
33. Beckstein, O., Sansom, M.S.P.: The influence of geometry, surface character, and flexibility on the permeation of ions and water through biological pores. Phys. Biol. **1**, 42–52 (2004). https://doi.org/10.1088/1478-3967/1/1/005
34. Rao, S., Lynch, C.I., Klesse, G., Oakley, G.E., Stansfeld, P.J., Tucker, S.J., Sansom, M.S.P.: Water and hydrophobic gates in ion channels and nanopores. Faraday Discuss. **209**, 231–247 (2018). https://doi.org/10.1039/c8fd00013a
35. Hummer, G., Rasaiah, J.C., Noworyta, J.P.: Water conduction through the hydrophobic channel of a carbon nanotube. Nature **414**, 188–190 (2001). https://doi.org/10.1038/35102535
36. Kalra, A., Garde, S., Hummer, G.: Osmotic water transport through carbon nanotube membranes. Proc. Natl. Acad. Sci. U.S.A. **100**, 10175–10180 (2003). https://doi.org/10.1073/pnas.1633354100
37. García-Fandiño, R., Sansom, M.S.P.: Designing biomimetic pores based on carbon nanotubes. Proc. Natl. Acad. Sci. U.S.A. **109**, 6939–6944 (2012). https://doi.org/10.1073/pnas.1119326109
38. Barboiu, M., Kumar, M., Baaden, M., Gale, P.A., Hinds, B.J.: Highlights from the faraday discussion on artificial water channels, Glasgow, UK. Chem. Commun. Camb. Engl. **55**, 3853–3858 (2019). https://doi.org/10.1039/c9cc90112d

39. Baaden, M., Barboiu, M., Bill, R.M., Chen, C.-L., Davis, J., Di Vincenzo, M., Freger, V., Fröba, M., Gale, P.A., Gong, B., Hélix-Nielsen, C., Hickey, R., Hinds, B., Hou, J.-L., Hummer, G., Kumar, M., Legrand, Y.-M., Lokesh, M., Mi, B., Murail, S., Pohl, P., Sansom, M., Song, Q., Song, W., Törnroth-Horsefield, S., Vashisth, H., Vögele, M.: Biomimetic water channels: general discussion. Faraday Discuss. **209**, 205–229 (2018). https://doi.org/10.1039/c8fd90020e
40. Baaden, M., Barboiu, M., Borthakur, M.P., Chen, C.-L., Coalson, R., Davis, J., Freger, V., Gong, B., Hélix-Nielsen, C., Hickey, R., Hinds, B., Hirunpinyopas, W., Horner, A., Hou, J.-L., Hummer, G., Iamprasertkun, P., Kazushi, K., Kumar, M., Legrand, Y.-M., Lokesh, M., Mi, B., Mitra, S., Murail, S., Noy, A., Nunes, S., Pohl, P., Song, Q., Song, W., Tornroth-Horsefield, S., Vashisth, H.: Applications to water transport systems: general discussion. Faraday Discuss. **209**, 389–414 (2018). https://doi.org/10.1039/c8fd90022a
41. Baaden, M., Barboiu, M., Bill, R.M., Casanova, S., Chen, C.-L., Conner, M., Freger, V., Gong, B., Góra, A., Hinds, B., Horner, A., Hummer, G., Kumar, M., Lokesh, M., Mitra, S., Noy, A., Pohl, P., Sadet, A., Sansom, M., Törnroth-Horsefield, S., Vashisth, H.: Structure and function of natural proteins for water transport: general discussion. Faraday Discuss. **209**, 83–95 (2018). https://doi.org/10.1039/c8fd90019a
42. Baaden, M., Borthakur, M.P., Casanova, S., Coalson, R., Freger, V., Gonzalez, M., Góra, A., Hinds, B., Hirunpinyopas, W., Hummer, G., Kumar, M., Lynch, C., Murail, S., Noy, A., Sansom, M., Song, Q., Vashisth, H., Vögele, M.: The modelling and enhancement of water hydrodynamics: general discussion. Faraday Discuss. **209**, 273–285 (2018). https://doi.org/10.1039/c8fd90021c
43. MacKinnon, R.: Potassium channels and the atomic basis of selective ion conduction (nobel lecture). Angew. Chem. Int. Ed. **43**, 4265–4277 (2004). https://doi.org/10.1002/anie.200400662
44. Roux, B.: Computational studies of the gramicidin channel. Acc. Chem. Res. **35**, 366–375 (2002). https://doi.org/10.1021/ar010028v
45. Saparov, S.M., Pohl, P.: Beyond the diffusion limit: Water flow through the empty bacterial potassium channel. Proc. Natl. Acad. Sci. **101**, 4805–4809 (2004). https://doi.org/10.1073/pnas.0308309101
46. Tang, C.Y., Zhao, Y., Wang, R., Hélix-Nielsen, C., Fane, A.G.: Desalination by biomimetic aquaporin membranes: review of status and prospects. Desalination **308**, 34–40 (2013). https://doi.org/10.1016/j.desal.2012.07.007
47. Barbour, L.J., Orr, G.W., Atwood, J.L.: An intermolecular $(H_2O)_{10}$ cluster in a solid-state supramolecular complex. Nature **393**, 671–673 (1998). https://doi.org/10.1038/31441
48. Yoshizawa, M., Kusukawa, T., Kawano, M., Ohhara, T., Tanaka, I., Kurihara, K., Niimura, N., Fujita, M.: Endohedral clusterization of ten water molecules into a "molecular ice" within the hydrophobic pocket of a self-assembled cage. J. Am. Chem. Soc. **127**, 2798–2799 (2005). https://doi.org/10.1021/ja043953w
49. Görbitz, C.H.: Microporous organic materials from hydrophobic dipeptides. Chem. – Eur. J. **13**, 1022–1031 (2007). https://doi.org/10.1002/chem.200601427
50. Müller, A., Bögge, H., Diemann, E.: Structure of a cavity-encapsulated nanodrop of water. Inorg. Chem. Commun. **6**, 52–53 (2003). https://doi.org/10.1016/S1387-7003(02)00679-2
51. Mitra, T., Miró, P., Tomsa, A.-R., Merca, A., Bögge, H., Ávalos, J.B., Poblet, J.M., Bo, C., Müller, A.: Gated and differently functionalized (new) porous capsules direct encapsulates' structures: higher and lower density water. Chem. – Eur. J. **15**, 1844–1852 (2009). https://doi.org/10.1002/chem.200801602
52. Esque, J., Sansom, M.S.P., Baaden, M., Oguey, C.: Analyzing protein topology based on Laguerre tessellation of a pore-traversing water network. Sci. Rep. **8**, 13540 (2018). https://doi.org/10.1038/s41598-018-31422-5

53. Horner, A., Pohl, P.: Single-file transport of water through membrane channels. Faraday Discuss. **209**, 9–33 (2018). https://doi.org/10.1039/c8fd00122g
54. Sauguet, L., Poitevin, F., Murail, S., Van Renterghem, C., Moraga-Cid, G., Malherbe, L., Thompson, A.W., Koehl, P., Corringer, P.-J., Baaden, M., Delarue, M.: Structural basis for ion permeation mechanism in pentameric ligand-gated ion channels. EMBO J. **32**, 728–741 (2013). https://doi.org/10.1038/emboj.2013.17
55. Baaden, M.: Visualizing biological membrane organization and dynamics. J. Mol. Biol. **431**, 1889–1919 (2019). https://doi.org/10.1016/j.jmb.2019.02.018
56. Hirst, J.D., Glowacki, D.R., Baaden, M.: Molecular simulations and visualization: introduction and overview. Faraday Discuss. **169**, 9–22 (2014). https://doi.org/10.1039/c4fd90024c
57. Sommer, B., Baaden, M., Krone, M., Woods, A.: From virtual reality to immersive analytics in bioinformatics. J. Integr. Bioinforma **15** (2018). https://doi.org/10.1515/jib-2018-0043
58. Hirst, J., Zheng, R., Zoppè, H., O'Donoghue, P., Glowacki, D., Brancale, B., Stone, B., Olson, C., Haag, K., Kouyoumdjian, L., Reiher, F., Raffin, B., Montes, D.: Virtual and augmented reality immersive molecular simulations: general discussion. Faraday Discuss. **169**, 143–166 (2014). https://doi.org/10.1039/c4fd90019g
59. Proctor, E., Petrov, O., Brooks, H., Stone, K., Zoppè, H., Baaden, R., Sommer, F., Rozmanov, G.: Advanced visualization and visual analytics: general discussion. Faraday Discuss. **169**, 245–264 (2014). https://doi.org/10.1039/c4fd90020k
60. Maes, A., Martinez, X., Druart, K., Laurent, B., Guégan, S., Marchand, C.H., Lemaire, S.D., Baaden, M.: MinOmics, an integrative and immersive tool for multi-omics analysis. J. Integr. Bioinforma **15** (2018). https://doi.org/10.1515/jib-2018-0006
61. Beckstein, O., Tai, K., Sansom, M.S.P.: Not ions alone: barriers to ion permeation in nanopores and channels. J. Am. Chem. Soc. **126**, 14694–14695 (2004). https://doi.org/10.1021/ja045271e
62. Lv, Z., Tek, A., Da Silva, F., Empereur-mot, C., Chavent, M., Baaden, M.: Game on, science - how video game technology may help biologists tackle visualization challenges. PLoS ONE **8**, e57990 (2013). https://doi.org/10.1371/journal.pone.0057990
63. Doutreligne, S., Cragnolini, T., Pasquali, S., Derreumaux, P., Baaden, M.: UnityMol: interactive scientific visualization for integrative biology. 2014 IEEE 4th Symposium on Large Data Analysis and Visualization (LDAV), pp. 109–110. IEEE, France (2014)
64. Laureanti, J., Brandi, J., Offor, E., Engel, D., Rallo, R., Ginovska, B., Martinez, X., Baaden, M., Baker, N.A.: Visualizing biomolecular electrostatics in virtual reality with UnityMol-APBS. Protein Sci. Publ. Protein Soc. **29**, 237–246 (2020). https://doi.org/10.1002/pro.3773

Modelling a Light Transducing Protocell Population

Fabio Mavelli[1,2], Paola Albanese[1], and Emiliano Altamura[1(✉)]

[1] Chemistry Department, University of Bari Aldo Moro, 70126 Bari, Italy
emiliano.altamura@uniba.it
[2] Institute of Nanotechnology, Italian National Research Council, 70126 Bari, Italy

Abstract. In this contribution, the approach to the modelling of a population of transducing light giant unilamellar vesicles will be addressed taking into account the problem of poly-dispersion in size and composition of the lipid compartments. A general approach is presented and discussed along with some preliminary results.

1 Introduction

Protocells are artificial compartments that mimic the behavior or some relevant features of modern living cells (Stano 2018). In the so-called semisynthetic approach (Luisi et al. 2006), they are mainly based on giant unilamellar vesicles (GUVs) that, for these purposes, entrap in the aqueous core or embed in the lipid membrane both synthetic compounds and biological extracts (enzymes, nucleic acids, cellular organelles, etc.) (Walde et al. 2010). In recent years, in our research group, we developed a research line focused on the preparation of GUVs able to transduce light energy in chemical energy with the aim of synthetizing protocells energetically autonomous (Stano et al. 2017). For this kind of reacting compartments, we coined the term of *Artificial Simplified-Autotroph Protocells* (ASAPs) referred to protocells able to sustain an internal metabolism by exploiting an external source of energy. In particular, our goal is to mimic the bacterial photosynthesis and to implement photo-active ASAPs following two different strategies: a single compartment and multi compartment approach respectively. Being inspired from the purple non-sulfur bacteria *Rhodobacter (R.) sphaeroides*, photosynthetic organisms that convert light energy in chemical energy stored in form of ATP molecules, we tried to reconstitute in GUVs all the enzymes engaged in this process. In the particular case of *R. sphaeroides*, there are three well-characterized integral membrane enzymes: the reaction center (RC), the ubiquinol:cytochrome c oxidoreductase (bc1) and the ATP synthase (ATPsyn). All these three enzymes are present in the bacterial cytoplasmic membrane and cooperate to perform the first step of the bacterial photosynthesis. The first two protein complexes are involved in a photo-induced cyclic electron transport and transmembrane proton translocation involving the ubiquinone/ubiquinol (Q/QH_2) and cytochrome c_2 (cyt^{2+}/cyt^{3+}) redox pools, while ATPsyn eventually converts the proton gradient into ATP molecules, via ADP phosphorylation (Fig. 1a).

S. Piotto et al. (Eds.): BIONAM 2019, LNBE, pp. 100–110, 2020.
https://doi.org/10.1007/978-3-030-47705-9_9

In the Single Compartment Approach (SCA), the three photosynthetic enzymes must be reconstituted in the giant vesicle membrane with the correct orientation, i.e. the same alignment they have in the cytoplasmic cellular membrane, to produce a photo-driven ATP inside the GUVs (Fig. 1b). Thanks to a suitable preparation procedure, we have been able to reconstitute RC in the membrane of GUVs keeping its photo-activity and 90% of its physiological orientation (Altamura et al. 2017b) and now we are going forward in this direction by reconstituting bc1 in the GUV lipid membrane (Altamura et al. 2020a). On the other hand, in the multi-compartment approach (MCA) artificial or biological organelles, producing ATP by light irradiation, can be entrapped in the internal lumen of GUVs in order to prepare artificial multi-compartment architectures for the light transduction. Following MCA, different authors have already presented example of ASAPs by encapsulating nanosized vesicles containing ATPsyn and Rhodopsin in GUVs. Indeed, under irradiation, these smaller vesicles work as organelles able to photo-produce ATP eventually exploited for sustaining protein expression (Berhanu et al. 2019) or a biological polymerization (Lee et al. 2018). In our group, we are exploring the possibility of extracting *chromatophores* directly from *R. sphaeroides* bacteria (Altamura et al. 2020b). Chromatophores are natural organelles containing all the photosynthetic protein complexes able to produce ATP driven by light (Altamura et al. 2018).

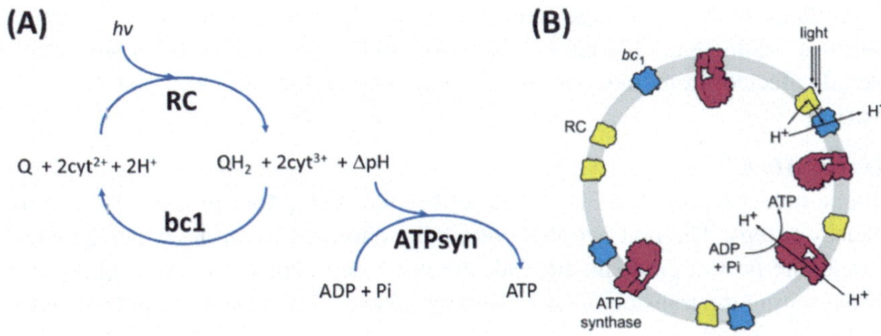

Fig. 1. (A) a simplified model of the photo-production of ATP, (B) a schematic drawing of the ASAP architecture prepared following the SCA procedures.

In this research framework, the aim of this contribution is to present a Monte Carlo hybrid simulation approach for kinetically modelling compartmentalized reacting systems. In particular, this approach allows to estimate the effect of the compartment poly-dispersity on the time course of the GUV population.

Indeed, the time behavior of compartmentalized reacting systems can be influenced by two sources of randomness: intrinsic and extrinsic stochasticity. The first kind of random source is due to reacting molecules that, being in energetically different conditions, can exhibit different reaction probabilities: intrinsic random effects (Mavelli and Piotto 2006, Mavelli and Ruiz-Mirazo 2010). These effects are more evident for lower numbers of reacting molecules in the system (Mavelli et al. 2014). For instance, in the case of giant vesicles, these effects are quite negligible, if the concentration is greater than 100 nM, due the large volume of the compartment. On the other hand, a second source of stochasticity can be ascribed to the poly-dispersity of a reacting compartment population in terms of morphology and composition. In particular, a population of giant spherical vesicles can exhibit both size and composition broad distributions that strongly depend on the preparation procedure (Stano et al. 2015). Moreover, the random composition of compartments and the presence of catalytic compounds can highly influence the compartment dynamic behavior. This can induce a strong displacement from the average time course of the population (Calviello et al. 2013). To have an evidence of this kind of effects, we used a simplified enzymatic model of light transduction for the photo-production of ATP as reported in Fig. 1(A) and a hybrid stochastic approach to simulate the protocell population (Mavelli and Stano 2015). The model describes the internal ATP production of a protocell of a given size and a defined composition prepared as a single compartment following the SCA procedures, Fig. 1(B). Moreover, to estimate the behavior of a vesicle population assumed monodispersed in size, uncorrelated Gaussian concentration distributions are assumed for the three photosynthetic enzymes. Thus, 1000 different GUVs are produced with a random enzymatic composition and for each compartment the deterministic time course is calculated. Thus, 1000 different time behaviors are collected and, from them, both the average time course and the standard deviation of the entire population are estimated. Of course, it is important to stress that no other substrates are considered affected from random distribution among the vesicles, just for sake of simplicity.

The Kinetic Model

The kinetic model reported in Fig. 1A describes the ATP photo-production in terms of three reaction steps. The first two that outline the photo-redox cycle, which generates a transmembrane proton gradient. Instead, the third step depicts the ADP phosphorylation by exploiting the transmembrane proton gradient. The rates of these three steps are assumed to be as follows:

$$R_{RC} = k_{RC}[RC]\left(\frac{[cyt^{2+}]_{ex}}{K_{cyt^{2+}} + [cyt^{2+}]_{ex}}\right)^{\gamma_1}\left(\frac{[Q]}{K_Q + [Q]}\right)\left(\frac{[H^+]}{K_{H^+} + [H^+]}\right)^2 \tag{1}$$

$$R_{bc1} = k_{bc1}[bc1]\left(\frac{[cyt^{3+}]_{ex}}{K_{cyt^{3+}} + [cyt^{3+}]_{ex}}\right)^{\gamma_2}\left(\frac{[QH_2]}{K_{QH_2} + [QH_2]}\right)\left(\frac{[H^+]}{K_{H^+} + [H^+]}\right)^2 \tag{2}$$

$$R_{ATP_{syn}} = k_{ATP_{syn}}[ATP_{syn}]\left(\frac{[ADP]}{K_{ADP} + [ADP]}\right)\left(\frac{[Pi]}{K_{Pi} + [Pi]}\right)\Delta[H^+] \tag{3}$$

where all the concentrations are calculated as volume concentrations in the internal aqueous core of GUVs, except for the concentrations of the redox pair cyt^{2+}/cyt^{3+} that are related to the external environment. Therefore, since the concentration of GUVs is very low, the total amount of both cyt^{2+} and cyt^{3+} is in excess compared to the total amount of entrapped substrates. As a consequence, the external concentrations $[cyt^{2+}]_{ex}$ and $[cyt^{3+}]_{ex}$ can be considered constant throughout the process.

Table 1. Parameters of the kinetic model: kinetic and Michaelis-Menten constants and surface enzyme concentrations.

Parameter	Value	Reference
C^S_{RC}	1200 molecule/μm^2	Altamura et al. (2017a)
C^S_{bc1}	200 molecule/μm^2	Senner et al. (2016)
C^S_{ATPsyn}	88 molecule/μm^2	Feniouk et al. (2002)
k_{RC}	2.74 10^6 min^{-1}	Altamura et al. (2017a)
k_{bc1}	8.32 10^4 min^{-1}	Altamura et al. (2017a)
k_{ATPsyn}	35 s^{-1}	Iino et al. (2009)
K_Q	1.19 10^2 mM	Altamura et al. (2017a)
$K_{cyt^{2+}}$	6.48 10^{-5} mM	Altamura et al. (2017a)
K_{QH_2}	8.88 10^{-1} mM	Altamura et al. (2017a)
$K_{cyt^{3+}}$	7.06 10^{-1} mM	Altamura et al. (2017a)
K_{H^+}	1.0 10^{-4} mM	Altamura et al. (2017b)
K_{ADP}	1.0 10^{-1} mM	Iino et al. (2009)
K_{Pi}	4.2 mM	Iino et al. (2009)
γ_1	4.8	Altamura et al. (2017a)
γ_2	3.5	Altamura et al. (2017a)
P_{H^+}	4.9e−5 cm/sec	Paula et al. (1996)

The three photosynthetic enzymes are transmembrane protein complexes and their expected surface concentrations C^S_E are reported in Table 1. C^S_E are converted in volume concentrations $[E]$ by multiplying them for the surface/volume ratio ($3/r_{GUV}$) of a spherical GUV of radius r_{GUV}. This allows to evidence better the contribution of each enzyme in the variation of the aqueous internal composition of GUVs of different sizes. In fact, the photosynthetic surface concentration of each enzyme is assumed constant in average on the GUV membrane independently of the vesicle radius.

As can be seen in the rate formulas, Eqs. (1)–(3), a Michaelis-Menten kinetic term, power to a stoichiometric or an optimized exponent, is introduced to consider the influence of each substrate on the enzymatic rate. Furthermore, $R_{ATP_{syn}}$ is assumed to depend linearly on the transmembrane proton gradient so coupling the ADP phosphorylation with the RC-bc1 photo-redox cycle. By assuming the presence of a pH chemical buffer in the outer aqueous solution, pH in the external environment is supposed to be constantly equal to the physiological value: pH = 7.4. No pH buffer is instead considered present in the internal aqueous core of vesicles; however, the internal proton concentration is initially set in equilibrium with the external proton concentration $[H^+]_0 = 10^{-7.4}$ mol/dm^3. As soon as the GUV suspension is irradiated by light, i.e. $t > 0$, the internal proton concentration decreases being dependent on the balance of three processes: the photo-redox cycle, that consumes 4 protons for each cycle from the internal lumen, the proton diffusion across the lipid membrane, driven by the concentration gradient, and the ATPsyn catalyzed phosphorylation.

In Table 1, all the kinetic parameters used for this model are reported along with the literature references from where the values have been derived.

Equations 4–9 show the complete ordinary differential equation set to be solved for obtaining the time course of the ATP production inside a GUV of a given radius r_{GUV}:

$$\frac{d[Q]}{dt} = (R_{bc1} - R_{RC}) \tag{4}$$

$$\frac{d[QH_2]}{dt} = -(R_{bc1} - R_{RC}) \tag{5}$$

$$\frac{d[H^+]}{dt} = -2(R_{bc1} + R_{RC}) + R_{ATP_{syn}} + \frac{P_{H^+}}{r_{GUV}}\left(10^{-7.4} - [H^+]\right) \tag{6}$$

$$\frac{d[ADP]}{dt} = -R_{ATP_{syn}} \tag{7}$$

$$\frac{d[ATP]}{dt} = +R_{ATP_{syn}} \tag{8}$$

$$\frac{d[Pi]}{dt} = -R_{ATP_{syn}} \tag{9}$$

As it can be seen, the internal species are the only species which concentrations are assumed to change in time. As already discussed, both the cyt^{2+} and cyt^{3+} pair and the external proton H^+ are considered constant due to an excess of molecules and/or to the presence of a pH buffer.

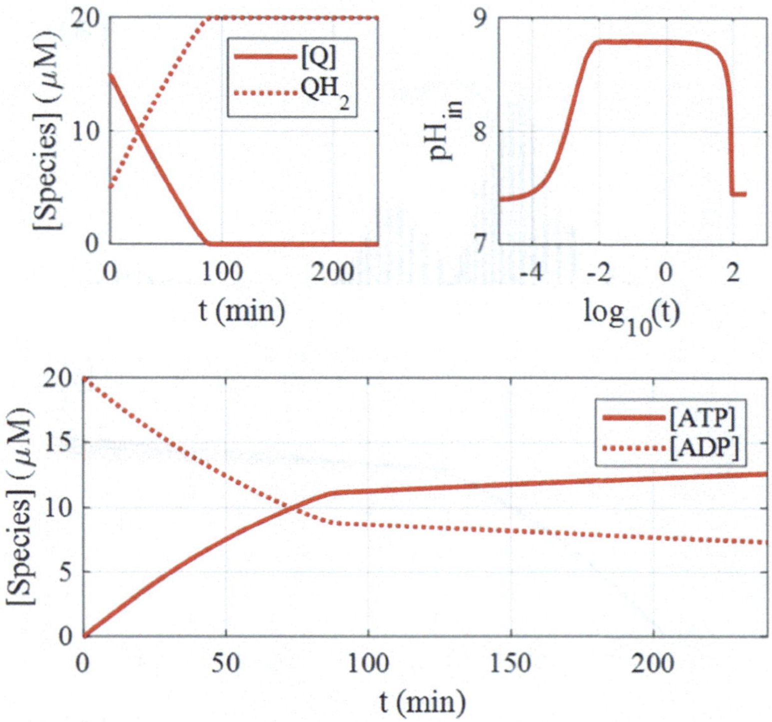

Fig. 2. Time course of concentration of the internal species, obtained by numerically solving ODEs (4)–(9) using the kinetic parameters reported in Table 1 with the following external concentrations: $[cyt^{2+}]_{ex} = [cyt^{3+}]_{ex} = 10$ μM, $pH_{ex} = 7.4$, that remain constant, and the initial concentrations: $[Q]_0 = 15$ μM, $[QH_2]_0 = 5$ μM, $[H]_0 = 10^{-7.4}$, $[ADP]_0 = 20$ μM, $[ATP] = 0$ μ M, $[Pi]_0 = 20$ mM of species that change in time.

2 Results

Figure 2 shows the time courses of species concentrations entrapped in a GUV of 15 μm radius. The concentration of phosphate [Pi] is not presented since it remains practically constant in the investigated time range (data not shown). In fact, the phosphate initial concentration is very high compared to ADP: $[Pi]_0 = 20$ mM $\gg$ $[ADP]_0 = 20$ μM.

At time 0, the GUV suspension is assumed to be irradiated by light, so that the photo-redox cycle starts working and an initial abrupt increase of the internal pH is observed. In about 90 min, all the amount of the quinone Q present in the GUV is converted in quinole QH_2. In fact, the RC is much more efficient than bc1 in these conditions, i.e. when the ratio RC/bc1 is 6/1 (Altamura 2017a). Therefore, in the first 90 min the ATP production is very fast, due to this abrupt pH increase. However, when the internal pH reaches a stationary value around pH = 7.5, the ATP production becomes slower, since the transmembrane proton gradient is now much decreased. In the stationary conditions, the rate remains constant for the next 2.5 h and the ATP concentration reaches 12.7 μM, with a yield of conversion about 64%, after four hours from the light irradiation start.

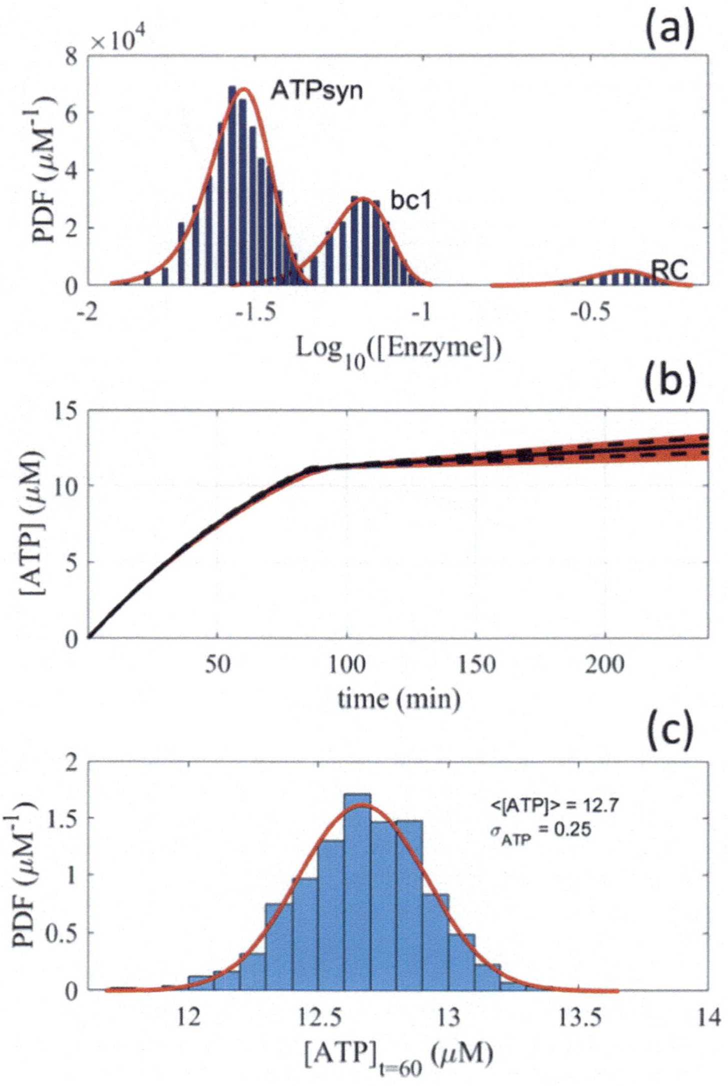

Fig. 3. Simulation outcomes of a transducing light vesicle population monodispersed in size: r_{GUV} = 15 μm. (a) Comparison between the simulated (bars) and the Gaussian (red line) enzyme concentration probability density functions: [RC] = 4.0e−01 ± 8.0e−02 μM, [bc1] = 6.64e −02 ± 1.3e−02 μM, and [ATP_{syn}] = 2.92e−02 ± 0.6e−02 μM; (b) ATP concentration time course in each GUV of the population (red line) along with the average <[ATP]> (continuous black line) and the random displacement <[ATP]> $\pm\sigma_{ATP}$ (dashed black line); (c) Comparison between the simulated (bars) and the Gaussian (red line) concentration probability density functions of ATP produced inside GUVs after 4 h.

In Fig. 3, the outcomes of hybrid stochastic simulations are shown. In particular, the average ATP production of a population of protocells randomly generated with the same size, but with a different internal composition is calculated. As a simplifying assumption, we consider that only the enzyme concentrations are randomly distributed

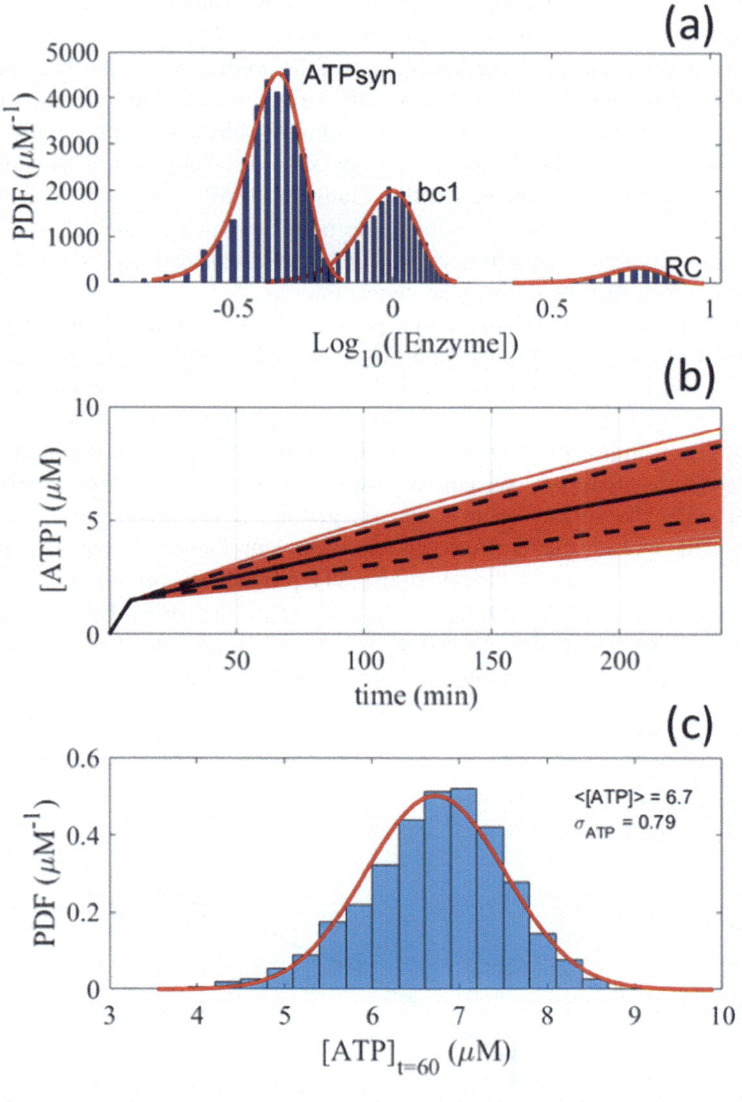

Fig. 4. Simulation outcomes of a transducing light vesicle population monodispersed in size: r_{GUV} = 1.0 μm. (a) Comparison between the simulated (bars) and the Gaussian (red line) enzyme concentration probability density functions: [RC] = 6.0 ± 1.2 μM, [bc1] = 1.0 ± 0.2 μM, and [ATP_{syn}] = 4.4e −01 ± 9e −02 μM; (b) ATP concentration time course in each GUV of the population (red line) along with the average < [ATP] > (continuous black line) and the random displacement < [ATP] > ± σ_{ATP} (dashed black line); (c) Comparison between the simulated (bars) and the Gaussian (red line) concentration probability density functions of ATP produced inside GUVs after 4 h.

among the compartments, while the concentrations of all the substrates are the same in all the protocells. A population of 1000 spherical GUVs with the same radius (r_{GUV} = 15.0 μm) is generated randomly with the photosynthetic enzymes distributed according to independent Gaussian probability density functions. In Fig. 3a, the

Gaussian distributions of the enzyme concentrations are shown by assuming for all the three enzymes a standard deviation equal to 20.0% of the expected average volume concentration. For each GUV, the time course of the produced ATP has been calculated solving the set of Eqs. (4)–(9) and reported in Fig. 3b (red lines), along with the < [ATP]> average population time course (continuous black line) and <[ATP]> $\pm_{ATP}$ (dashed black lines). In Fig. 2c, the GUV population distribution of the produced ATP after 4 h is reported and compared with a Gaussian distribution. Note that the $_{ATP}$ /< [ATP]> is 1.9%, much lower than the same ratio of each enzyme set to 20%. This means that a sort of compensation on the population kinetic activity takes place among the non-correlated distributed enzyme concentrations.

In Fig. 4, the outcomes of simulations performed starting from a population of 1000 GUVs of radius r_{GUV} = 1.0 μm are reported. As it can be seen, the final average value of the ATP concentration is lower than what obtained in the case of the larger size protocell simulation, compare Fig. 3c and Fig. 4c. Indeed, in smaller vesicles the stationary state is reached faster, in less than 10 min. In fact, although the surface concentration of the three enzymes is assumed constant in average, the enzyme volume concentration increases, since the surface/volume ratio is getting higher when the vesicle radius decreases. This makes smaller protocells more efficient than the larger ones, but when the stationary state is reached, the rate of the ATP production abruptly decreases in any case. Moreover, it is worthwhile to note that the dispersion around the average behaviour is also higher for smaller compartments. In fact, 1.0 μm radius protocells exhibit a broad dispersion around the average concentration of ATP produced after 4 h.

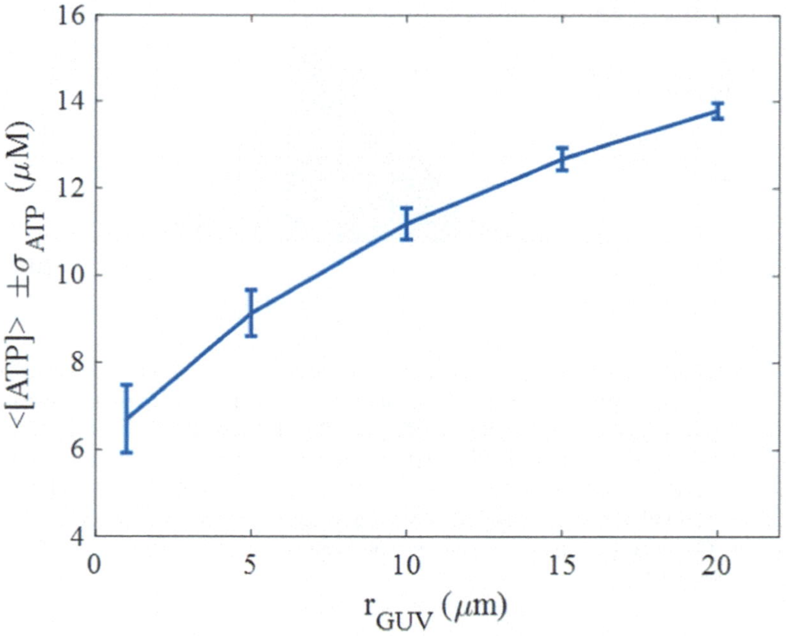

Fig. 5. Average ATP concentration produced by 4 h of light irradiation against the radius of the GUV population. Outcomes obtained by statistically independent hybrid simulations on 1000 vesicles.

The average ATP concentration produced after 4 h of light irradiation obtained by outcomes of hybrid simulations is reported in Fig. 5 against the radius of the vesicles. This plot shows as the behaviour of GUVs can be quite different since the smaller vesicles exhibit a percentage of ADP conversion around 34% after 4 h. However, it is worthwhile to remark that the model indicates that when the stationary state is reached, smaller protocells are more efficient than the larger ones. Indeed, they exhibit a stationary rate of ATP production that is faster, as clearly shown by plots in Figs. 3b and 4b.

3 Conclusions

In this contribution the problem of describing the time course of a population of giant unilamellar vesicles transducing light energy is addressed. In particular, a hybrid stochastic approach is presented suitable to take into account the extrinsic stochastic effects on the time behaviour of protocells that is the effect of the dispersion in size and composition of a GUV population. To get this goal, a simplified deterministic model is presented and discussed. This model describes the kinetic coupling of the photo-redox cycle, producing the transmembrane proton gradient, with the ADP phosphorylation. Moreover, by assuming uncorrelated Gaussian distributions of the three photosynthetic enzymes: RC, bc1 and ATPsyn among protocells, a populations of 1000 GUVs with the same radius, but with an enzymatic random composition has been produced. For each GUV, the enzymatic composition has been then randomly generated according to uncorrelated Gaussian distributions and the time course of the ATP concentration calculated by solving numerically the deterministic differential equation set. Thus, for each monodispersed population the average ATP time course and the standard deviation have been also obtained showing as the ADP conversion after 4 h of light irradiation increases with the size of the vesicles, see Fig. 4. By assuming the surface concentration of the three enzymes constant at least in average, this approach has shown that after 4 h the yield of ADP-ATP conversion is unpredictably higher in larger vesicles, although the smaller protocells are more efficient in longer time, due to a favourable surface/volume ratio, i.e. since they have higher enzyme volume concentrations.

Finally, we would like to remark how this approach has clearly shown that protocells can exhibit different kinetic behaviour due to size and composition polydispersions. Therefore, hybrid stochastic simulations can be suitable tools to investigate the time behaviour of a reacting protocell population in presence of detailed information on the vesicle composition and size distribution.

References

Altamura, E., Albanese, P., Milano, F., et al.: Single compartment approach for assembling photosynthetic protocells. In: Piotto, S., Rossi, F., Concilio, S., et al. (eds.) Advances in Bionanomaterials II. Lecture Notes in Bioengineering, pp. 1–14. Springer, Cham (2020a)

Altamura, E., Albanese, P., Marotta, R., et al.: Light-driven ATP production promotes mRNA biosynthesis inside hybrid multi-compartment artificial protocells, bioRxiv 2020.02.05.933846 (2020b). https://doi.org/10.1101/2020.02.05.933846

Altamura, E., Mavelli, F., Milano, F., Trotta, M.: Photosynthesis without the organisms: the bacterial chromatophores. In: Piotto, S., Rossi, F., Concilio, S., Reverchon, E., Cattaneo, G. (eds.) Advances in Bionanomaterials. Lecture Notes in Bioengineering, pp. 165–175. Springer, Cham (2018)

Altamura, E., Fiorentino, R., Milano, F., et al.: First moves towards photoautotrophic synthetic cells: in vitro study of photosynthetic reaction centre and cytochrome BC1 complex interactions. **229**, 46–56 (2017a)

Altamura, E., Milano, F., Tangorra, R.R., et al.: Highly oriented photosynthetic reaction centers generate a proton gradient in synthetic protocells. PNAS **14**(15), 3837–3842 (2017b)

Berhanu, S., Ueda, T., Kuruma, Y.: Artificial photosynthetic cell producing energy for protein synthesis. Nat. Commun. **10**, 1325–1332 (2019)

Calviello, L., Stano, P., Mavelli, F., et al.: Quasi-cellular systems: stochastic simulation analysis at nanoscale range. BMC Bioinf. **14**, S7 (2013)

Iino, R., Hasegawa, R., Tabata, K.V., Noji, H.: Mechanism of inhibition by c-terminal alpha-helices of the epsilon subunit of escherichia coli FoF1-ATP synthase. J. Biol. Chem. **284**, 17457–17464 (2009)

Feniouk, B.A., Cherepanov, D.A., Voskoboynikova, N.E., et al.: Chromatophore vesicles of rhodobacter capsulatus contain on average one F0F1-ATP synthase each. Biophys. J. **82**, 1115–1122 (2002)

Lee, K.Y., Park, S.J., Lee, K.A., et al.: Photosynthetic artificial organelles sustain and control ATP-dependent reactions in a protocellular system. Nat. Biotechnol. **36**, 530–535 (2018)

Luisi, P.L., Ferri, F., Stano, P.: Approaches to semi-synthetic minimal cells: a review. Naturwissenschaften **93**(1), 1–13 (2006)

Mavelli, F., Piotto, S.: Stochastic simulations of homogeneous chemically reacting systems. J. Mol. Struct. **771**, 55–64 (2006)

Mavelli, F., Ruiz-Mirazo, K.: ENVIRONMENT: a computational platform to stochastically simulate reacting and self-reproducing lipid compartments. Phys. Biol. **7**(036002), 13 p. (2010)

Mavelli, F., Altamura, E., Cassidei, L., Stano, P.: Recent Theoretical Approaches to Minimal Artificial Cells. Entropy, **16**(5), 2488–2511 (2014)

Mavelli, F., Stano, P.: Experiments on and numerical modeling of the capture and concentration of transcription-translation machinery inside vesicles. Artif. Life **21**(4), 445–463 (2015)

Paula, S., Volkov, A.G., Van Hoek, A.N., et al.: Permeation of protons, potassium ions, and small polar molecules through phospholipid bilayers as a function of membrane thickness. Biophys. J. **70**, 339–348 (1996)

Sener, M., Strumpfer, J., Singharoy, A., et al.: Overall energy conversion efficiency of a photosynthetic vesicle. eLife **5**, e09541 (2016)

Stano, P., Altamura, E., Mavelli, F.: Novel directions in molecular systems design: the case of light-transducing synthetic cells. Commun. Integr. Biol. **10**(1–7), e1365993 (2017)

Stano, P., De Souza, T.P., Carrara, P., et al.: Recent biophysical issues about the preparation of solute-filled lipid vesicles. Mech. Adv. Mater. Struct. **22**, 748–759 (2015)

Stano, P.: Is research on "Synthetic Cells" moving to the next level? Life **9**, 29 (2018). https://doi.org/10.3390/life9010003

Walde, P., Cosentino, K., Engel, H., Stano, P.: Giant vesicles: preparations and applications. Chem. Bio. Chem. **11**, 848–865 (2010)

Molecular Dynamics Simulation of Antimicrobial Permeable PVC-Based Films

Lucia Sessa(✉), Anna Maria Nardiello, Miriam Di Martino, Francesco Marrafino, and Pio Iannelli

Department of Pharmacy, University of Salerno, Via Giovanni Paolo II, 132, 84084 Fisciano, SA, Italy
lucsessa@unisa.it

Abstract. Polymeric materials are widely used in different fields. Due to their versatility and properties, they can be used as chemically stable matrices for the non-covalent incorporation of low molecular weight antimicrobial agents. Polyvinyl chloride (PVC) is one of the polymers widely used in the medical field and in food packaging. We realized antimicrobial PVC based films by adding a different percentage of antimicrobial agents containing an azo group. The antimicrobial azo-compounds blended in PVC matrices showed good inhibition of *S. aureus* and *C. albicans* biofilm formation. Compared to glass or metal packaging materials, polymeric packaging is permeable to gas molecules, such as water vapor and organic vapor, or to flavors and additives. Knowledge of the diffusion of gases through polymeric films is crucial, for polymers used in the food industry and the biomedical field. In addition, the migration of small molecules blended in the polymeric materials could affect the quality of the device and, the potential release of these molecules into the body is a severe public health concern. The diffusion of antimicrobial agent, named A4, in Polyvinyl chloride (PVC) film were investigated by molecular dynamics (MD) simulation studies. The effects of the presence of the azo compound on the diffusion of O_2, H_2O, and CH_4 in PVC matrices were also studied by MD simulation.

Keywords: PVC · Diffusion · Azo dye · MD

1 Introduction

Polymeric materials, thanks to their versatility and properties, have become an essential product in several fields. They are widely used for food packaging, transport, or to produce medical devices. Due to its chemical resistance, long term stability, optical clarity, and weldability, as well as its meager cost, Polyvinyl chloride (PVC) is widely used in the medical field for making blood bags and flexible tubing for infusers and it is the second most commonly used synthetic polymer in food packaging [1]. PVC based films are hard and rigid, but the addition of plasticizers permits the preparation of soft and flexible plastic packaging [2]. To prevent the development of bacterial infections, the use of specific and powerful antimicrobial materials helps to fight and eliminate infections. In general, plastic materials do not show intrinsic antibacterial properties. However, their use plays a crucial role, as they can be used as a chemically stable matrix for the non-covalent incorporation of low molecular weight antimicrobial

S. Piotto et al. (Eds.): BIONAM 2019, LNBE, pp. 111–119, 2020.
https://doi.org/10.1007/978-3-030-47705-9_10

agents. Several antimicrobial materials are prepared by adding triclosan, chlorhexidine, and azobenzene, into different polymeric matrices such as polystyrene, polycaprolactone, polypropylene and polyvinyl chloride [3–6].

In our previous work [7], using the solvent casting technique, we realized several polymeric films formed by PVC and Polyoxyethylene sorbitan monostearate (PS60) as plasticizer (Fig. 1). We used this kind of plasticizer to improve the mechanical properties and to avoid crystallization during the peeling step [8]. Solvent casting is one of the oldest techniques in the production of polymeric films. After complete evaporation of the organic solvent, the film is peeled off the support. The dried films are then cut and packaged appropriately.

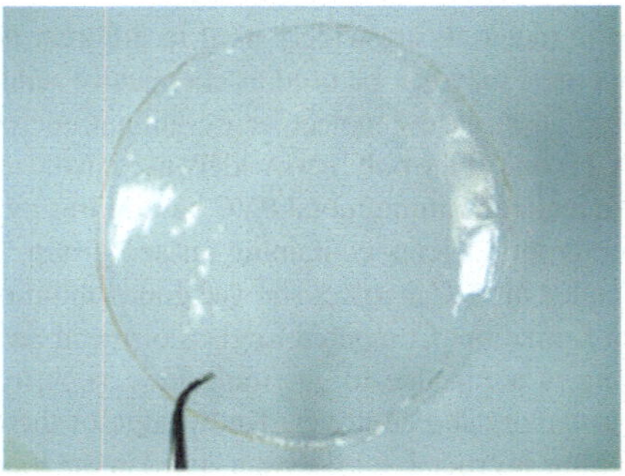

Fig. 1. PVC/PS60 based film realized by the solvent casting technique

We also prepared antimicrobial polymer films by adding a different percentage of three antimicrobial agents containing an azo moiety [9–11]. We reached the final concentration of 0.01%, 0.05%, and 0.1% (w/w) of antimicrobial agent. Homogeneous and elastic films were obtained, approximately 600 mm thick, as measured by a thickness gauge.

X-ray analysis of dried films confirmed the amorphous nature of the materials and the absence of crystals of azo compounds in the matrix. Thermal characterizations indicated that the presence of the highest percentage of azo dye in the matrix (0.1%) does not influence the thermal behavior of the material. The release of azo molecules from the polymer films reached a plateau within 10 h. After that time, the polymer films appeared undamaged and with the same initial mechanical properties. The antimicrobial azo compounds blended in PVC matrices showed good biocide activity. PVC films loaded with 0.01% of azo compounds were able to completely inhibit the biofilm formation of *S. aureus* bacterium and *C. albicans* fungus. This aspect confirmed that the realized materials might be interesting for biomedical tools, antibacterial surfaces, and films for active packaging.

Compared to glass or metal packaging materials, polymeric packaging is permeable to gas molecules, such as water vapor and organic vapor, or to flavors and additives. Knowledge of the diffusion of gases through polymeric films is crucial, especially for polymers used in the food industry, where contamination from the external environment must be prevented. The study of the diffusion rate is also essential to evaluate the lifetime of biomedical devices. In addition, the migration of small molecules blended in the polymeric materials could affect the quality of the device and, the potential release of these molecules into the body is a severe public health concern [12].

The molecular dynamics simulation (MD) is a method widely used to analyze the microscopic movement of molecules in a polymeric matrix [13]. This technique has been widely used to study the diffusion of oxygen in membrane polymers of polyethylene, polypropylene, and PVC [14], and again to study the diffusion of small gas molecules of CO_2 in Polyethylene terephthalate/ Polylactic acid mixtures [15]. Data from the literature confirmed that the simulation results had the same trend as the experimental results [16, 17]. This is encouraging for the application of MD as an effective method to study the diffusion behavior of gas molecules in the polymers.

In this work, we used MD simulations at the atomic scale to study the diffusion of gas molecules (O_2, H_2O, and CH_4) in PVC matrices. We also investigated the molecular diffusion of the antimicrobial agent (A4) in plasticized PVC film and its effect on the diffusion of gases in the materials from the microscopic view.

2 Materials and Methods

2.1 Building of the Molecular and the Polymer Structures

The molecular models of O_2, H_2O, CH_4, and 4'-Hydroxy-(4-hydroxy-3,5-dimethyl)-azobenzene (A4) were built with the Materials Visualizer module of Materials Studio 7.0 BIOVIA Materials Studio [18]. The PVC polymer was built as a homopolymer formed by 20 repeat units of vinyl chloride monomer. The structure of the plasticizer Polysorbate 60 (PS60) was retrieved from the library PubChem [19]. Models were geometrically optimized by using Forcite module with the "Smart Minimizer" method, under the Dreiding force field [20, 21]. The force field we used, is an old and generic force field that well parametrizes the azo compounds.

The amorphous cells containing PVC, the plasticizer and the azo compound were built to simulate the experimental condition of our previous study [7]. To analyze the diffusion of O_2, H_2O, and CH_4 in PVC films, the molecules of gas were placed into the cell structure of PVC. We fixed the value of 2% for gas molecules in the matrixes. We also analyzed the matrix without the plasticizer to study the behavior of the azo compound in different systems. We generated ten amorphous cells with an initial density of 1.4 g/cm^3 according to the density of PVC film [22]. The compositions in terms of weight percentage and number of molecules of each component are listed in Table 1.

The energy of the amorphous cells was optimized using COMPASS as Forcefield. The nonbonded electrostatic and van der Waals forces were controlled by "group-based" summation and a cutoff value of 15.5 Å. A short molecular dynamics simulation (MD) at constant energy (NVE) was performed to equilibrate the systems. Five annealing cycles were performed with the initial temperature of 300 K and the mid-cycle temperature of 500 K. Then, the equilibration was performed at constant temperature (NVT) to ensure that the system is at the correct temperature. It was carried out using Velocity Scale as the thermostat, for a total time of 50 ps. This time was enough to equilibrate the cell correctly. We monitored the equilibration progress by looking up the energies in the live update chart, which was constant apart from small fluctuations and we could assume that the systems were indeed in the thermal equilibrium condition.

Table 1. Composition of the amorphous cells

Model	Composition	Weight %	Molecules loaded
PVC/O_2	PVC	98.0	10
	O_2	2.0	8
PVC/PS60/O_2	PVC	88.8	10
	PS60	9.2	2
	O_2	2.0	9
PVC/PS60/A4/O_2	PVC	86.0	14
	PS60	9.6	3
	A4	2.4	2
	O_2	2.0	13
PVC/H_2O	PVC	98.0	10
	H_2O	2.0	14
PVC/PS60/H_2O	PVC	88.8	10
	PS60	9.2	2
	H_2O	2.0	16
PVC/PS60/A4/H_2O	PVC	86.0	14
	PS60	9.6	3
	A4	2.4	2
	H_2O	2.0	23
PVC/CH_4	PVC	98.0	10
	CH_4	16	2.0
PVC/PS60/CH_4	PVC	88.8	10
	PS60	9.2	2
	CH_4	2.0	18
PVC/PS60/A4/CH_4	PVC	86.0	14
	PS60	9.6	3
	A4	2.4	2
	CH_4	2.0	25
PVC/A4	PVC	97.3	14
	A4	2.4	2.0
PVC/PS60/A4	PVC	87.8	14
	PS60	9.8	3
	A4	2.4	2

2.2 Simulation Method

All simulations were performed using the Forcite module in Materials studio 8.0. MD simulations at constant energy and at the temperature of 300 K were performed for 2500 ps using Dreiding as force field and timestep of 1.0 fs. The nonbonded electrostatic and van der Waals forces were controlled by "group-based" summation and a cutoff value of 15.5 Å. The trajectories were collected at an interval of 250 ps. We ran dynamics at constant energy (NVE) because some thermostats can interfere with the dynamics of the system, and potentially affect the diffusion coefficient.

3 Results and Discussion

3.1 Fractional Free Volume (FFV)

Molecular dynamics is an essential tool to evaluate the molecular packing of the polymeric system. The spaces between molecules of the blend systems are called free volume. The size and shape of these holes in the polymeric matrix affect the diffusion behavior of the gas molecules. In general, high values of diffusion coefficients are caused by large holes in the system.

The free space within the simulation cell can be obtained using the Connolly surface method [23]. It can be represented by the volume on the side of the Connolly surface without atoms and can be calculated by rolling a spherical probe over the van der Waals surface. We selected a probe radius of 1.5 A and this value was enough to define the values of free volume and volume occupied by the polymeric matrix. To compare the free volume between the simulated PVC cells, we calculated the fractional free volume (FFV) by using the following equation [24]:

$$\mathrm{FFV} = \frac{V_f}{V_f + V_0} \times 100\% \tag{1}$$

where V_f is the volume of the holes, and V_0 is the volume occupied by the materials.

In Table 2, we report the percentage of FFV for the 11 analysed systems.

3.2 Gas Diffusion in Polymeric Matrixes

The diffusivity of a gas in a polymeric matrix can be calculated by running a molecular dynamics simulation and determining the mean square displacement (MSD) of the gas in the material. The Einstein equation [25] is a useful method to correlate the diffusion coefficient (D) with the increase of MSD during the simulation time. The MSD was calculated using the Forcite module available in MS 7.0, applying the Einstein Eq. (2):

$$D = \frac{1}{6N_\alpha} \lim_{t\to\infty} \frac{d}{dt} \sum_{i=1}^{N_\alpha} \left[r_i(t) - r_i(0)\right]^2 \tag{2}$$

where N_α is the number of diffusive molecule *i* in the system, *t* is the diffusion time, and $r_i(t)$ and $r_i(0)$ are the diffusion position vector of the diffusion molecule *i* at time *t* and *0* respectively.

Figure 2 shows the MSD curves for the diffusion of the three chosen gases (O_2, H_2O, and CH_4) in the simulated cells of PVC before and after the inclusion of the azo compound. We calculated the coefficient D (cm^2/ps) according to the above Eq. (2). We performed a linear regression by analyzing the mean square displacement (MSD) curve. We deleted the data corresponding to the initial, non-linear, part of the graph, and we abstracted the slope *a* from each straight line. In Table 2 are shown the values of D listed in descending order of diffusion.

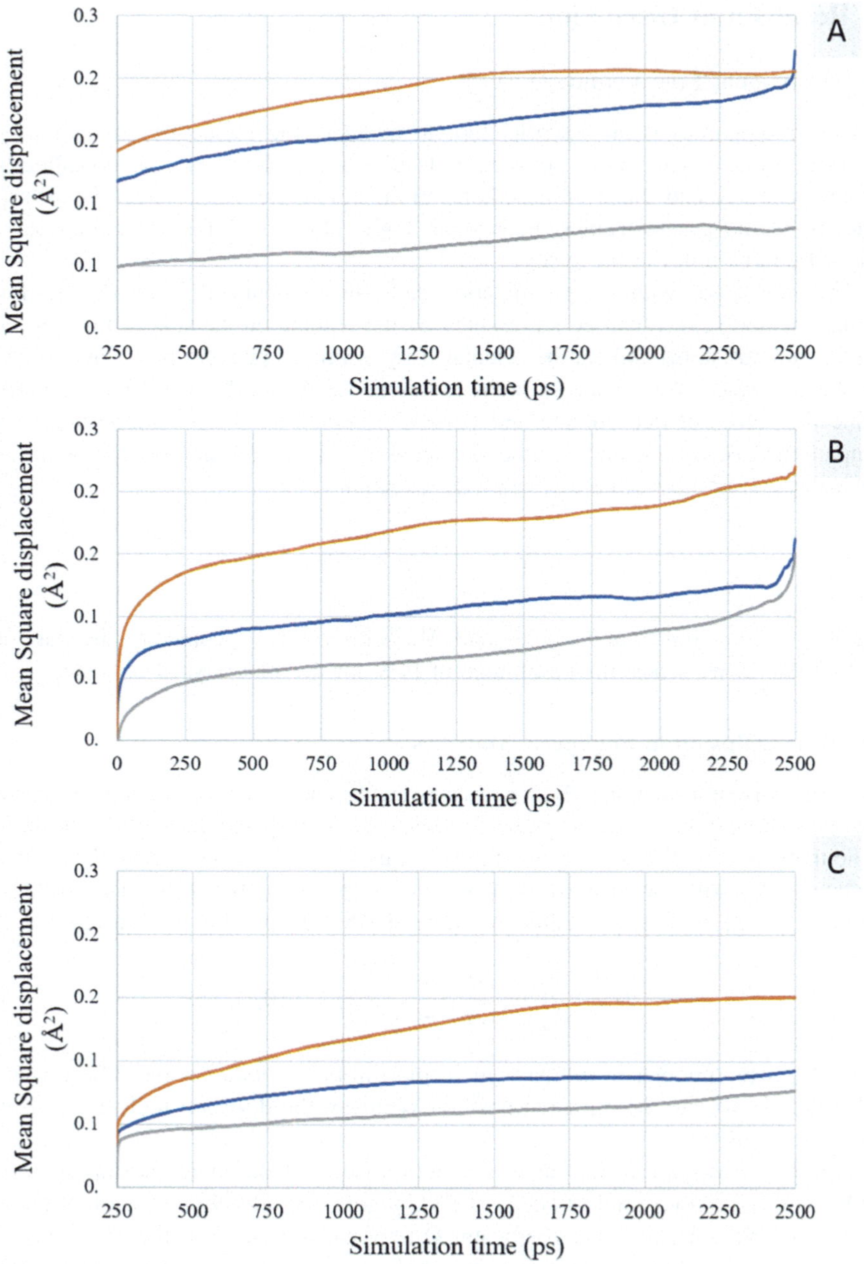

Fig. 2. Linear fitting of Log_{10} curves of MSD for (A) O_2 molecules in PVC (blue), in PVC/PS60 (orange) and in PVC/PS60/A4 (gray); (B) H_2O molecules in PVC (blue), in PVC/PS60 (orange) and in PVC/PS60/A4 (gray); (C) CH_4 molecules in PVC (blue), in PVC/PS60 (orange) and in PVC/PS60/A4 (gray).

PVC based materials are very permeable to molecules of gas of oxygen and water. As we can observe from the predicted diffusion values shown in Table 2, the addition of the plasticizer in the matrix increases the diffusion to these gases, increasing the volume of voids in the cell. This reduction of gas barrier properties can be neglected when considering the advantages of the mechanical properties that plasticizers give to the final material. In contrast, the simulated systems show a good ability to provide an excellent barrier property to methane gas molecules.

Table 2. Fractional free volume comparison and Diffusion coefficient (D) for gases

Model	FFV	D (cm^2/ps)
PVC/PS60/H_2O	2.55%	6.93E−06
PVC/PS60/O_2	2.33%	6.86E−06
PVC/O_2	1.88%	4.57E−06
PVC/PS60/CH_4	1.64%	2.11E−06
PVC/H_2O	1.46%	9.27E−07
PVC/PS60/A4/H_2O	1.23%	7.01E−07
PVC/CH_4	1.11%	3.61E−07
PVC/PS60/A4/CH_4	0.88%	2.80E−07
PVC/PS60/A4/O_2	0.83%	2.78E−07

In this study, it was found that the inclusion of the azo dye in a polymer matrix has several advantages. The main one is the opportunity to prepare an antimicrobial material by mixing the antimicrobial agent with the inert polymer. The second one is the ability of the antimicrobial molecule to increase the barrier property of gas.

3.3 Azo Compound Diffusion in Polymeric Matrixes

We simulated the diffusion of the antimicrobial azo compound (A4) in a polymeric matrix before and after the inclusion of three gases (O_2, H_2O, and CH_4). We calculated the coefficient D (cm^2/ps) according to the Einstein Eq. (1).

Table 3. Diffusion coefficient (D) for A4 azo compound

System	D (cm^2/ps)	FFV
PVC/PS60/A4	1.53E−07	2.10%
PVC/A4	1.47E−07	1.85%
PVC/PS60/A4/H_2O	8.38E−08	1.23%
PVC/PS60/A4/O_2	6.82E−08	0.83%
PVC/PS60/A4/CH_4	3.84E−08	0.88%

In Table 3, we compared the diffusion of the A4 compound in the five simulated systems. We observed that A4 has higher mobility in PVC and PVC/PS60 systems. In this last case, the presence of the plasticizer PS60 does not affect the diffusion process of the antimicrobial agent. The simulation data showed that the presence of water vapor increased the diffusion of the A4 molecule, which is instead reduced when methane is added in the cell.

4 Conclusion

We simulated the systems formed by PVC, Polysorbate 60, and the azo molecules A4, as a representative structure for the antimicrobial agents. We applied the molecular dynamics technique to analyze the effect of the azo compound on the diffusion of O_2, H_2O, and CH_4 in PVC cells. From the predicted coefficient diffusion, we found that the PVC based systems are permeable to molecules of oxygen and water gas, but they provide a suitable barrier property to methane gas molecules. Compared to the pure matrices, systems with plasticizers show a higher gas permeability. The higher water vapor permeability limits the application of PVC based materials in hydrated conditions. The molecule A4 significantly reduces the diffusion of gas molecules of oxygen, water, and methane by competing for free holes in PVC. The azo dye dispersed in the polymeric matrix provides a gas barrier function. The higher free volume in pure PVC matrix and PVC/PS60 systems, facilities the diffusion of the azo dye. In contrast, the migration of A4 molecule is drastically reduced by the presence of gas. Among the three different gases considered, we observed a slight increase migration of the azo dye in the hydrated film, back to the limitations of PVC applications.

The molecular dynamics has been an efficient method for the *in-silico* study of the films. We can conclude that the insertion of azo molecules in polymeric matrix allows both the production of materials with antimicrobial property and with high gas barrier properties.

References

1. Robertson, G.L.: Food Packaging: Principles and Practice. CRC Press, Boca Raton (2005)
2. Villanueva, M.E., González, J.A., Rodríguez-Castellón, E., Teves, S., Copello, G.J.: Antimicrobial surface functionalization of PVC by a guanidine based antimicrobial polymer. Mater. Sci. Eng. C **67**, 214–220 (2016)
3. Chung, D., Papadakis, S.E., Yam, K.L.: Evaluation of a polymer coating containing triclosan as the antimicrobial layer for packaging materials. Int. J. Food Sci. Technol. **38**, 165–169 (2003)
4. Petersen, R.C.: Triclosan antimicrobial polymers. AIMS Mol. Sci. **3**, 88 (2016)
5. Piotto, S., Concilio, S., Sessa, L., Iannelli, P., Porta, A., Calabrese, E.C., Galdi, M.R., Incarnato, L.: Novel antimicrobial polymer films active against bacteria and fungi. Polym. Compos. **34**, 1489–1492 (2013)
6. Panunzi, B., Borbone, F., Capobianco, A., Concilio, S., Diana, R., Peluso, A., Piotto, S., Tuzi, A., Velardo, A., Caruso, U.: Synthesis, spectroscopic properties and DFT calculations of a novel multipolar azo dye and its zinc(II) complex. Inorg. Chem. Commun. **84**, 103–108 (2017)
7. Sessa, L., Concilio, S., Iannelli, P., De Santis, F., Porta, A., Piotto, S.: Antimicrobial azobenzene compounds and their potential use in biomaterials. American Institute of Physics Inc. (2016)

8. Vieira, M.G.A., da Silva, M.A., dos Santos, L.O., Beppu, M.M.: Natural-based plasticizers and biopolymer films: a review. Eur. Polymer J. **47**, 254–263 (2011)
9. Piotto, S., Concilio, S., Sessa, L., Porta, A., Calabrese, E.C., Zanfardino, A., Varcamonti, M., Iannelli, P.: Small azobenzene derivatives active against bacteria and fungi. Eur. J. Med. Chem. **68**, 178–184 (2013)
10. Concilio, S., Piotto, S., Sessa, L., Iannelli, P., Porta, A., Calabrese, E.C., Galdi, M.R., Incarnato, L.: Antimicrobial polymer films for food packaging. In: AIP Conference Proceedings, pp. 256–258. AIP Conference Proceedings (2012)
11. Concilio, S., Sessa, L., Petrone, A.M., Porta, A., Diana, R., Iannelli, P., Piotto, S.: Structure modification of an active azo-compound as a route to new antimicrobial compounds. Molecules **22**, 875 (2017)
12. Hahladakis, J.N., Velis, C.A., Weber, R., Iacovidou, E., Purnell, P.: An overview of chemical additives present in plastics: migration, release, fate and environmental impact during their use, disposal and recycling. J. Hazard. Mater. **344**, 179–199 (2018)
13. Friedman, R., Khalid, S., Aponte-Santamaría, C., Arutyunova, E., Becker, M., Boyd, K.J., Christensen, M., Coimbra, J.T.S., Concilio, S., Daday, C., van Eerden, F.J., Fernandes, P.A., Gräter, F., Hakobyan, D., Heuer, A., Karathanou, K., Keller, F., Lemieux, M.J., Marrink, S. J., May, E.R., Mazumdar, A., Naftalin, R., Pickholz, M., Piotto, S., Pohl, P., Quinn, P., Ramos, M.J., Schiøtt, B., Sengupta, D., Sessa, L., Vanni, S., Zeppelin, T., Zoni, V., Bondar, A.N., Domene, C.: Understanding conformational dynamics of complex lipid mixtures relevant to biology. J. Membr. Biol. **251**, 609–631 (2018)
14. Börjesson, A., Erdtman, E., Ahlström, P., Berlin, M., Andersson, T., Bolton, K.: Molecular modelling of oxygen and water permeation in polyethylene. Polymer **54**, 2988–2998 (2013)
15. Pavel, D., Shanks, R.: Molecular dynamics simulation of diffusion of O2 and CO2 in blends of amorphous poly (ethylene terephthalate) and related polyesters. Polymer **46**, 6135–6147 (2005)
16. Sun, B., Lu, L., Zhu, Y.: Molecular dynamics simulation on the diffusion of flavor, O2 and H2O molecules in LDPE film. Materials **12**, 3515 (2019)
17. Gårdebjer, S., Larsson, M., Gebäck, T., Skepö, M., Larsson, A.: An overview of the transport of liquid molecules through structured polymer films, barriers and composites–experiments correlated to structure-based simulations. Adv. Coll. Interface. Sci. **256**, 48–64 (2018)
18. Biovia, D.S.m.: Materials Studio 8.0. San Diego: Dassault Systèmes (2014)
19. https://pubchem.ncbi.nlm.nih.gov/compound/24832100
20. Mayo, S.L., Olafson, B.D., Goddard, W.A.: DREIDING: a generic force field for molecular simulations. J. Phys. Chem. **94**, 8897–8909 (1990)
21. Landi, G., Fahrner, W.R., Concilio, S., Sessa, L., Neitzert, H.C.: Electrical hole transport properties of an ambipolar organic compound with Zn-Atoms on a crystalline silicon heterostructure. IEEE J. Electron Devices Soc. **2**, 179–181 (2014)
22. Bhunia, K., Sablani, S.S., Tang, J., Rasco, B.: Migration of chemical compounds from packaging polymers during microwave, conventional heat treatment, and storage. Compr. Rev. Food Sci. Food Saf. **12**, 523–545 (2013)
23. Connolly, M.L.: Analytical molecular surface calculation. J. Appl. Crystallogr. **16**, 548–558 (1983)
24. Peng, F., Jiang, Z., Hoek, E.M.: Tuning the molecular structure, separation performance and interfacial properties of poly (vinyl alcohol)–polysulfone interfacial composite membranes. J. Membr. Sci. **368**, 26–33 (2011)
25. Rowley, R.L., Painter, M.M.: Diffusion and viscosity equations of state for a Lennard-Jones fluid obtained from molecular dynamics simulations. Int. J. Thermophys. **18**, 1109–1121 (1997)

Pseudo-semantic Approach to Study Model Membranes

Anna Maria Nardiello(✉), Stefano Piotto, Luigi Di Biasi, and Lucia Sessa

Department of Pharmacy, University of Salerno, Via Giovanni Paolo II, 132, 84084 Fisciano, SA, Italy
annardiello@unisa.it

Abstract. It is well known that, during rapid temperature variations, lipid membranes are sensitive structures within cells. It is, therefore, not surprising that membranes are one of the most critical cellular targets for temperature adaptation. Many organisms adapt to the different temperature changing the degree of unsaturation of the lipids in the membrane. In this study, we describe a pseudo-semantic analysis approach applied to molecular dynamics. This approach is based on the encoding of data into strings and on the calculation of similarities. The described approach is universally applicable and, in this work, the fluidity of a POPC (palmitoyl-oleoyl-glycerophosphocholine) membrane under different conditions was monitored. In three simulations we varied the temperature above and below the phase transition temperature (T_m). In a fourth simulation we added an external molecule as fluidifier. The results of the clustering, obtained by similarity values, were consistent with the experimental data.

Keywords: Molecular dynamics · Semantic encoding · Membrane

1 Introduction

Phospholipid membranes play a critical role in compartmentalization, as a medium for proteins to functionality and as structural scaffolding for cells. At the same time, the protein activity is influenced and, in some cases, modulated by the physical state of membranes [1]. Atomistic simulations have been made to reproduce and predict many fundamental properties of phospholipid membranes. During the last years, molecular dynamics (MD) simulations have increased in length and time scales by orders of magnitude. In an MD simulation, all atoms in the system are treated classically, and the chemical bonds between atoms are generally modeled with harmonic expressions. For atom pairs that are not chemically bonded, potential energy expressions include coulombic '6–12 interactions' that consists of a repulsive part, which decays as $1/r^{12}$, and an attractive part, which decays as $1/r^6$ [2]. This permits to model interactions due to polarization effects between atomic electron clouds. Torsional potentials model the interactions between next-nearest neighbor atoms on the same molecule. The set of functions, and the parameters characterizing the strengths of the various interactions, is commonly referred to as 'force field'. This approach has severe limitations. The first

S. Piotto et al. (Eds.): BIONAM 2019, LNBE, pp. 120–127, 2020.
https://doi.org/10.1007/978-3-030-47705-9_11

one concerns the parameters. They can be calculated by *ab initio* computations or derived by experiments. Of course, it is not possible to forecast the evolution of a system if a simplified description of the system, like the one provided by the force field, cannot capture the essential characteristics of the aggregate. The second limitation is the maximum time step for which the integration of the equations of motion remains stable. A typical value is 2 fs ($2 \cdot 10^{-15}$ s). This means that $5 \cdot 10^5$ integration steps are necessary to calculate the dynamics of a system during 1 ns. With current computers and algorithms, the size of systems and the time scales for which phospholipid bilayers can be studied preclude the accurate examination of phase transitions, the interaction of proteins with the membrane, or the flip flop of lipids.

Successful MD simulations of the gel phase of phospholipids appeared in recent years [3]. The simulations demonstrate convincingly that all-atom models are necessary for the study of the gel phase, as is the use of constant pressure, rather than constant volume simulation constraints. For more complex membranes, with several lipid components, many reports have appeared recently [4–6]. One of the essential molecules to be considered is cholesterol. Cholesterol is capable of modifying the mechanical, thermophysical, and lateral organizational properties of membranes [7].

MD simulations start to offer the prospect of analysis of the interactions between membrane proteins and surrounding lipid molecules. This is extremely important, since only limited data on lipid-protein interactions may be obtained by examining interactions in those crystal structures where lipid molecules are present. The lipid membrane physical state influences the conformational space explored by a protein. The secondary structure of a transmembrane peptide is found to deviate from its structure in aqueous solution through reversible conformational transitions. The peptide moderately alters the lateral compressibility of the bilayer by changing the equilibrium area of the membrane. This results in an alteration of the chemical-physical parameters of the membrane, such as fluidity, spontaneous curvature, and bending rigidity coefficients. The distribution of local pressure inside a bilayer, the so-called lateral pressure profile (LPP) [8], is a fascinating but still poorly understood property. The lateral pressure profile is related to many relevant macroscopic and measurable quantities, such as surface tension, surface free energy, and spontaneous curvature. Changes in the lateral pressure profile may affect the functionality of mechanosensitive proteins in cell membranes, which could explain, for example, the action of general anesthetics and the coupling between protein functionality and lipid content. Also signaling via receptors coupled to the G protein and the activity of channels and/or transporters involve structural rearrangements and such structural changes are also influenced by changes in the lateral pressure profile of the membrane induced by anaesthetics.

Anything capable to modulate the membranes physical state can modulate G protein binding to bilayers [9]. The presence of unsaturations in lipids induce changes in the binding of G proteins that depend on the type of G protein (monomeric, dimeric or trimeric) and the membrane composition (PC:PE molar ratio). These results imply that the fatty acids associated with G_α subunits not only participate in the attachment of G proteins to membranes but also, in the interactions and sorting of G proteins to different membrane domains. Geranylgeraniol (GG) moieties induce the most relevant changes in the binding of the different forms of G proteins to lipid bilayers. In non-lamellar prone membranes, GG significantly reduces the binding of $G_{\alpha i}$ protein to membranes with a PC:PE content similar to that found at the inner leaflet of the plasma membrane [9].

The state of a membrane is often interpreted in terms of fluidity. Unfortunately, for a membrane, an anisotropic fluid, the concept of fluidity is poorly defined. The LPP is sometimes used. LPP, however, has the high disadvantage of not being measurable with an instrument, is difficult to apply because the membrane is heterogeneous, and it returns the state of the membrane in a specific position.

The classic methods for the evaluation of membrane fluidity include an average over time of the order parameter (SCD) of the individual carbons composing the phospholipid chains. This causes an approximation and inaccuracy of the obtained results. Despite its popularity, the concept of fluidity and the related concept of viscosity is not directly applicable to membranes because it is not isotropic media. Observations are accumulating that challenge the dominant hypothesis that membrane fluidity serves as a measure of lipid saturation regulation [10].

The LPP and, more generally, the membrane physical state can be tuned by temperature changes and by the presence of endogenous molecules. For example, benzyl alcohol (BA) has a well-known fluidizing effect on both artificial and cellular membranes [11]. In the context of membrane biology, BA is generally referred to as a membrane fluidifier due to its ability in increasing the fluidity and decreasing the order of bilayer membranes, such as artificial lipid bilayers and cellular membranes. The BA-induced disordering of lipid bilayers has been measured by several techniques including capacitance and conductance [12], electron spin resonance [13, 14], NMR [15], differential scanning calorimetry [16] and fluorescence polarization [16]. X-ray studies have shown that BA can induce an interdigitated phase in bilayers [17], and the compound has been reported to localize in the interfacial region of a bilayer with the hydroxyl group oriented in the polar head group region and the benzene ring towards the membrane hydrophobic core. In addition to increasing disorder among lipids in both model and biological membranes, BA can affect also other properties of biological membranes. BA has been reported to induce destabilization of sterol-rich domains formed by sphingomyelin and cholesterol in artificial bilayers. Hence, the ability of BA to affect the activity of specific membrane proteins could depend on the lateral localization of the proteins within a membrane as well as on the ability of BA to affect the interactions between the proteins and the surrounding lipids in the domains.

BA will probably also modulate the activities of specific membrane proteins and reduce membrane order. This phenomenon is probably related to BA ability to disrupt the interactions between membrane proteins and surrounding lipids by fluidizing the lipid nozzle. The components of biological membranes are laterally diverse in transient compositions of varying content and order, and many proteins are activated or inactivated by their location within or outside the membrane domains with different physical phases.

The pseudo-semantic approach is able to consider the full state of the membrane with an atomistic spatial resolution. This type of analysis permits to evaluate if the addition of an exogenous molecule can reproduce the physical state conditions of the membrane that we obtained by varying the temperature.

This approach is valid for different types of MD simulations analysis, allowing to reduce the computational time needed for the analysis by 3–4 orders of magnitude. In general, any parameter that can be converted into a text string can be analyzed with this method, for example the analysis of phase transitions, or the prediction of agonist/antagonist mechanism of action of a protein receptor.

To prove the applicability of this method to complex biosystems such as biological membranes, the fluidity of the membrane has been analyzed. All lipids have a characteristic temperature (T_m) at which they undergo a transition from gel (solid) to liquid phase. In the liquid phase the lipids have a greater fluidity and greater freedom of movement within the bilayer [18].

In this work, we applied a pseudo-semantic approach for the MD analysis of a palmitoyl-oleoyl-glycerophosphocholine (POPC) model membrane [19]. The method, described in detail below, was used to analyze the variation of membrane fluidity by changing the temperature and by adding exogenous fluidifying molecules to the simulation environment.

2 Materials and Methods

2.1 Molecular Dynamics

Molecular dynamics simulations were performed on four different systems, for 10 ns each. Three systems consist of POPC model membranes at different temperature. The fourth system is a membrane of POPC at 270 K upon addition of benzyl alcohol (BA) as a fluidifier. The first 250 ps of each simulation were not used for analysis and used to balance the system and. This time is necessary to let the POPC molecules to compact properly. The simulated systems are composed by 56 lipids. The system is fully hydrated (water density 0.997 g/mL). Only the final 5 ns of the simulation were analyzed from the trajectory files. The phase transition temperature (T_m) of POPC is 271 K [20]. Therefore, two simulations were performed at a temperature than lower T_m (250 K–270 K), one was performed above it (290 K) and the last one at 270 K with the addition of BA as fluidifier to evaluate its effect on the system. MD simulations were performed with Yasara Structure software package [21]. The used force-field was AMBER14 under NPT conditions. The Particle Mesh Ewald (PME) algorithm was used for electrostatic interactions with a 8Å cut-off. The SCD (Deuterium Order Parameter) was calculated using the MEMBPLUGIN plugin of VMD program [22] to obtain information on the chain orientation and to predict possible alterations of the lipophilic chains of phospholipids.

2.2 Strings Preparation

The analyzed parameter is the SCD, calculated using VMD MembPlugin [22]. The 21 frames, corresponding to the final 5 ns of the simulation have been analyzed. For each phospholipid the order parameter is obtained for both the saturated and unsaturated chain. Two files are generated for each system, corresponding to the two aliphatic chains. Before the conversion, the two files are concatenated. For each SCD value a letter is assigned. A total of 84 alphabetical strings are produced (21 for each of the 4 systems). In each string the SCD values of each carbon of the saturated and unsaturated aliphatic chain of phospholipids are reported in letters.

2.3 Strings Analysis

The strings are processed with ProtComp program, an online software package able to analyze the similarities among strings (www.yadamp.unisa.it/protcomp). ProtComp is a software that pre-processes input files to eliminate redundant data and then extracts sub-strings of length k (k-mer), defined by the user. The input files are read through a sliding window, creating dictionaries (one for each k-mer) that share some common information. When all input files have been processed, they are merged into a single MEF (Most Expressed Features) dictionary. It is possible to define an Expressed Features (EF) as a string of length *l* and a number of occurrence of the strings (o_i). The dictionary MEF contains all EFs in descending order starting from the sequences with larger o_i [23].

Protcomp calculates distances among strings (Fig. 1). The metric used to calculate the distances is the Easy Similarity Score (ESS), as defined in Eq. 1.

ESS corresponds to the overlap of the two dictionaries divided by the smallest. k-mers with a length between 6 and 8 were used.

$$ESS_{lmin,lmax}(X,Y) = \frac{\sum_{lmin}^{lmax} |MEF_i(X) \cap MEF_i(Y)|}{MIN(X,Y)} \tag{1}$$

Where:

$ESS_{lmin,lmax}$ (X, Y) = Easy similarity score between two sets (X, Y);

$MEF_i(X)$ = MEF dictionary of the set X with string length i;

$MIN(X, Y)$ is the smallest dictionary.

To provide an effortless display of the calculated distances between the 44 trajectories (11 for each type of system), we have applied t-SNE (t-distributed stochastic neighbor embedding) [24]. t-SNE is a nonlinear dimensionality reduction technique. The data was displayed using a dispersion graph (Fig. 2).

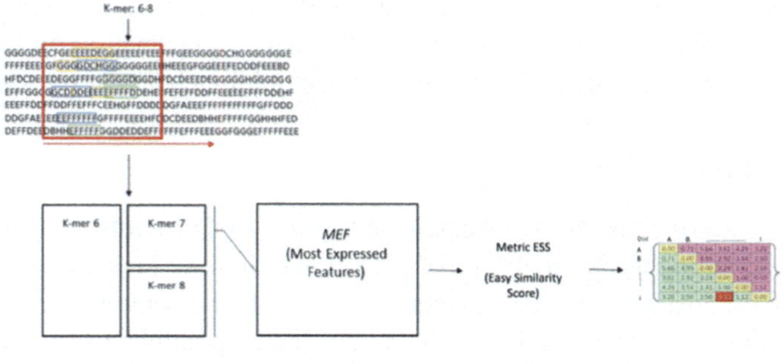

Fig. 1. Example of the Protcomp workflow. Protcomp pre-processes input files and extracts sub-strings of length *k* (*k*-mer). The input files are read through a sliding window (in red), creating dictionaries of the most abundant sub-strings MEF (Most Expressed Features) dictionary. The Easy Similarity Score (ESS) metric is pairwise applied to the MEFs to generate a matrix of distances.

3 Results

The analysis proposed here represents an alternative to traditional studies on the physical state of membranes. All the most commonly used methods suffer from some severe problems. For example, computational investigations using techniques such as the SCD order parameter calculation necessarily offer averaged values. In contrast, those carried out with LPP provide a fundamental analysis but without the possibility of experimental verification. Finally, the most common experimental methods, such as the use of fluorescent probes, hardly discriminate between membranes of similar composition and inevitably alter the same physical state they claim to measure. The method described in this paper does not extract a numerical value to be attributed to the membrane physical state, but can determine the extent of perturbation of a membrane by calculating the similarity among bilayers.

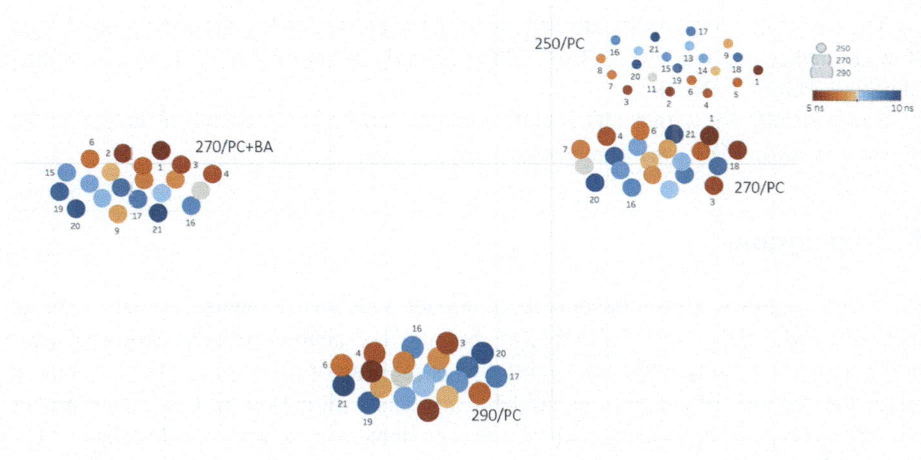

Fig. 2. t-SNE dispersion graph. Each point corresponds to a frame of molecular dynamics. Each dynamic generated 21 frames, indicated by a time scale colored from red to blue. The figure shows the four systems analyzed at different temperatures.

In Fig. 2, each point indicates a snapshot of MD simulations. Each simulation of MD generates 21 snapshots, whose color indicates the time (ns). The size of the points signifies the temperature applied; the smaller dots indicate a lower temperature.

The figure shows the four analyzed systems at different temperatures: at the top right, the two systems are consisting of the POPC membrane before the phase transition temperature (250–270 K). At the center bottom, there is the 290 K POPC system, and at the left, there is the system membrane+BA.

The systems simulated below the T_m temperature (250–270 K) show high similarity, as evident by the extreme proximity of the points in the graph. This indicates a similar physical state of the membrane.

For the 270 K BA system, there is a difference compared to the 270 K membrane alone. The two clusters are distant from each other. This shows that the BA has modified the fluidity of the membrane as already experimentally reported in ref. [25].

The system with BA at 270 K is more similar to POPC at 290 K rather than POPC at 270 K without a fluidifier. This is even more evident from the tree obtained from the distance matrices of the 4 systems.

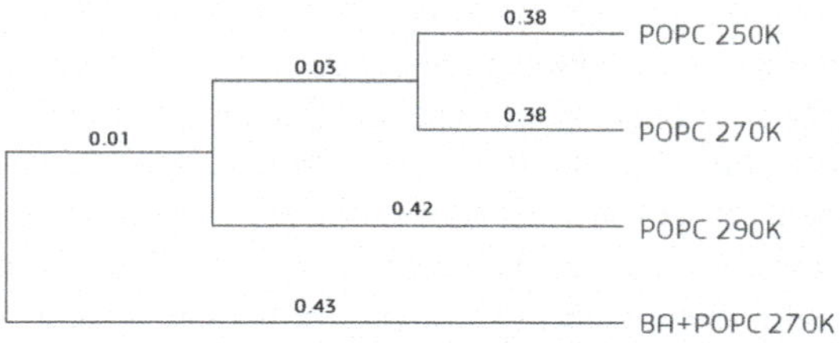

Fig. 3. The tree obtained from the distance matrix. The numbers correspond to the distance calculated with ESS.

The tree (Fig. 3) shows the distance among the clusters. We can see from the length of the branches that the 270 K BA+POPC system is closer to the 290 K system, rather than pure POPC at 270 K.

These results show that this pseudo semantic analysis is capable of detecting the physical state of POPC membranes correctly.

4 Conclusions

This work proposes a pseudo-semantic approach that uses a newly defined metric to analyze an MD trajectory of a POPC membrane. The analysis offered here represents an alternative to traditional studies on the physical state of membranes. In particular, it solves the problem of creating unavoidable artifacts with the use of fluorescent probes and offers a type of holistic analysis that is complementary to traditional methods. The method described can be used to investigate the state of membranes and how it modulates the dynamics of embedded proteins. However, it is important to underline how this approach has its strength in its applicability. The method can be used for the analysis of any system that can be encoded in strings. This allows the analysis of protein structures, crystals, and even molecular dynamics. The application of the method for the study of protein folding is currently being validated.

References

1. Friedman, R., et al.: Understanding conformational dynamics of complex lipid mixtures relevant to biology. J. Membrane Biol. **251**(5–6), 609–631 (2018)
2. Scott, H.L.: Modeling the lipid component of membranes. Curr. Opin. Struct. Biol. **12**(4), 495–502 (2002)
3. Török, Z., et al.: Plasma membranes as heat stress sensors: from lipid-controlled molecular switches to therapeutic applications. Biochim. Biophys. Acta (BBA)-Biomembr. **1838**(6), 1594–1618 (2014)

4. Polley, A., Mayor, S., Rao, M.: Bilayer registry in a multicomponent asymmetric membrane: dependence on lipid composition and chain length. J. Chem. Phys. **141**(6), 064903 (2014)
5. Marquardt, D., Geier, B., Pabst, G.: Asymmetric lipid membranes: towards more realistic model systems. Membranes **5**(2), 180–196 (2015)
6. Lee, I.-H., et al.: Live cell plasma membranes do not exhibit a miscibility phase transition over a wide range of temperatures. J. Phys. Chem. B **119**(12), 4450–4459 (2015)
7. Finegold, L.: Cholesterol in model membranes. CRC Press, Boca Raton (1993)
8. Ollila, S., Hyvönen, M.T., Vattulainen, I.: Polyunsaturation in lipid membranes: dynamic properties and lateral pressure profiles. J. Phys. Chem. B **111**(12), 3139–3150 (2007)
9. Casas, J., et al.: G protein-membrane interactions II: effect of G protein-linked lipids on membrane structure and G protein-membrane interactions. Biochim. Biophys. Acta (BBA)-Biomembr. **1859**(9), 1526–1535 (2017)
10. Ballweg, S., et al.: Regulation of lipid saturation without sensing membrane fluidity. Nat. Commun. **11**(1), 1–13 (2020)
11. Maula, T., Westerlund, B., Slotte, J.P.: Differential ability of cholesterol-enriched and gel phase domains to resist benzyl alcohol-induced fluidization in multilamellar lipid vesicles. Biochim. Biophys. Acta (BBA)-Biomembr. **1788**(11), 2454–2461 (2009)
12. Reyes, J., Latorre, R.: Effect of the anesthetics benzyl alcohol and chloroform on bilayers made from monolayers. Biophys. J. **28**(2), 259 (1979)
13. Sauerheber, R.D., Esgate, J.A., Kuhn, C.E.: Alcohols inhibit adipocyte basal and insulin-stimulated glucose uptake and increase the membrane lipid fluidity. Biochim. Biophys. Acta (BBA)-Biomembr. **691**(1), 115–124 (1982)
14. Gordon, L., et al.: The increase in bilayer fluidity of rat liver plasma membranes achieved by the local anesthetic benzyl alcohol affects the activity of intrinsic membrane enzymes. J. Biol. Chem. **255**(10), 4519–4527 (1980)
15. Pope, J., et al.: The ordering of benzyl alcohol and its influence on phospholipid order in bilayer membranes. J. Am. Chem. Soc. **108**(18), 5426–5433 (1986)
16. Chen, C.-H., Hoye, K., Roth, L.G.: Thermodynamic and fluorescence studies of the underlying factors in benzyl alcohol-induced lipid interdigitated phase. Arch. Biochem. Biophys. **333**(2), 401–406 (1996)
17. McIntosh, T., McDaniel, R., Simon, S.: Induction of an interdigitated gel phase in fully hydrated phosphatidylcholine bilayers. Biochim. Biophys. Acta (BBA)-Biomembr. **731**(1), 109–114 (1983)
18. Chapman, D.: Fluidity and phase transitions of cell membranes. In: Eisenberg, H., Katchalski-Katzir, E., Manson, L.A. (eds.) Aharon Katzir Memorial Volume, pp. 1–9. Springer, Boston (1975)
19. Concilio, S., et al.: A novel fluorescent solvatochromic probe for lipid bilayers. Supramol. Chem. **29**(11), 887–895 (2017)
20. Silvius, J.R.: Thermotropic phase transitions of pure lipids in model membranes and their modifications by membrane proteins. Lipid-protein Interact. **2**, 239–281 (1982)
21. Krieger, E., Vriend, G., Spronk, C.: YASARA–yet another scientific artificial reality application, vol. 993. YASARA. org (2013)
22. Guixa-González, R., et al.: MEMBPLUGIN: studying membrane complexity in VMD. Bioinformatics **30**(10), 1478–1480 (2014)
23. Piotto, S., et al.: Transmembrane peptides as sensors of the membrane physical state. Front. Phys. **6**, 48 (2018)
24. Maaten, L.V.D., Hinton, G.: Visualizing data using t-SNE. J. Mach. Learn. Res. **9**(Nov), 2579–2605 (2008)
25. Vígh, L., et al.: Membrane-regulated stress response. In: Molecular Aspects of the Stress Response: Chaperones, Membranes and Networks, pp. 114–131. Springer (2007)

Encoding Materials Dynamics for Machine Learning Applications

Stefano Piotto(✉), Anna Maria Nardiello, Luigi Di Biasi, and Lucia Sessa

Department of Pharmacy, University of Salerno, Via Giovanni Paolo II, 132, 84084 Fisciano, SA, Italy
piotto@unisa.it

Abstract. Recent years witnessed an explosion of machine learning methods in all sectors. However, in the materials sector, and even more specifically in the biomaterials sector, although there have been numerous attempts at generalization, there has been a severe problem of coding the problem. The reason lies mainly in the temporal and spatial dimensions of the materials and their intrinsic complexity. In this contribution, we wish to suggest a possible universal coding of materials. This coding exploits a pseudo-semantic analysis and can be particularly advantageous in the study of polymeric biomaterials.

Keywords: Pseudo-semantic analysis · MD · ProtComp

1 Introduction

The process of discovering new materials can be significantly accelerated and simplified if we can learn effectively from existing data. Several contributions have shown that an efficient and accurate prediction of material properties is possible using automated (or statistical) learning methods trained on the results of quantum mechanics calculations in combination with concepts of chemical similarity [1–6]. For example, a general formalism for a family of one-dimensional chain systems has been presented, allowing us to discover decision-making rules that establish a mapping between the easily accessible attributes of a system and its properties. It has been shown that fingerprints based on chemostructural information (composition and configuration information) or electronic charge distribution can be used to make ultra-fast but accurate property predictions [7]. The use of such learning paradigms expands recent efforts to explore and exploit vast chemical spaces systematically. These methodologies can significantly accelerate the discovery of new application-specific materials. Forecasts are typically interpolating, where first a fingerprint material is captured numerically, and then a mapping (determined by a learning algorithm) between the fingerprint and the property of interest is followed. Fingerprints, also called "descriptors," can be of many types and sizes, depending on the application and needs. Forecasts can also be extrapolated - the extension to new materials - spaces provided that predictive uncertainties are adequately considered.

S. Piotto et al. (Eds.): BIONAM 2019, LNBE, pp. 128–136, 2020.
https://doi.org/10.1007/978-3-030-47705-9_12

Extracting or learning from these resources or other existing reliable data can lead to the identification of previously unknown relationships between properties and to the discovery of qualitative and quantitative rules - also called surrogate models - that can predict the magnitude of material properties faster, more cheaply and with less human effort than the benchmark simulation or experimental methods used to generate data.

Challenges are accompanied by great opportunities. The question always arises as to what kind of material science problems are best suited for a data-based approach. A clear understanding of this aspect is essential before deciding to use automatic learning to address a problem of interest.

Perhaps the most critical aspect of the automatic learning paradigm is that it deals with the numerical representation of cases or input materials [8, 9]. A numerical representation is indispensable to design the concept of prediction quantitatively. The selection of the numerical representation can be made valid only with sufficient knowledge of the problem and the objectives (i.e., competence or experience) and is usually iterative, taking into account the aspects of the material with which the real property can be correlated [10].

In materials science, particularly in polymer science, the concepts underlying QSAR/QSPR finally led to the success of group contribution methods [11]. Van Krevelen and his colleagues studied the properties of polymers and found that they were strongly related to chemical structure (e.g., type of polymer repetition unit, end groups, etc.) and molecular weight distribution. They observed that the properties of polymers, such as glass transition temperature, solubility parameters, and mass modulus, which were and are difficult to calculate with conventional calculation methods, were and are related to the presence of different chemical groups and group combinations in the repetitive unit. Based on a purely data-based approach, they developed an "atomic group contribution method" to express the different properties as a weighted linear sum of the contribution (called atomic group parameters) from each atomic group that formed the repetitive unit. These groups can be units such as CH_2, C_6H_4, CH_2-CO, etc. that make up the polymer.

However, the full application of these replacement models is still limited because (1) the definition of the atomic groups is somehow ad hoc, and (2) it is assumed that the target properties are linearly linked to the parameters of the group.

This method provides a uniform and seamless pathway to represent all polymers within this class, and the method can be generalized indefinitely considering fragments of a higher-order (i.e., quadruplets, quintuples, etc. of the types of atoms). The strength of these modern, data-based learning approaches based on molecular fragments (such as its group contribution predecessor) lies in the knowledge that any property related to the molecular structure - whether calculable via DFT [12, 13] (e.g., bandgap, dielectric constant), with all atoms or coarse grain molecular dynamics [5, 14] or experimentally measurable (e.g., glass transition temperature, dielectric loss) - can be learned and predicted. The representation based on molecular fragments is not limited to polymeric materials.

MoA analysis of a ligand is a complex task on a multi-state receptor such as GPCR. It is commonly accepted that GPCR receptors oscillate between (at least) two states, referred to as open and closed conformation [15, 16]. The simple binding of a ligand is not enough to establish the biological response of the molecule and distinguish between the agonist and antagonist. This aspect is usually expressed by evaluating the final effect of the entire path initiated by the receptor and not by a mechanical interpretation of the system.

To compare the effect of different molecules on the same receptor is necessary to identify a suitable metric to compare the dynamics of the systems. A dynamic state is identified with a string of characters, each representing a residue of the receptor. *n* strings, 1 for each frame, embed the information of the whole dynamics of the system.

The adopted method to transform the information contained in the molecular dynamics calculations into processable data can be easily used to train a Neural Network. The final aim is to verify the possible correlation between the agonist/antagonist activity of a ligand and the conformational changes that occur in the receptor.

A key point is the calculation of the similarity between the various strings. A transition between α-helix and β leaflet should be observable as a transition from one point to another, without oscillations. The effect of a ligand on the dynamics of the systems can be quickly evaluated. If a ligand, for example, blocked the conformation in a defined way or denatured the receptor, a defined and irreversible variation would be observed in the strings. Having a periodic oscillation is an indication that the ligand does not block the system but alters the frequency of movement and not the mechanics of the system itself.

2 Methods

MD trajectories must be converted in strings and submitted to the ProtComp software (www.yadamp.unisa.it/protcomp), able to analyze the similarities between them.

ProtComp is a method that counts the number of substrings of length k (k-mer) in each string. The input files are read by sliding-window scrolling, forming dictionaries. Dictionaries share common information. When all input files have been processed, they are merged into a single MEF (Most Expressed Features) dictionary. You can define MEF as an l-length dictionary as a pair (ci, oi) where there is an l-length segment that occurs oi times in a string X with oi $\geq$ 1.

ProtComp uses the Easy Similarity Score (ESS) metric to calculate distances. This metric considers the intersections between dictionaries by dividing them by the smallest dictionary [17].

The hydrogen bonds formed during the simulation were monitored for each water molecule in the system. The number of H-bonds varies between 0 and 4, so the state of the system in a given instant is represented by a string with length equal to the number of water molecules.

In the case of protein receptors, molecular dynamics calculations were performed for the following systems:

1. free receptor (derived from the structure with pdb code: 6A93)
2. receptor bound to LSD (agonist)
3. receptor bound to 8NU (risperidone, antagonist)

Each system was inserted in a membrane model (POPC 100%), and a simulation of molecular dynamics long 30 ns was carried out. The system was fully hydrated, and an initial minimization of 250 ps was performed. The force-field used was AMBER15FB; the simulation cell is 80 × 80 × 80. At the end of the simulations, 120 simulation snapshots were produced for each system.

Subsequently, for each frame and for each residue, the RMSD (Root-Mean-Square Deviation) was calculated between the receptor in the presence of the different ligands and the only receptor present in the membrane. Moreover, for each residue, the values of the dihedral angles (φ e ψ) of the receptor scene with the various ligands were calculated, compared to the values of the dihedral angles of the receptor alone.

For the dynamics of the systems 2, 3, and 4, were calculated, for each residue and for each snapshot (Receptor = R, Ligand = L):

- the values of RMSD for each residue between the system R + L and the free receptor
- the values of the dihedral angles (φ e ψ) for each pair of residues of the system R + L and the free receptor

The obtained angles were first converted from degrees to radians, and of these, the sin(rad) was calculated. This step was necessary to avoid large fluctuations in value in case of a change of sign. The conversion in sin (rad) allows us to obtain all positive values, contained in a less wide range, also obtaining values that provide, if reported in the graph, a clean baseline (smooth) and fluctuations clearly visible in the presence of a significant difference. The differences between the two have been calculated:

- $\sin\varphi(R+L) - \sin\varphi\, R$
- $\sin\psi(R+L) - \sin\psi\, R$

For each of the 2–4 complexes, a file containing the RMSD values was obtained $\Delta \sin \varphi$ and $\Delta\sin\psi$ for each residue of the receptor.

The receptor 5HT2A [18] has 370 amino acid residues. For each file, R + L/free R, 370 RMSD values for 121 simulations (table 370 × 121), and 740 Δsin values for 121 simulations (720 × 121) are given (the extra string represents the initial state of the system).

Each simulation is then converted into an alphabetical string, where each letter corresponds to a range of values; the first ten letters of the alphabet (A - J) were used.

The values of RMSD are divided into 10 ranges, with the last letter (J) corresponding to values higher than 10 Å. The values of Δsin are divided into 10 ranges, between 0 and 2 (A - J).

For each complex R + L/free R, two files were obtained: one containing 121 strings of 370 characters, corresponding to RMSD values, and one containing 121 strings of 740 characters, corresponding to $\Delta \sin \varphi$ e $\Delta\sin\psi$ values (for each residue).

The similarity matrix was initially calculated on 10 blocks of 12 strings each, for each complex - the dynamics in 3 ns blocks are therefore considered. The similarity between these strings is then measured. The strings are then subjected to semantic analysis with ProtComp to evaluate the similarity between the strings [17].

3 Results

3.1 Analysis of Water Phase Transitions

Once the molecular dynamics were performed, and the hydrogen bonds of each individual water molecule were calculated over time, this information was transformed into a string to be submitted to ProtComp [17].

ProtComp generated a distance matrix comparing every single snapshot of the dynamics performed. From the distance matrix, a heatmap was derived for more efficient and faster data visualization.

The heatmap in Fig. 1 shows the similarity between the snapshots of the dynamics, indicating with the number on the axes the different temperatures used. The color scale shows that the more dissimilar the two snapshots are (tending to 1), the more the color tends to yellow, while close distances are represented with blue (tending to 0).

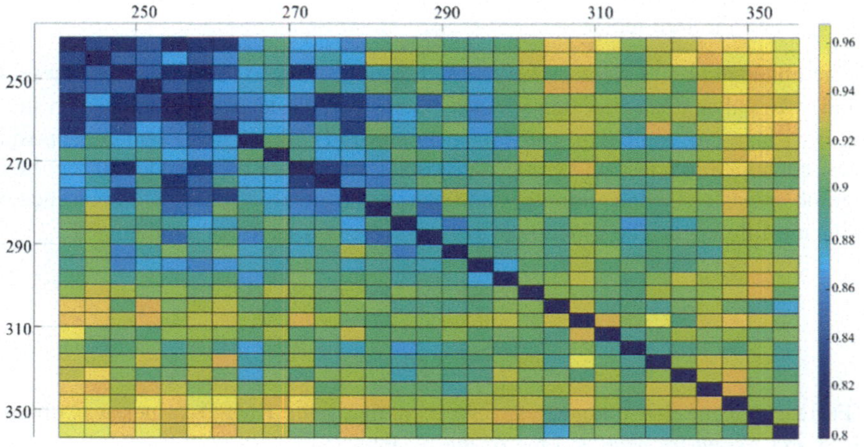

Fig. 1. Heatmap representation of the similarity distances among MD snapshots. Blue corresponds to high similarity, yellow to low similarity.

As shown in the figure, the greater distance between the number of hydrogen bonds is between 250 and 350K. Less distance between closer temperatures. This shows how, through an alphabetical string, we can discriminate the different water transition states.

3.2 Analysis of the Trajectory of a Receptor-Ligand System

A more complex and much more relevant example is the following.

There are many known protein receptors whose activity is regulated after interaction with the natural substrate.

If a molecule after interaction induces an increase in receptor activity, we have an agonist. If not, we have an antagonist. It is essential to say that the difference between agonist and antagonist cannot be deduced from the differences in binding. In fact, in many cases, the portion of the protein affected by binding is the same, and there are no reliable predictive models for the design of an agonist or an antagonist (Fig. 2). It is only possible to design a suitable binder, or at best, an inhibitor if the binding results in a block of enzymatic activity.

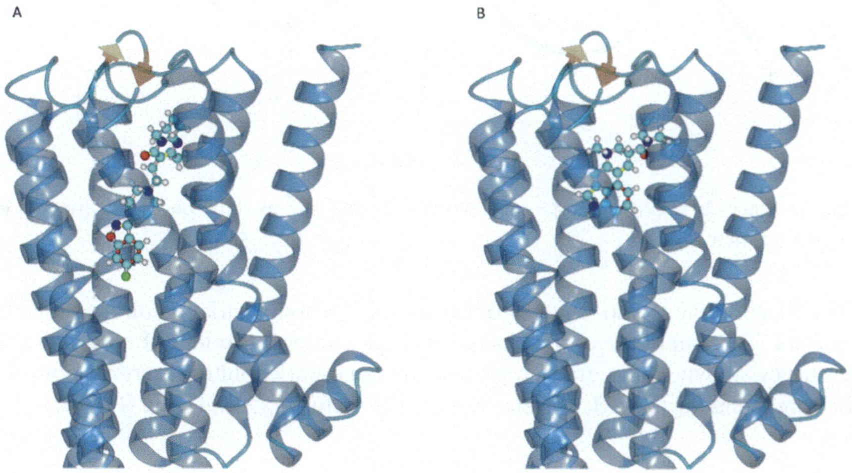

Fig. 2. Models of the binding site of serotonin 2A receptor in complex with the antagonist (A) and the agonist (B)

Depending on the type of enzyme, and in particular, for GPCR, conformational changes affect the receptor domains, resulting in a different interaction with the G protein and activation of the signal cascade. In the case of the serotonin 2A receptor (5HT2AR), the interaction with a ligand involves different stability of the open or closed form of the receptor, resulting in greater or lesser interaction with the coupled G protein.

The conversion into strings allows us to quickly and efficiently compare the changes to the dihedral angles in the 3 complexes considered. The rotation of the dihedral angles in a peptide sequence reflects the movement of the lateral chains in the same; only the angles φ and ψ have been considered. The difference of the angles is used for the analysis with ProtComp, to obtain the graphs of similarity. Based on RMSD, the graphs give us an indication of the overall response of the system, the entire trajectory in response to the three different effectors.

If we plot the distances between the systems R/L at different time and the system R/L at an arbitrary time, we obtained two dimensional plots of global geometrical variation.

This type of graph shows the overall trend of the receptor with ligand with respect to the free receptor over time; in other words, it is possible to see the temporal trend of the oscillations of similarity (Fig. 3).

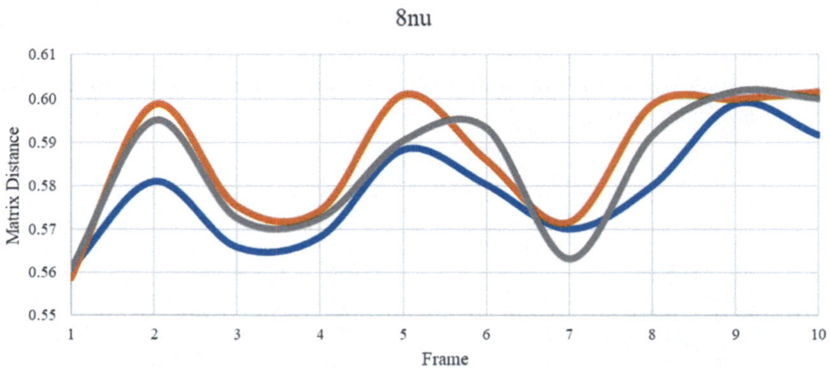

Fig. 3. Distances between the system receptor/8nu ligand at 1 ns (blue line), 2 ns (orange line) and 3 ns (grey line)

The Risperidone (8 nu), an antagonist, shows a periodic variation of the phi and psi parameters. The amplitude of the changes is high compared to that of an agonist. The three curves shown in the graph represent the temporal evolution starting from three distinct moments of the R-L system, respectively, after 1, 2 and 3 ns (Fig. 4).

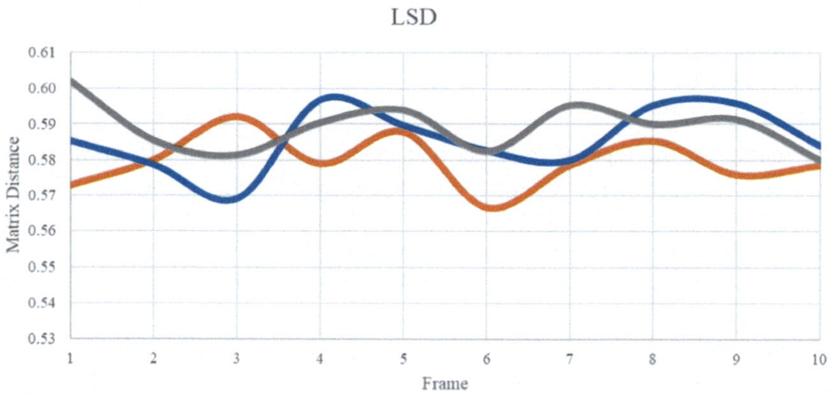

Fig. 4. Distances between the system receptor/LSD and the geometries at 1 ns (blue line), 2 ns (orange line) and 3 ns (grey line)

The Lysergic acid diethylamide (LSD) shows a small variation of fluctuations. The 3 signals, relative to 1, 2, and 3 ns, show a counter phase trend.

The interaction between the ligand and receptor in the binding pocket leads to changes in the shape of the residues in the binding pocket itself. Depending on the enzyme type, and especially for GPCRs, conformational changes affect the receptor domains, resulting in a different interaction with the G protein and in the activation of the signaling cascade. In the case of 5HT2AR, the interaction with a ligand involves different stability of the open or closed form of the receptor, resulting in greater or lesser interaction with the coupled G protein.

The conversion into strings allows us to quickly and efficiently compare the changes to the dihedral angles in the 3 considered systems. The rotation of the dihedral angles in a peptide sequence reflects the movement of its side chains—only the angles φ and ψ have been considered. The difference of the angles is used for the analysis with ProtComp, to obtain the similarity graphs. Based on RMSD, the graphs give us an indication of the overall response of the system, the entire trajectory in response to the three different effectors.

4 Conclusions

In this work, two cases were shown: the first, extremely simple, showed how the appropriate coding of the dynamics of water molecules allows the detection of phase transitions. The second, much more sophisticated, showed how to distinguish the agonist or antagonist effect of a binder on a protein. The method described finds countless applications in materials science, physiology, solid-state chemistry, and pharmacology. It allows the investigation of phase transition in crystalline systems, or to monitor the effect of local anesthetics on lipid membranes. It has an immense advantage over conventional methods of being applicable in areas where there is no unique and accepted method and is also enormously faster than traditional methods.

References

1. Diana, R., et al.: Solid-state fluorescence of two zinc coordination polymers from bulky dicyano-phenylenevinylene and bis-azobenzene cores. Inorg. Chem. Commun. **110** (2019)
2. Piotto, S., et al.: Novel antimicrobial polymer films active against bacteria and fungi. Polym. Compos. **34**(9), 1489–1492 (2013)
3. Piotto, S., et al.: Computer simulations of natural and synthetic polymers in confined systems. In: Macromolecular Symposia. Wiley Online Library (2009)
4. Piotto, S., Nesper, R.: CURVIS: a program to study and analyse crystallographic structures and phase transitions. J. Appl. Crystallogr. **38**(1), 223–227 (2005)
5. Piotto, S., et al.: DDFT simulations of the assembly of block copolymers in confined systems. In: AIP Conference Proceedings. AIP (2008)
6. Friedman, R., et al.: Understanding conformational dynamics of complex lipid mixtures relevant to biology. J. Membr. Biol. **251**(5–6), 609–631 (2018)
7. Ramprasad, R., et al.: Machine learning in materials informatics: recent applications and prospects. npj Comput. Mater. **3**(1), 1–13 (2017)
8. Mueller, T., Kusne, A.G., Ramprasad, R.: Machine learning in materials science: recent progress and emerging applications. Rev. Comput. Chem. **29**, 186–273 (2016)
9. Nakajima, K., et al.: Exploiting the dynamics of soft materials for machine learning. Soft Robot. **5**(3), 339–347 (2018)
10. Goh, G.B., Hodas, N.O., Vishnu, A.: Deep learning for computational chemistry. J. Comput. Chem. **38**(16), 1291–1307 (2017)
11. Van Krevelen, D., Te Nijenhuis, K.: Properties determining mass transfer in polymeric systems. Prop. Polym. **4** (1990)

12. Panunzi, B., et al.: Synthesis, spectroscopic properties and DFT calculations of a novel multipolar azo dye and its zinc (II) complex. Inorg. Chem. Commun. **84**, 103–108 (2017)
13. Concilio, S., et al.: A novel fluorescent solvatochromic probe for lipid bilayers. Supramol. Chem. **29**(11), 887–895 (2017)
14. Piotto, S., et al.: Dissipative particle dynamics study of alginate/gelatin aerogels obtained by supercritical drying. In: Advances in Bionanomaterials, pp. 75–84. Springer, Cham (2018)
15. Gurevich, V.V., Gurevich, E.V.: Molecular mechanisms of GPCR signaling: a structural perspective. Int. J. Mol. Sci. **18**(12), 2519 (2017)
16. Mozumder, S., et al.: Comprehensive structural modeling and preparation of human 5-HT2A G-protein coupled receptor in functionally active form. Biopolymers **11**(1), e23329 (2019)
17. Piotto, S., et al.: Transmembrane peptides as sensors of the membrane physical state. Front. Phys. **6**, 48 (2018)
18. Kimura, K.T., et al.: Structures of the 5-HT 2A receptor in complex with the antipsychotics risperidone and zotepine. Nat. Struct. Mol. Biol. **26**(2), 121–128 (2019)

Applications of Bio and Nanomaterials

Eco-Friendly Lipid Nanoformulations: Toward Greener Alternatives for Organic Extraction Solvents

Ilaria Clemente[1,2(✉)], Felicia Menicucci[2], Giulia Sautariello[2], and Sandra Ristori[2]

[1] Department of Biotechnology, Chemistry and Pharmacy, University of Siena, via Aldo Moro 2, 53100 Siena, Italy
ilaria.clemente@student.unisi.it
[2] Department of Chemistry "Ugo Schiff" & CSGI, University of Florence, via Della Lastruccia 3, 50019 Sesto Fiorentino, Italy

Abstract. Dealing with lipid extraction procedures, the use of organic solvents, i.e. primarily chloroform and methanol, is still very popular, despite of the proved toxicity of these solvents. In the last decades, several contributions reported the possibility of employing more eco-friendly solvents as extraction agents, but the efficacy of chloroform and methanol has not been attained yet, making their substitution one of the big challenges of green chemistry.

In this work, we compared four alternative solvent mixtures for the extraction of the lipid component from olive pomace, the exhausted waste of olive oil processing used as resource to build eco-sustainable nanoformulations for agrotechnological applications. Specifically, the mixtures tested were ethyl acetate/ethanol (EtAc/EtOH), ethyl acetate/methanol (EtAc/MeOH), acetonitrile/ethanol (ACN/EtOH), and acetonitrile/methanol (ACN/MeOH) to compare with the standard one. The lipid extracts thus obtained were analyzed by Gas Chromatography-Mass Spectrometry and then employed for the manufacturing of nanoformulations.

The obtained lipid nanovectors were then fully characterized by Dynamic Light Scattering and Small Angle X-ray Scattering. The comparison of the collected data shows that among the mixtures tested, EtAc/EtOH and ACN/MeOH resulted to be the most suitable alternatives to the standard method, thus representing a possible different choice that better matches with green chemistry requirements.

1 Introduction

Lipid carriers currently represent a widely studied and applied nanotechnology area, which allows to fabricate nanoformulations for biomedical, pharmacological, agricultural and industrial interest [1]. Such widespread use on a large-scale level would involve the preparation of massive amounts of product, with serious issues on the safety and toxicity of the process, both for the operators and the environment. Indeed, concerns have been risen particularly about the chemicals employed, in the effort of developing harmless and eco-friendly procedures. Among the existing alternatives, the

S. Piotto et al. (Eds.): BIONAM 2019, LNBE, pp. 139–150, 2020.
https://doi.org/10.1007/978-3-030-47705-9_13

extraction of natural lipids from agricultural waste represents a valuable option to enhance biocompatibility, and at the same time to allow the recycling of otherwise useless materials. Since the extraction of lipids from natural matrices involves the use of organic solvents, another step towards a totally eco-friendly procedure would require their substitution. Even though the first lipid extraction methods proposed by Folch [2] and by Bligh and Dyer [3] date back to the 1950s, they are still considered the gold standards for such procedures. Indeed, the solvents used, i.e. chloroform and methanol, are very effective to retrieve both polar and non-polar compounds. However, according to the green chemistry approach, hazardous chemicals should be replaced by greener, possibly bio-sourced substitutes, and recently many attempts have been made in this sense [4]. Numerous solvents and mixtures have been proposed, but none proved to be quite as effective as the original ones to recover lipids from biological matrices. In a recent study by Breil et al., [4] the use of both computational and experimental techniques allowed a detailed screening of alternatives on the basis of multiple solvation parameters, to simulate the relative solubility of lipids in these solvents. From the initial screening, the most appropriate substitutes were found to be ethyl acetate for chloroform and ethanol for methanol; these were then experimentally tested to perform extraction of lipids from microorganisms. Comparing this mixture with the original one, it was shown that the performances were quantitatively similar, opening up a valuable option for improving the standard extraction methods.

In this work, we propose an evaluation of four solvent mixtures to extract lipids from agricultural waste, to be then employed to prepare green lipid nanoformulations. We previously developed a protocol to prepare unconventional and biocompatible lipid nanocarriers for delivery purposes in agronomy [5], by exploiting an agricultural waste (i.e. olive pomace) as a natural extraction material. In order to make the procedure totally eco-friendly, we compared the efficacy of greener solvents to extract amphiphilic compounds suitable for the preparation of nanovectors. Based on toxicity parameters [6] and on their physico-chemical properties, the chosen mixtures were: (i) ethyl acetate/ethanol (EtAc/EtOH), (ii) ethyl acetate/methanol (EtAc/MeOH), (iii) acetonitrile/ethanol (ACN/EtOH), and (iv) acetonitrile/methanol (ACN/MeOH) (Table 1). Acetonitrile was chosen as a polar solvent with less toxicity with respect to chloroform. The lipid composition of the obtained extracts was analyzed through Gas Chromatography-Mass Spectrometry, then these extracts were employed to prepare lipid nanovectors. The structural characterization of the nanoformulations was carried out by means of Dynamic Light Scattering and Small Angle X-ray Scattering, that allowed to define the best performing extract to obtain nanovectors with respect to the one from Folch procedure. Our results show that acetonitrile/methanol are the solvents with more overall similarities to the standard method, in terms of both type of extracted compounds and obtained lipid nanovectors. These findings are meant to contribute to the development of a green and eco-friendly protocol for large-scale production of lipid nanosystems.

Table 1. TLV-TWA (Threshold Limit Value - Time Weighted Average) for traditional and alternative solvents

Solvent	TLV-TWA (ppm)
chloroform	10
acetonitrile	40
ethyl acetate	400
methanol	200
ethanol	1000

2 Materials and Methods

2.1 Lipid Extraction and Preparation of Lipid Nanovectors

Olive pomace from *Olea europaea* was stored in freezer at −4 °C to prevent oxidation and de-frozen right before use. Then 200 mg/ml were extracted in each solution: (a) chloroform/methanol 2:1 v/v (Folch solution); (b) acetonitrile/methanol 2:1 v/v; (c) acetonitrile/ethanol 2:1 v/v; (d) ethyl acetate/methanol 2:1 v/v; (e) ethyl acetate/ethanol. The extraction was carried out under magnetic stirring for 24 h at room temperature. Afterwards the residues of olive pomace were eliminated, and the remaining solvent containing the extracted lipids was evaporated under ventilation overnight, leading to the formation of a lipid film. Then, a second extensive (under-vacuum) evaporation at 30 °C for about 2 h was carried out, in order to obtain a completely dried lipid film. After rehydration with MilliQ water, 8 cycles of freeze-and-thaw were performed. Finally, to obtain unilamellar vesicles and homogenize the size of nanovectors, 5 cycles of 4 min each of high-power sonication were carried out.

2.2 Gas Chromatography-Mass Spectrometry (GC-MS)

GC-MS analysis was carried out on a Varian Saturn Ion Trap 2200 GC/MS/MS System with CP-3800 GC.

2.3 Esterification

GC-MS analysis was performed to verify the type of extracted lipids and the possible differences among the various extraction solutions. An aliquot of the extracts obtained from the previous step was taken for analysis and derivatized. Indeed, the fatty acids need to be esterified to be less polar and more volatile, therefore detectable by the instrument. A 2.5% acidic methanol solution was prepared, adding dropwise 2.5 ml H_2SO_4 to MeOH. Then 500 µl of sample diluted 1:5 (v/v) + 500 µl of acidic methanol were mixed in vials and placed in a ~60 °C bath for one hour to favor the reaction. The esterification was then stopped by adding 1 ml of n-hexane, in order to obtain a phase separation and extract only the fatty acids. Finally, the upper phase, containing the esterified fatty acids was taken and transferred into a suitable vial for gas-chromatography.

For samples in Folch, the chloroform was evaporated, then the formed lipid film was rehydrated in acetone and subsequently the esterification followed as above.

2.4 Standard Preparation

As a reference, appropriate lipid standards with known hydrocarbon chains were analyzed at GC-MS, i.e. DOPE (dioleoyl phosphatidylethanolamine), DOPC (dioleoyl phosphatidylcholine), POPE (palmitoyl-oleoyl-phosphatidylethanolamine) and POPC (palmitoyl-oleoyl-phosphatidylcholine). These compounds were also derivatized before being inserted in the instrument. Stock solutions of 10 mg/ml of DOPE, POPE and POPC in acetone, and DOPC in chloroform, were prepared, then esterification with methanol 1:1 was carried out as previously described.

2.5 Dynamic Light Scattering (DLS)

DLS measurements were carried out on a Malvern Zetasizer (Nano ZS) equipped with a He-Ne 633 nm, 4 mW laser with backscattering optics, after 1:20 dilution for all samples to obtain a suitable optical turbidity.

2.6 Small Angle X-ray Scattering (SAXS)

Small Angle X-ray Scattering experiments were performed at the ID02 beamline of the European Synchrotron Radiation Facility (Grenoble, France). The wavelength of the incoming beam was 1 μm, and the sample detector distance was 1 m, which covered a q range 0.103–6.5 nm^{-1} ($q = (4\pi/\lambda) \sin\theta$, where 2θ is the scattering angle). The 2D SAXS patterns initially recorded were normalized to the absolute scale using a standard procedure reported elsewhere [7, 8]. Samples were loaded on a flow through capillary of 2 mm diameter to ensure accurate background subtraction.

3 Results and Discussion

3.1 GC-MS Analysis

GC-MS is a powerful analytical technique commonly used for the separation and identification of volatile compounds [9]. In the present study GC-MS was performed on all the extracts, to compare qualitatively the classes of lipid molecules in the different solvent mixtures with respect to the Folch extract. The control sample itself was correlated to the profiles obtained for the standard purified phospholipids (Fig. 1), for a better identification of m/z fragments. In the Folch control (Fig. 2) the main peaks were at 193 m/z, 208 m/z, 282 m/z (linoleic acid) (Table 2), which corresponded to fragments present in all the standard phospholipid, at 287 m/z, a fragment found in the POPC profile, and at 361 m/z, corresponding to oleuropein aglycone, the main polyphenol commonly found in olive tree leaves and fruits [10]. Consequently, the components of Folch extract were found to be mainly phospholipids and polyphenols. The extract obtained from ACN/MeOH solvent mixture presented a similar composition, made up of phospholipid and polyphenol fragments at the same m/z of the control and two additional peaks found in all the phospholipid standards. The substitution of the alcoholic solvent in ACN/EtOH shifted the extraction yield towards bigger fragments over 400 m/z, related to compounds of polyphenolic nature (e.g. secoiridoids and lignans), though the intensity was lower. Both ethyl acetate mixtures showed a partial affinity to ACN/MeOH. Indeed, EtAc/MeOH extracted mainly phospholipid compounds, though all the peaks had quite low intensity, and apart for a couple of peaks, the same could be observed for EtAc/EtOH. On the contrary, neither mixture extracted polyphenols. It is also possible that such combinations were able to extract predominantly lipidic classes such as mono-, di- and triglycerides, that are not detectable by means of GC [11]. Moreover, it was not possible to identify fragments with m/z over $\sim$435, to avoid excessively long elution times. All the peaks below 80 m/z were attributed to the solvent and/or to fragments with scarcely diagnostic character.

3.2 DLS Measurements

DLS is a technique that exploits the light-matter interaction to derive information about a dispersed system subjected to Brownian motion. The time-dependent fluctuation of the scattered intensity can be correlated to the motion of particles with submicron size, subsequently, a Laplace inversion or cumulant expansion allows to obtain the distribution of dimensions and polydispersity of the sample [12]. The Folch control sample showed the presence of two populations, at 50 nm and 250 nm, with a polydispersity index (PI) of 0.49, but quite reproducible. Comparing the four alternative mixtures (Table 3), the ACN/MeOH sample showed rather similar distributions with respect to the control (Fig. 3), with two peaks at 90 nm and 290 nm, respectively, even though with very high PI. The substitution of methanol with ethanol in the ACN/EtOH mixture brought some relevant changes in the type of nanovectors obtained; these displayed less reproducible, but still controlled, distributions with generally larger size. On the contrary, both ethyl acetate mixtures showed quite wide and sometimes complex distributions, with poor reproducibility and high polydispersity (Fig. 4).

3.3 SAXS Measurements

SAXS is an advanced structural technique to obtain supramolecular information on dispersed samples at the nanoscale [13]. The scattered intensity of a monochromatic and well collimated incident x-ray beam gives a pattern which is the Fourier transform of the electron density profile within the sample. This technique allows to obtain structural detail at high resolution ($\sim$1 nm) over a wide range of size (typically hundreds of nm) in the direct space [14]. In this work SAXS analysis was performed on samples prepared with lipids extracted with ACN/MeOH and ACN/EtOH only, since these samples were identified from DLS results as suitable for this kind of measurements. On the contrary, as inferred from DLS data, the nanovectors prepared employing both ethyl acetate extracts were too polydisperse to be studied by SAXS. Indeed, such mixtures apparently extracted classes of lipids able to assemble into very large and poorly reproducible aggregates, that made the samples too heterogeneous for in depth structural analysis. Lipids extracted with ACN mixtures self-assembled into nanostructures similar to those obtained with the Folch procedure in both cases

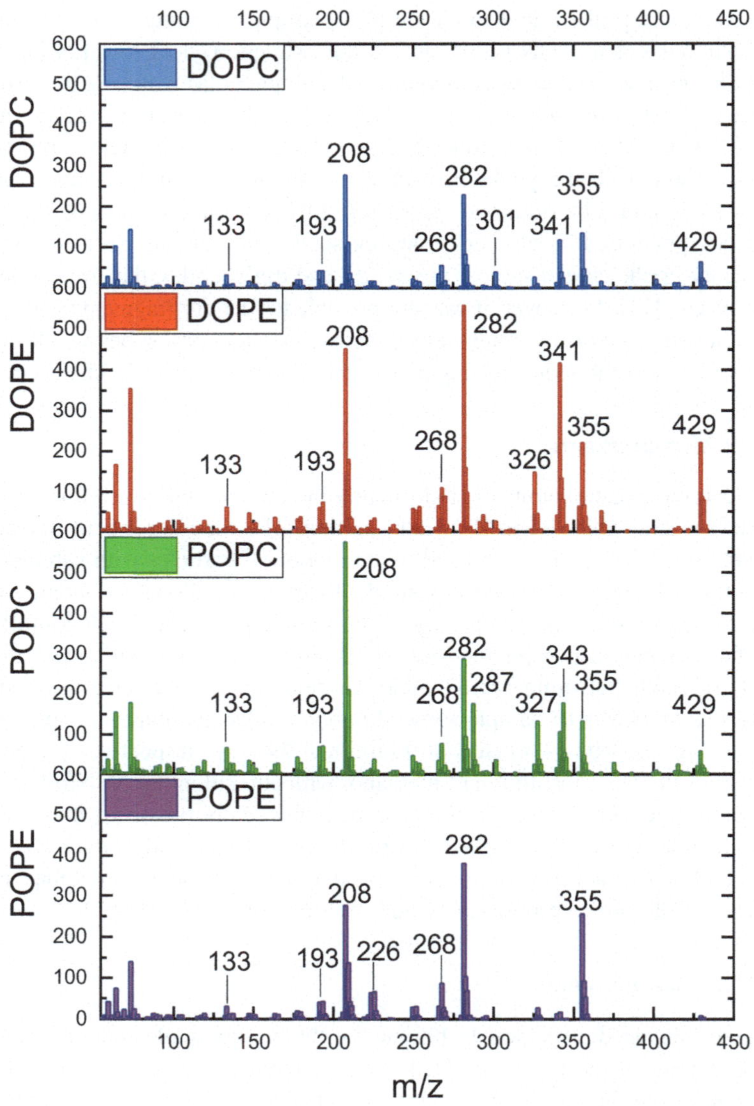

Fig. 1. GC-MS spectra in the 50–450 m/z region of chosen reference lipids (dissolved in Folch solution)

(ACN/MeOH and ACN/EtOH). Indeed, as it can be seen from Fig. 5, all the three curves present a typical pattern of mostly unilamellar aggregates without any superior order, indicated by the absence of intercorrelation peaks. Also, a diffused background that could indicate the presence of another, more polydisperse distribution is shown by all the samples. This is in line with the complex composition of the extracts evidenced by GC-MS analysis. Three more samples with DOPE addition were analyzed by means of SAXS (Fig. 6). Based on our previous findings, we found interesting the comparison

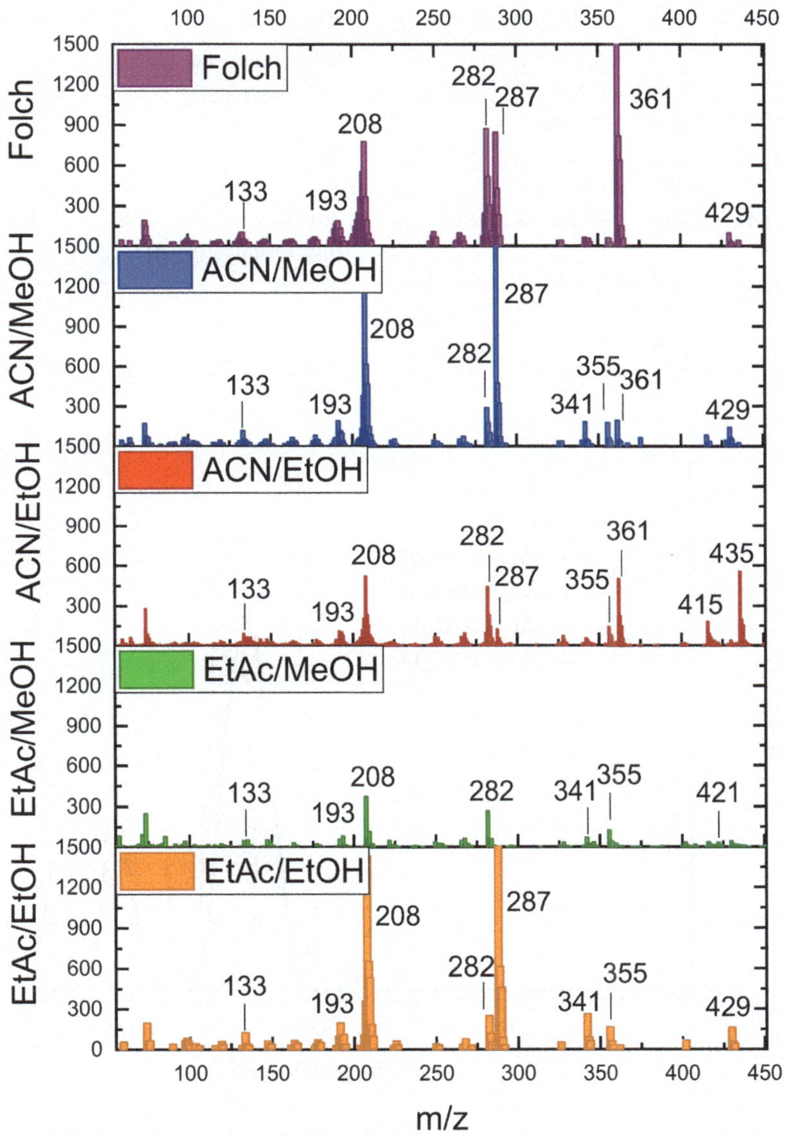

Fig. 2. GC-MS spectra in the 50–450 m/z region of the lipids extracted with different solvent mixture (see text for details)

of the structuring abilities of such systems when DOPE was added to the different extracts [15]. Indeed, as previously seen, the addition of a small amount of this purified phospholipid was able to increase the correlation among bilayers. Even though the Bragg peaks indicating the presence of a nonlamellar hexagonal arrangement were more defined in the Folch sample, they could be evidenced also for ACN samples. These data confirmed the higher polydispersity of such systems suggested by DLS

Table 2. List of significant fragments taken into account for each solvent mixture

Sample	Phospholipid fragments (m/z)	Other fragments
Control (Folch)	133, 193, 208, 282 (linoleic acid), 287, 429	361 (oleuropein)
ACN/MeOH	133, 193, 208, 282, 287, 341, 355, 429	361
ACN/EtOH	133, 193, 208, 282, 287, 355	361, 415 (acetoxypinoresinol), 435 (ligstroside aglycon)
EtAc/MeOH	133, 193, 208, 282, 341, 355	421
EtAc/EtOH	133, 193, 208, 282, 287, 341, 355, 429	–

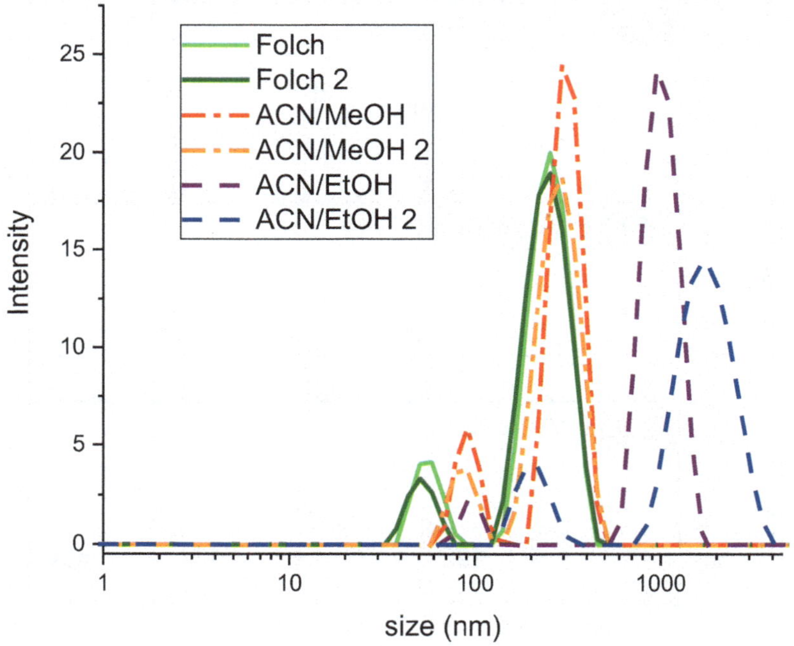

Fig. 3. Size distribution of the nanostructures obtained from lipids extracted with Folch and different ACN mixtures from olive pomace. For each sample the results of two different runs are reported.

results. As already seen in our previous work [15], the most relevant parameter to obtain different lipid structures was the addition of DOPE to the lipid mixtures, regardless of the solvent used.

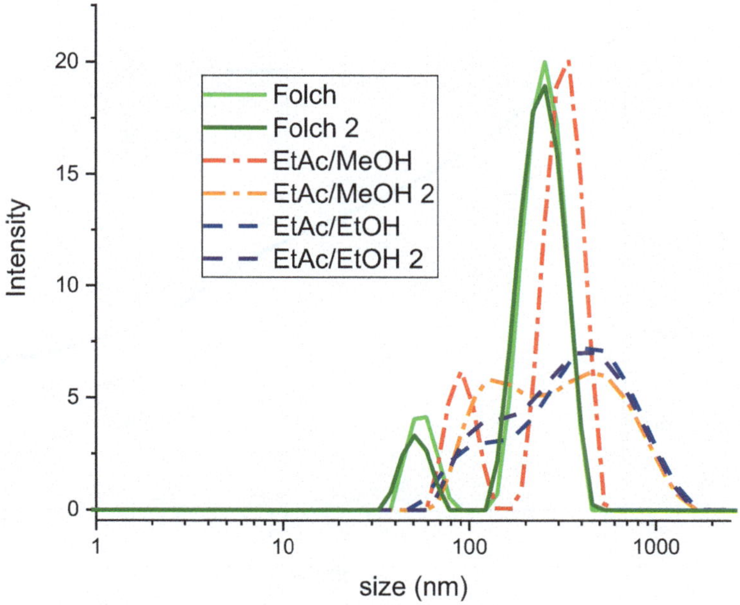

Fig. 4. Size distribution of the nanostructures obtained from lipids extracted with Folch and different EtAc mixtures from olive pomace. For each sample the results of two different runs are reported.

Table 3. Mean values and polydispersity indexes (P.I.) for the mono- and bi-modal size distributions observed in nanosystems formulated with lipids extracted by using different solvents

Sample	Peak 1	Peak 2	Convoluted peak	P.I.	Dilution
Folch control	50 nm	250 nm	–	0.49	1:20
ACN/MeOH	90 nm	290 nm	–	>0.9	1:20
ACN/EtOH	180 nm	1580 nm	–	0.44	1:40
EtAc/MeOH	230 nm	2000 nm	360 nm	0.47–0.59	1:20
EtAc/EtOH	–	–	410 nm	0.5–0.6	1:20

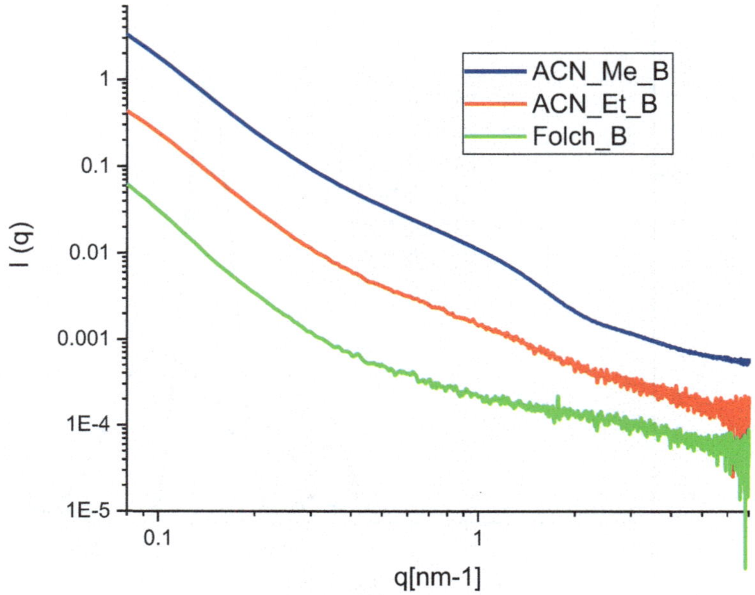

Fig. 5. One-dimensional SAXS intensity plot comparing nanovectors obtained from ACN/MeOH, ACN/EtOH and Folch lipid extraction

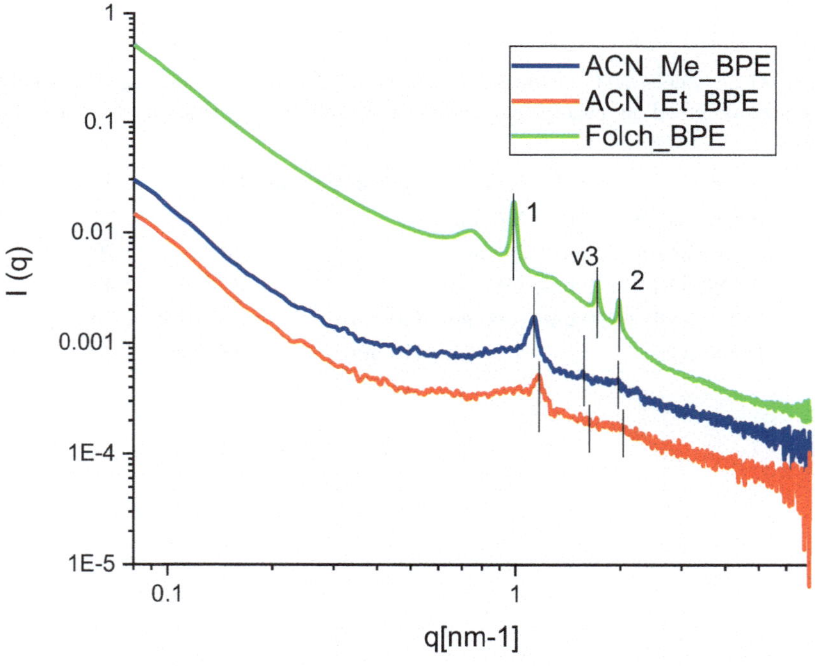

Fig. 6. One-dimensional SAXS intensity plot comparing nanovectors obtained from ACN/MeOH, ACN/EtOH and Folch lipid extraction with addition of a purified phospholipid adjuvant (DOPE)

4 Conclusions

In this work, an eco-friendly approach for the extraction of lipids from natural raw materials was devised and critically evaluated. The traditional Folch solution, considered as a gold standard for the extraction of biomolecules from natural tissues, was replaced by four less toxic alternative mixtures, i.e. ethyl acetate/ethanol (EtAc/EtOH), ethyl acetate/methanol (EtAc/MeOH), acetonitrile/ethanol (ACN/EtOH), and acetonitrile/methanol (ACN/MeOH). Then, the efficacy of these greener solutions was investigated on the basis of their physico-chemical characteristics and their performance in the preparation of lipid nanovectors, comparing the obtained results with those from of Folch-extracted lipids. GC-MS analysis showed that different classes of lipids could be extracted, depending on the mixture used. Specifically, phospholipids and polyphenols were obtained in the case of both Folch and ACN/MeOH, which proved to be the most similar to the control (i.e.) traditional Folch mixture, while the other three either extracted larger fragments (e.g. ACN/EtOH) or did not extract polyphenols, as in the case of both EtAc mixtures. Concerning the nanovectors obtained from the lipids extracted with different solvents, DLS confirmed a similarity to Folch sample for ACN/MeOH, even if high polydispersity values were obtained, whereas the other three alternatives gave nanoformulations with bigger sizes of aggregates and higher polydispersity. SAXS measurements valuably complemented these results, evidencing a similar type of aggregates regardless of the solvent used, even though ACN samples contained less defined and less ordered scattering objects with respect to the nanoformulations obtained through the classical procedure. Among the four proposed alternatives, ACN/MeOH was the best performing one compared to control, so from this point of view acetonitrile can be a suitable, less toxic alternative to chloroform.

Acknowledgement. We thank the ESRF (Grenoble, France) for beamtime allocation at the ID02 beamline.

References

1. Akbarzadeh, A., et al.: Liposome: classification, preparation, and applications. Nanoscale Res. Lett. **8**, 102 (2013). https://doi.org/10.1186/1556-276X-8-102
2. Folch, J., Lees, M., Sloane-Stanley, G.H.: A simple method for the isolation and purification of total lipids from animal tissues. J. Biol. Chem. **226**, 497 (1957)
3. Bligh, E.G., Dyer, W.J.: A rapid method of total lipid extraction and purification. J. Biochem. Physiol. **37**, 911 (1959)
4. Breil, C., Abert Vian, M., Zemb, T., Kunz, W., Chemat, F.: "Bligh and Dyer" and Folch methods for solid–liquid–liquid extraction of lipids from microorganisms. Comprehension of solvatation mechanisms and towards substitution with alternative solvents. Int. J. Mol. Sci. **18**, 708 (2017)
5. Clemente, I., Menicucci, F., Colzi, I., Sbraci, L., Benelli, C., Cristiana, C., Gonnelli, C., Ristori, S., Petruccelli, F.: Unconventional and sustainable nanovectors for phytohormone delivery: insights on Olea Europaea. ACS Sustain. Chem. Eng. **6**, 15022 (2018)

6. Sullivan, J.B., Krieger, G.R. (eds.): Clinical Environmental Health and Toxic Exposures, 2nd edn. Lippincott Williams & Wilkins Publisher, Philadelphia (2001)
7. Lindner, P., Zemb, T. (eds.): Neutron, X-Rays and Light Scattering Methods Applied to Soft Condensed Matter. North Holland Press, Amsterdam (2002)
8. Boesecke, P.: Reduction of two-dimensional small- and wide-angle X-ray scattering data. J. Appl. Cryst. **40**, S423 (2007)
9. Watson, J.T., Sparkman, O.D.: Introduction to Mass Spectrometry: Instrumentation, Applications, and Strategies for Data Interpretation. Wiley, Hoboken (2007)
10. Perri, E., Raffaelli, A., Sindona, G.: Quantitation of oleuropein in virgin olive oil by ionspray mass spectrometry – selected reaction monitoring. J. Agric. Food Chem. **47**(10), 4156 (1999)
11. La Nasa, J., Degano, I., Brandolini, L., Modugno, F., Bonaduce, I.: A novel HPLC-ESI-Q-ToF approach for the determination of fatty acids and acylglycerols in food samples. Anal. Chim. Acta **1013**, 98 (2018)
12. Eiser, E.: Dynamic light scattering. In: Multi Length-Scale Characterisation, vol. 1, p. 233 (2014)
13. Glatter, O.: Scattering Methods and their Application in Colloid and Interface Science. Elsevier, Amsterdam (2018)
14. Narayanan, T., Sztucki, M., Van Vaerenbergh, P., Leonardon, J., Gorini, J., Claustre, L., Sever, F., Morse, J., Boesecke, P.: A multipurpose instrument for time-resolved ultra-small-angle and coherent X-ray scattering. J. Appl. Cryst. **51**, 1511 (2018)
15. Clemente, I., Falsini, S., Di Cola, E., Fadda, G.C., Gonnelli, C., Spinozzi, F., Bacia-Verloop, M., Grillo, I., Ristori, S.: Green nanovectors for phytodrug delivery: in-depth structural and morphological characterization. ACS Sustain. Chem. Eng. **7**, 12838 (2019)

Controlling Drug Release of Anti-inflammatory Molecules Through a pH-Sensitive, Bactericidal Polymer Matrix: Towards a Synergic and Combined Therapy

Lorella Izzo[1(✉)], Giuliana Gorrasi[2], Andrea Sorrentino[3], Andrea Tagliabue[4], and Massimo Mella[4]

[1] Dipartimento di Biotecnologie e Scienze della Vita, Università degli Studi dell'Insubria, Via J. H. Dunant, 3, 21100 Varese, Italy
lorella.izzo@uninsubria.it
[2] Dipartimento di Ingegneria Industriale, Università degli Studi di Salerno, Via Giovanni Paolo II, 132, 84084 Fisciano, SA, Italy
[3] IPCB-CNR, P.le Enrico Fermi, 1, 80055 Portici, NA, Italy
[4] Dipartimento di Scienza e Alta Tecnologia, Università degli Studi dell'Insubria, Via Valleggio, 11, 22100 Como, Italy

Abstract. This study aimed to tune the release of non-steroidal, antiinflammatory and antimicrobial diclofenac sodium salt from a bactericidal, pH-sensitive polymer matrix in view of the development of a synergic and combined therapy.

The polymer matrix was represented by $A(BC)_n$ linear or branched copolymers (n = 1, 2) synthesized by Atom Transfer Radical coPolymerization (ATRP). Manufactures were obtained and analysed in film form, using solvent casting technique.

2-(dimethylamino) ethyl methacrylate (DMAEMA), 2-(diethylamino) ethyl methacrylate (DEAEMA) and 2-(diisopropylamino) ethyl methacrylate (DIPAEMA) have been used as comonomer bearing the bactericidal amino-groups and to modulate the pH-sensitivity of the material. The electrolytic and structural factors of the amino-groups, juxtaposed with the copolymer structure, gave raise to different interactions with the negatively charged diclofenac and imparted a finer control on the drug release kinetics. Data evidenced that copolymers with DMAEMA released diclofenac faster than those with DEAEMA and DIPAEMA; thus, the lower was the amine K_b, the faster was the release. The correlation between copolymer structure and kinetic/equilibrium of the drug release was analysed and correlated with the T_g of copolymers when diclofenac was loaded.

1 Introduction

In few applications of antimicrobial materials such as biomedical devices, antibacterial properties must be conserved over an extended timeframes, thus non-active materials releasing bactericidal molecules may not be the most appropriate choice.

In addition to this issue, scientists are aware that bacteria are developing resistance toward antibiotics, generally achieved through antibiotic inactivation, alternative routes

S. Piotto et al. (Eds.): BIONAM 2019, LNBE, pp. 151–163, 2020.
https://doi.org/10.1007/978-3-030-47705-9_14

to the methabolic pathway, target modification or altered permeability of the outer envelope; the latter is due to inhibition of the cell wall synthesis. In this framework, alternative strategies to small molecule-based treatments may need to exploit "weak spots" that are less likely to change due to bacteria rapid mutation.

To this end, antimicrobial mechanism of action of soluble polycations, characterized by a certain degree of hydrophobicity, appears rather intriguing. They cause permeabilization of cell through the displacement of Ca^{2+} and Mg^{2+} ions, which stabilize the negative charges on the outer envelope (negative charges of teichoic and lipoteichoic acids of the Gram-positive cell wall or lipopolysaccharides and phospholipids of the Gram-negative outer membrane) [1–6]. The consequence of this action is the adhesion of the polymer, followed by its penetration into the cell. It is worth noting that polyelectrolyte adsorption onto charged nanoparticles have been studied by a theoretical approach, and it was found that effects such as spatial partitioning of ions may be influenced by the polyelectrolyte structure (linear or star–like) and size [7]. This is important to understand the bactericidal activity of a polymer when in contact with the outer envelope of a bacteria cell.

Although the mechanism adopted by polyelectrolyte is probably less suitable for inducing bacteria modification with respect to the ones used by antibiotics, soluble polyelectrolytes can result hemolytic and consequently their application is limited (see e.g. ref [8]).

However, a similar mechanism of action can be extended to surfaces if coated with short polycations or to thermoplastic polymers if these contain charged groups. Evidently, such a kind of material would result not soluble in water but able to act as antimicrobial by contact. Furthermore, it has been demonstrated that in such a case, no hydrophobicity is needed since polymers do not penetrate the cells but, when in contact, simply make them more permeable due to the displacement of divalent cations that neutralize the negative charges of the envelope [9, 10].

Considering the potentiality of inherently antimicrobial materials, some of us recently developed and studied different types of thermoplastic copolymers capable of killing bacteria by contact. The bactericidal activity was given by the presence of protonable (substituted amines) [11, 12] or cations coordinating (crown ethers) pedant moieties [13]. The former materials also resulted to be pH-sensitive and with potential application in controlled drug delivery of negative charged molecules.

In this framework, we report here a preliminary study on the kinetic release of diclofenac sodium salt from few films obtained by casting bactericidal, pH-sensitive copolymer with different architecture (branched and linear) with the aim of finely tune molecule release.

Diclofenac is non-steroidal, anti-inflammatory drug (NSAID), widely used for treatment of inflammation, pain and fever caused by bacterial infections.

Recently, has been demonstrated that few NSAIDs also show antibacterial activity probably through mechanisms different from those of antibiotics. For this reason they may be represent a further alternative to the latter substance family.

So far, NSAIDs are used in combination with antibiotics, particularly in veterinary medicine, producing synergistic effect [14, 15].

However, the approach proposed in this paper, meaning the use of systems capable of controlling release of NSAID by tuning the pK_b of amino group in bactericidal polymer matrix, can pave the way to a novel synergic and combined therapy for *e.g.* burn wound or health-care associated infection.

2 Experimental Section

2.1 Materials

Poly(ethylene glycol) monomethylether (mPEG) (M_n = 2000 Da, M_w/M_n = 1.16), benzaldehyde dimethyl acetal, 2,2-Bis(hydroxymethyl)propionic acid, p-toluenesulfonic acid monohydrate (TsOH), acetone, N,N'-dicyclohexylcarbodiimide (DCC), 4-(dimethylamino)pyridine (DMAP), methanol, Pd/C 10%, 2-bromoisobutyryl bromide (BMPB), triethylamine (TEA), diethyl ether, ethanol, CuBr, 2,2'-bipyridine (bpy), chloroform, fluorescein sodium salt, cetyltrimethylammonium chloride (25 wt%) and Al_2O_3 were purchased from Aldrich and used without further purification.

All manipulations involving air-sensitive compounds were carried out under nitrogen atmosphere using Schlenk or drybox techniques. Toluene (Aldrich) was dried over sodium and distilled before use. CH_2Cl_2 (Carlo Erba) was dried over CaH_2 and then distilled. Methyl methacrylate (MMA), and the 2-(alkylamino)ethyl methacrylate (AAEMA) monomers such as 2-(dimethylamino)ethyl methacrylate (DMAEMA), 2-(diethylamino)ethyl methacrylate (DEAEMA), 2-(diisopropylamino)ethyl methacrylate (DIPAEMA), (Aldrich) were dried over CaH_2 and then distilled under reduced pressure of nitrogen.

2.2 Synthesis of mPEG-(PMMA-PAAEMA)$_n$ Copolymers

2.2.1 Synthesis of mPEG-Br Linear and mPEG-Br$_2$ Branched Macroinitiators

Linear and branched macroinitiators were synthesized according to the literature procedure [12].

2.2.2 Synthesis of mPEG-(PMMA-*ran*-PAAEMA) Linear and mPEG-(PMMA-*ran*-PAAEMA)$_2$ Copolymers by ATRP [12]

mPEG-(PMMA-*ran*-PAAEMA) linear copolymers were synthesized in toluene at 90 ° C. The reaction was carried out in a 100 mL glass flask charged, under nitrogen atmosphere, with 0.1 g of mPEG-Br linear macroinitiator in 15 mL of dry toluene. After the dissolution of the macroinitiator, 0.013 g of CuBr, 0.03 g of bpy, 5 mL of MMA and 1.0 or 4.0 mL of AAEMA were added (0.5, 1.0, 2.5, or 4.0 mL were used in the case of DIPAEMA). The mixture was thermostated at 90 °C and magnetically stirred. The reaction was stopped with *n*-hexane after 18 h. The copolymer was recovered, dissolved in the minimum amount of chloroform and passed over a column of activated Al_2O_3 to remove the catalyst. The solution was dried in vacuum, the polymer was washed with cold methanol and then dried.

mPEG-(PMMA-*ran*-PAAEMA)$_2$ copolymers were synthesized using a molar ratio mPEG-Br$_2$/CuBr/bpy = 1/4/8.

^{1}H NMR (400 MHz, $CDCl_3$): □ δ 0.83–1.10 (CH_3 main chain), 1.79–1.87 (CH_2 main chain), 2.27 (-N(CH_3)$_2$), 2.56 (-O-CH_2-CH_2-N(CH_3)$_2$), 3.57 (-OCH_3), 3.61

(-OCH_2CH_2-), 4.06 (-O-CH_2-CH_2-N$(CH_3)_2$). ^{13}C-NMR (400 MHz, $CDCl_3$): □ 16.6-18.9 (CH_3 main chain), 44.9 (quaternary carbon in the main chain), 46.0 (-N$(CH_3)_2$), 52.0 (-OCH_3, MMA), 54.5 (CH_2 main chain), 57.2 (-O-CH_2-CH_2-N$(CH_3)_2$), 63.2 (-O-CH_2-CH_2-N$(CH_3)_2$), 70.7 (-OCH_2CH_2-), 176.3–178.2 (-*C* = O).

The molar mass of copolymers was determined by the degree of polymerization (DP) as evaluated from ^{13}C NMR, by the molar mass of monomers (MM) and by the signal intensities (I):

$$M_n = DP_{mPEG}(MM_{EO}) + DP_{MMA}(MM_{MMA}) + DP_{AAEMA}(MM_{AAEMA}) \quad (1)$$

where:

$$DP_{mPEG} = M_{n(mPEG)}/44; \cdot DP_{MMA} = DP_{mPEG} \cdot (2I_{MMA}/I_{mPEG});$$
$$DP_{AAEMA} = DP_{mPEG} \cdot (2I_{AAEMA}/nI_{mPEG})$$

44 is the molecular weight of monomeric units of mPEG; $(I_{m\text{-}PEG})/2$ is half the integration of the signal relative to mPEG units: –OCH_2CH_2-; I_{MMA} is the integration of the signal relative to MMA units: -OCH_3; I_{AAEMA} is the integration of methyl group of the signal relative to AAEMA units: -N$(CH_3)_2$, -N$(CH_2CH_3)_2$, -N$[CH(CH_3)_2]_2$ and –NHC$(CH_3)_3$ respectively; n indicates the number of methyl carbons of AAEMAs (2 for DMAEMA and DEAEMA, and 4 for DIPAEMA respectively).

Analogously, the monomers content in the copolymers were calculated using the following equations:

$$X_{mPEG} \cong \frac{I_{mPEG}/2}{I_{mPEG}/2 + I_{MMA} + I_{AAEMA}/n}; \quad (2)$$

$$X_{MMA} \cong \frac{I_{MMA}}{I_{mPEG}/2 + I_{MMA} + I_{AAEMA}/n}; \quad (3)$$

$$X_{AAEMA} \cong \frac{I_{AAEMA}/n}{I_{mPEG}/2 + I_{MMA} + I_{AAEMA}/n} \quad (4)$$

The polydispersity index (PDI = M_w/M_n) was evaluated by GPC.

2.3 NMR Analysis

Spectra were recorded on a Bruker Avance 400 MHz spectrometer at 25 °C with D1 = 5 s. The samples were prepared by introducing 20 mg of sample in 0.5 mL of $CDCl_3$ into a tube (5 mm outer diameter). TMS was used as internal reference.

2.4 GPC Measurements

The molecular weights (M_n) and the polydispersity index (PDI = M_w/M_n) of polymer samples were measured by GPC at 30 °C, using THF as solvent, flow rate of eluant 1.0 mLmin^{-1}, and narrow polystyrene standards as reference. The measurements were performed on a Waters 1525 binary system equipped with a Waters 2414 RI detector using four Styragel columns (range 1000–1,000,000 Å). Every value was the average of two independent measurements.

2.5 Preparation of Films by Casting

Thin films were prepared by dissolving 200 mg of polymer in 50 mL of $CHCl_3$ at 25 ° C. The solution was cast in a teflon petri dish (diameter 6 cm) and the solvent evaporated at room temperature. Films containing 3% wt/wt of diclofenac sodium salt were prepared using the same procedure adopted for the unfilled samples. All casted films were further dried in vacuum at 40 °C for three days before performing thermal analysis and studies on kinetic release. The resulting film is insoluble in water and does not appear to leach material.

2.6 Release Kinetic Studies

The release kinetics of diclofenac were obtained by ultraviolet spectrometric measurement at ambient temperature, using a Spectrometer UV-2401 PC Shimadzu (Japan). The tests were performed using rectangular specimens of 4 cm^2 and same thickness (about 120 μm), placed into 25 mL of physiological solution at 100 rpm in an orbital shaker (VDRL MOD. 711+, Asal S.r.l.). The release medium was withdrawn at fixed time intervals and replenished with fresh medium. The considered absorption band was at 275 nm.

2.7 Differential Scanning Calorimetry (DSC)

DSC was carried out on the films of about 10 mg using a DTA Mettler Toledo (DSC 30) under nitrogen atmosphere. Films were submitted to the following thermal cycles: cooling from 25 °C to −50 °C at 50 °C/min, keeping at −50 °C for 5 min, heating from −50 °C to 120 °C at 10 °C/min. The Tg (°C) of the materials was extracted from the third thermal scan.

3 Results and Discussion

pH-sensitive copolymers: synthesis, characterization and preparation of films.

Polymer matrixes were prepared from $A(BC)_n$ block copolymers with linear and branched architecture (n = 1, 2). Chemically, copolymers were made of a block of monomethyl ether polyethylene glycol (mPEG, A) and one or two random copolymer chains (BC) based on methylmethacrylate (MMA) and an alkylamminoethyl methacrylate (AAEMA).

The mPEG block confers to the material the hydrophilicity needed for the protonation of AAEMA both on the surface than in the bulk, while the block containing the protonable monomer, whose solubility in water would change with pH, is definitely made hydrophobic by the presence of MMA, so that the copolymers can be classified as not soluble in water and thermoplastic.

The effect of two parameters over the diclofenac release was studied: the copolymer structure and the chemical substituents on the AAEMA amino-groups. As for the structure, it was varied basing on the fact that two of us estimated the chances of interchain interactions between an ammonium and an amino group (charged hydrogen bonds) to be roughly 2.5 times higher than for intrachain ones, a factor that may, *de facto*, increases the polymer surface basicity. Thus, we aimed to verify if the amount of charged hydrogen bonds, and so the stabilization of the positive charge of the copolymer, can affect the interaction with the negatively charged diclofenac and consequently its release.

The latter process may also be affected by a change in the amine pK_b, i.e. by the chemical nature of AAEMA amino-pedant, if diclofenac release requires the development of a charge on the material to take place. To investigate this possibility, 2-(dimethylamino) ethyl methacrylate (DMAEMA), 2-(diethylamino) ethyl methacrylate (DEAEMA) and 2-(diisopropylamino) ethyl methacrylate (DIPAEMA) were used as the protonable comonomers to impart a change in pH-sensitivity to the materials, knowing that they also modulate the bactericidal activity.

Copolymers were synthesized by Atom Transfer Radical coPolymerization (ATRP) using the classic CuBr/bpy system in toluene at 90 °C in presence of a macroinitiator based on mPEG with a terminal chain having one or two functionalities, respectively for the synthesis of linear and branched copolymers (Fig. 1).

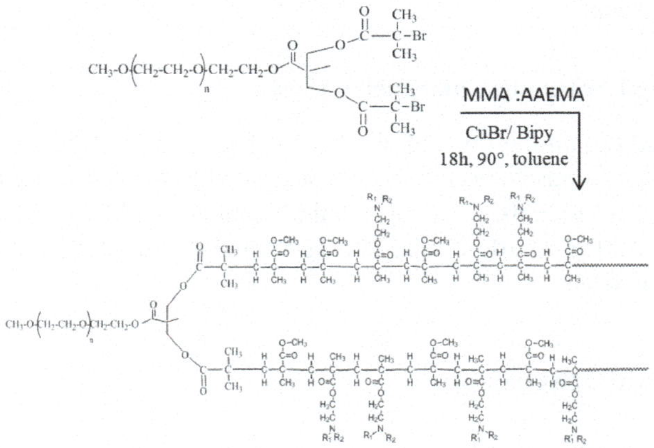

Fig. 1. Schematic representation of the copolymers synthesis

Copolymers obtained, their composition and molecular masses are reported in Table 1.

Kinetic Release Studies

Experimental data show that both the alkyl substituents on the amino-groups and the copolymer architecture strongly influence diclofenac release over the range of time considered (1200 h), as reported in Table 2.

Table 1. Chemical composition and molecular masses of copolymers used for the formation of films

Sample	Structure	AAEMA amino groups	%mPEG (in mol)	%MMA (in mol)	%AAEMA (in mol)	M_n[a] (kDa)
$PMDMe_1$	A(BC)	$N(CH_3)_2$	7	51	42	76
$PMDMe_2$	$A(BC)_2$	$N(CH_3)_2$	7	59	34	78
$PMDEt_1$	A(BC)	$N(CH_2CH_3)_2$	9	49	42	70
$PMDEt_2$	$A(BC)_2$	$N(CH_2CH_3)_2$	7	53	40	87
$PMDiPr_1$	A(BC)	$N[CH(CH_3)_2]_2$	8	53	39	80
$PMDiPr_2$	$A(BC)_2$	$N[CH(CH_3)_2]_2$	7	53	40	103

[a] Determined by ^{13}C-NMR

Table 2. Diclofenac release from copolymer matrixes

Sample	Structure	Groups on N	Release (%) 1 h	Release (%) 2 h	Release (%) 5 h	Release (%) 72 h	Release (%) 1200 h
$PMDMe_1$	A(BC)	Me	2.8	3.7	10.0	29.4	93.8
$PMDMe_2$	$A(BC)_2$	Me	15	16.7	35.1	54.6	100
$PMDEt_1$	A(BC)	Et	1.8	2.2	3.2	7.3	32.2
$PMDEt_2$	$A(BC)_2$	Et	6.5	10.8	17.0	32.7	50.8
$PMDiPr_1$	A(BC)	iPr	10.4	14.9	23.1	38.3	49.5
$PMDiPr_2$	$A(BC)_2$	iPr	2.3	3.0	7.9	10.1	15.6

The behavior of linear copolymers with different AAEMAs at 25 °C is shown in Fig. 2. Interestingly, the differences in release kinetics are already evident after the first few hours of soaking (Fig. 2, bottom panel). In fact, $PMDiPr_1$ released more than the others just after 1 h, and in 5 h it has already released more than 20% of the molecule; $PMDMe_1$ and $PMDEt_1$-based matrix released, instead, only 10% and 3%, respectively, in the same timeframe.

After 24 h, $PMDiPr_1$ released about 30% of diclofenac, while $PMDMe_1$ only 17%; these percentages became 40% and 30%, respectively, after 72 h. Somewhat surprisingly, the relative performance of $PMDiPr_1$ and $PMDMe_1$ changed radically at longer times, the former not releasing further, while the latter retro-dissolved 93% of the loaded diclofenac over 50 days.

As in the first few hours, $PMDEt_1$ released slower than the others: the diclofenac passed in solution was slightly below 5% after 24 h and 7% after 72 h (Fig. 2, middle panel). Even after 50 days, the amount of molecules released is only 32% (Fig. 2, top panel).

For branched copolymers (Fig. 3), the release kinetic of $PMDMe_2$-based matrix is faster compared to the corresponding linear one; in fact, the amount in solution was already about 30% after the first 5 h, half of the loading was re-emitted after 72 h, and the rest of diclofenac was found in solution after 47 days.

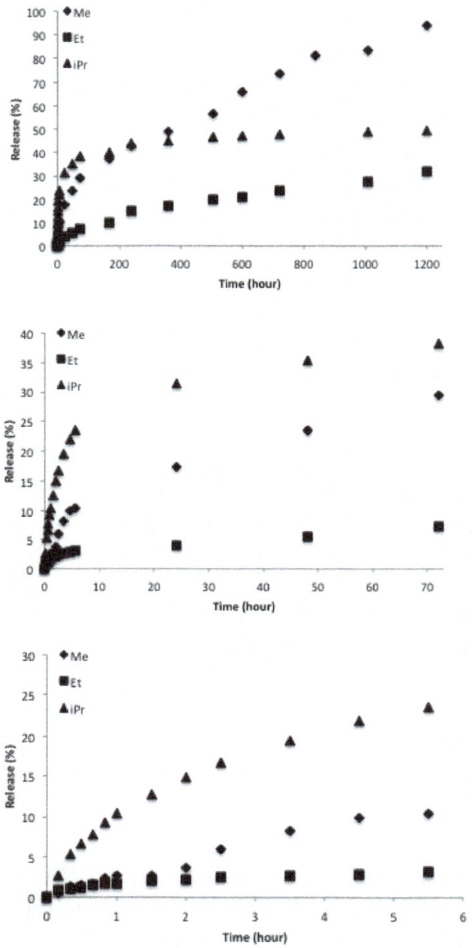

Fig. 2. Amount percentage of diclofenac sodium salt released after 5.5 h (bottom panel), 72 h (middle panel), and 1200 h (top panel) from a polymer matrix made of linear copolymers with different AAEMA immersed a physiological solution at 25 °C.

Analogously, the release kinetic of $PMDEt_2$ sped up with respect to the linear-based matrix: about 17% of diclofenac was found in solution after 5 h, 33% after 72 h, and about the 50% after 50 days.

In contrast, release from $PMDiPr_2$ was slower compared to the corresponding linear copolymer-based matrix. Thus, it released just about 7% of diclofenac after 5 h, a percentage that increased to 10% in 72 h, and to 16% in 50 days. Noteworthy, this polymer shows the slowest kinetics among the branched copolymers releasing diclofenac.

As for a possible rationalization of the differences in release kinetics and characteristics for the various matrices, we notice that there is no apparent correlation with any of the chemical variables changed in this work. In this respect, the impact of

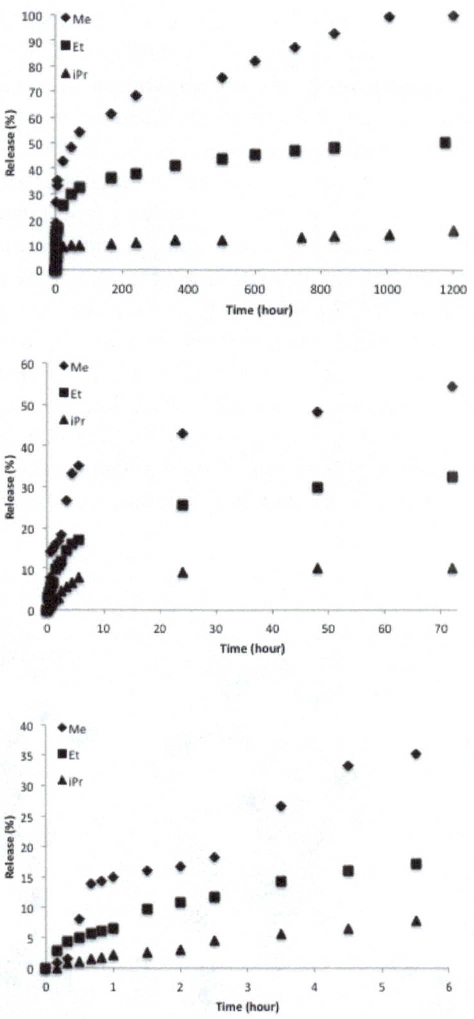

Fig. 3. Amount percentage of diclofenac sodium salt released after 5.5 h (bottom panel), 72 h (middle panel), and 1200 h (top panel) from a polymer matrix made of branched $A(BC)_2$ copolymers with different AAEMA immersed a physiological solution at 25 °C.

varying pK_b ought to be expected small in the release condition employed by us, as the pH of the medium is commonly characterized by a pH $\sim$ 5.5 due to the presence of carbon dioxide. In this situation, we would expect all water exposed amino groups to be fully protonated (e.g. with a $pK_b = 5.7$, DMAEMA would have a protonated/neutral ratio of roughly 640) irrespectively of the alkyl groups bound to the nitrogen atom. As the molar fraction of AAEMA monomers in the six polymers is quite similar, this implies that also the total charge per unit mass imposed to the various matrices by the acid-base equilibrium should be analogous, unless there are morphological or structural peculiarities for each material that impact on water access to the ionizable groups.

Another possible aspect to consider is represented by the fact that the diclofenac (pK_a = 4.15), which in the saltine conditions is largely in the anionic form, needs to escape the attractive interactions with the formed ammonium cations to be released in solution, bearing in mind that the concentration of cations inside the materials may be enhanced by formation of strong hydrogen bonds [7, 16–19]. To test whether or not different alkyl substituents may impact on the interactions between the mentioned moieties, we have computed their binding energies and found them to differ by limited amounts (in the range 51–54 kcal/mol). Importantly, the optimized geometry for the ammonium-anion dimers differ only slightly (see Fig. 4 for the optimized structures), thus suggesting that the size of the alkyl groups does not impact on the electrostatic component of the interactions. It may, instead, modify the dispersion contribution, strengthening it as the size of the substituent increases. Being characterized by a hydrogen bond between ammonium and the carboxylate group, the dimer structures suggest, however, that dispersion may not play a massive role, as the bulkiest part of diclofenac is far away from the alkyl substituents. Interestingly, this idea seems to be also supported by the lack of a trend shown by our data as a function of the substituent size.

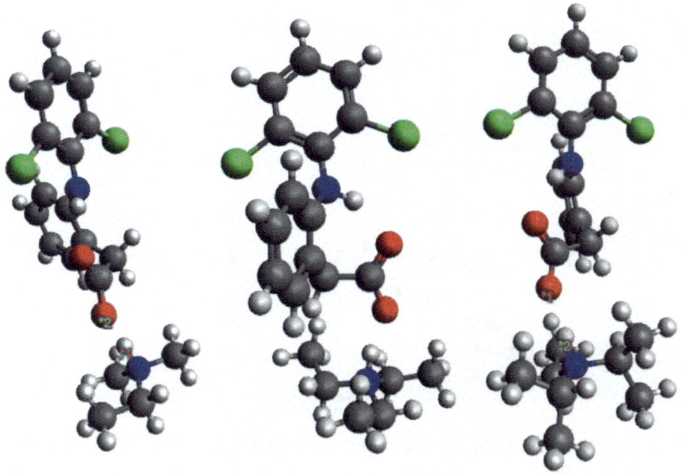

Fig. 4. Optimized geometries for the ammonium-diclofenac anion; from left to right, dimethyl ethyl ammonium, triethyl ammonium, and di-isopropyl ethyl ammonium. Energies and structures obtained at the B3LYP/6-31++G(d, p) employing the PCM model to represent water.

From what we presented so far, one may thus be led to conclude that the different release behaviours indicated in Table 2 could descend by differences in the overall structure, morphology, and, possibly, chain mobility of the films either unloaded or loaded with diclofenac. DSC measurements were then performed. As reported in Table 3, diclofenac would affect differently the T_g values of the thermoplastic matrix depending on AAEMA amino pedant groups. In fact, it is capable of increasing copolymer T_g in presence of ethyl-amino substituents, and this effect is more evident in the branched architecture rather than in the linear one. As consequence of this, one would suggest diclofenac to act as a sort of "linker" between chains, and such a linker is more effective in the branched architecture, the one characterized by a lower T_g when unloaded.

Table 3. T_g of loaded and unloaded copolymer matrices

Sample	Structure	Groups on N	Tg (°C) Unloaded matrix	Tg (°C) Loaded matrix	%AAEMA (in mol)	M_n^a (kDa)
$PMDMe_1$	A(BC)	Me	71.2	71.4	42	76
$PMDMe_2$	$A(BC)_2$	Me	71.0	70.3	34	78
$PMDEt_1$	A(BC)	Et	56.5	60.0	42	70
$PMDEt_2$	$A(BC)_2$	Et	45.0	67.5	40	87
$PMDiPr_1$	A(BC)	iPr	82.2	65.0	39	80
$PMDiPr_2$	$A(BC)_2$	iPr	90.0	72.4	40	103

At variance with the latter finding, the impact on polymers with the isopropyl groups is exactly the opposite compared to PMDEt. In fact, the presence of diclofenac dramatically decreases the T_g of the linear that in branched structures; so, the molecule works as a powerful "plasticizer".

Finally, the copolymer matrices with the faster release, namely the ones containing DMAEMA, appear not to be affected by the presence of diclofenac in terms of modification of thermal properties and consequently of inter- and intra-chains interactions among PMMA/AAEMA blocks. Such an aspect needs further investigation and will be deeply considered in a forthcoming paper.

4 Conclusions

In this paper, we report a kinetic study on the release of an NSAID such as diclofenac sodium salt from matrices made of inherently biocidal, thermoplastic copolymers. Matrices chemical composition was varied introducing into the back-bone amino-pedant groups with different pK_b; two different architectures, meaning linear and branched, were used to enlarge the material parameter space explored.

Interestingly, diclofenac release was differently controlled by both the amino-groups and the copolymer structure. In particular, the polymers containing the dimethyl amino pendants released the complete load over 1200 h, while the other species appeared to release slower. This notwithstanding, all matrices demonstrated a controlled re-dissolution of the model anti-inflammatory drug chosen in this work. At the moment, it appears, however, complicate to provide a unifying rationale for the set of described behaviors. To improve on such aspect, work in the direction of experimentally probing the acid-base properties of the materials ought to be undertaken; the latter may somewhat control the amount of swelling induced in the matrices.

References

1. Tashiro, T.: Antibacterial and bacterium adsorbing macromolecules. Macromol. Mater. Eng. **286**, 63–87 (2001)
2. Grapski, J.A., Cooper, S.L.: Synthesis and characterization of non-leaching biocidal polyurethanes. Biomaterials **22**, 2239–2246 (2001)
3. Chen, C.Z., Cooper, S.L.: Interactions between dendrimer biocides and bacterial membranes. Biomaterials **23**, 3359–3368 (2002)
4. Kenawy, E.-R., Mahmoud, Y.A.-G.: Biological active polymers, synthesis and antimicrobial activity of some linear copolymers with quaternary ammonium and phosphonium groups. Macromol. Biosci. **3**, 107–116 (2003)
5. Kenawy, E.-R., Worley, S.D., Broughton, R.: The chemistry and applications and antimicrobial polymers: a state-of-the-art review. Biomacromolecules **8**, 1359–1384 (2007)
6. Waschinski, C.J., Barnert, S., Theobald, A., Schubert, R., Kleinschmidt, F., Hoffmann, A., Saalwächter, K., Tiller, J.C.: Insights in the antibacterial action of poly(methyloxazoline)s with a biocidal end group and varying satellite groups. Biomacromolecules **9**, 1764–1771 (2008)
7. Mella, M., Tagliabue, A., Mollica, L., Izzo, L.: Monte Carlo study of the effects of macroion charge distribution on the ionization and adsorption of weak polyelectrolytes and concurrent counterion release. J. Colloid Interface Sci. **560**, 667–680 (2020)
8. Duncan, R., Izzo, L.: Dendrimer biocompatibility and toxicity. Adv. Drug Deliv. Rev. **57**, 2215–2237 (2005)
9. Kügler, R., Bouloussa, O., Rondelez, F.: Evidence of a chargedensity threshold for optimum efficiency of biocidal cationic surfaces. Microbiology **151**, 1341–1348 (2005)
10. Murata, H., Koepsel, R.R., Matyjaszewski, K., Russell, A.J.: Permanent, non-leaching antibacterial surfaces-2: how high density cationic surfaces kill bacterial cells. Biomaterials **28**, 4870–4879 (2007)
11. Vigliotta, G., Mella, M., Rega, D., Izzo, L.: Modulating antimicrobial activity by synthesis: dendritic copolymers based on nonquaternized 2-(Dimethylamino)ethyl methacrylate by Cu-mediated ATRP. Biomacromol **13**, 833–841 (2012)
12. Matrella, S., Vitiello, C., Mella, M., Vigliotta, G., Izzo, L.: The role of charge density and hydrophobicity on the biocidal properties of self-protonable polymeric materials. Macromol. Biosci. **15**, 927–940 (2015)
13. Izzo, L., Matrella, S., Mella, M., Benvenuto, G., Vigliotta, G.: Escherichia coli as a model for the description of the antimicrobial mechanism of a cationic polymer surface: cellular target and bacterial contrast response. ACS Appl. Mater. Interfaces **11**, 15332–15343 (2019)
14. De Rosa, M., Vigliotta, G., Soriente, A., Capaccio, V., Gorrasi, G., Adami, R., Reverchon, E., Mella, M., Izzo, L.: Leaching or not leaching: an alternative approach to antimicrobial materials via copolymers containing crown ethers as active groups. Biomater. Sci. **5**, 741–751 (2017)
15. Yin, Z., Wang, Y., Whittel, L.R., Jergic, S., Liu, M., Harry, E., Dixon, N.E., Kelso, M.J., Beck, J.L., Oakley, A.J.: DNA replication is the target for the antibacterial effects of nonsteroidal anti-inflammatory drugs. Chem. Biol. **21**, 481–487 (2014)
16. Mella, M., Mollica, L., Izzo, L.: Influence of charged intramolecular hydrogen bonds in weaks polyelectrolytes: a monte carlo study of flexible and extendible polymeric chains in solution and near charged sphere. J. Polym. Sci. Part B: Polym. Phys. **53**, 650–663 (2015)
17. Mella, M., Izzo, L.: Modulation of ionization and structural properties of weak polyelectrolytes due to 1D, 2D and 3D confinement. J. Polym. Sci. Part B: Polym. Phys. **55**, 1088–1102 (2017)

18. Tagliabue, A., Izzo, L., Mella, M.: Absorbed weak polyelectrolytes: impact of confinement, topology, and chemically specific interactions on ionization, conformation free energy, counterion condensation, and absorption equilibrium. J. Polym. Sci. Part B: Polym. Phys. **57**, 491–510 (2019)
19. Tagliabue, A., Izzo, L., Mella, M.: Impact of charge correlation, chain rigidity, and chemical specific interactions on the behavior of weak polyelectrolytes in solution. J. Phys. Chem. B **123**, 8872–8888 (2019)

Oral Fast and Topical Controlled Ketoprofen Release Through Supercritical Fluids Based Processes

Paola Franco and Iolanda De Marco(✉)

Department of Industrial Engineering, University of Salerno, Via Giovanni Paolo II, 132–84084 Fisciano, SA, Italy
idemarco@unisa.it

Abstract. Ketoprofen (KET) is a non-steroidal anti-inflammatory drug (NSAID) widely used for different phlogistic diseases of rheumatoid and non-rheumatoid origin. When a fast release is required, KET is orally administered in form of capsules, tablets or granulates. In this case, due to KET poor solubility in water, large drug doses with consequent side effects, mainly gastrointestinal one are required. KET bioavailability can be enhanced through its coprecipitation with a hydrophilic carrier, such as polyvinylpyrrolidone (PVP). Another way to reduce the dosing frequency and avoid gastrointestinal irritation is the transdermal drug delivery with a controlled release. In this work, two different supercritical carbon dioxide ($scCO_2$) based processes were used to modify KET dissolution rate: the supercritical antisolvent technique to coprecipitate PVP and KET in form of controlled dimensions microparticles for an oral delivery, and the supercritical adsorption to impregnate KET in alginate aerogel for a topical delivery. In the case of oral KET, composite spherical microparticles with controlled diameters were successfully produced, leading to a faster NSAID dissolution rate than unprocessed KET. In the case of topical KET, alginate aerogel was successfully impregnated with KET; it promotes a controlled release, suitable for transdermal anti-inflammatory patches, reducing frequency of administration and side effects. Supercritical techniques allow to obtain a fast or controlled release of the NSAID, according to the specific therapy desired.

Keywords: Supercritical carbon dioxide · Ketoprofen · Oral fast release · Topical controlled release

1 Introduction

The inflammation is a complex biological protective response of body tissues against pathogens, damaged cells or irritants. From a clinical point of view, a sequence of five classical symptoms is associated to the inflammation: *calor, dolor, rubor, tumor and functio laesa*; i.e. heat, pain, redness, swelling and loss of function.

Non-Steroidal Anti-Inflammatory Drugs (NSAIDs) are generally prescribed to relieve the acute ache caused by different phlogistic diseases, such as headache, toothache, migraine, back pain, postoperative ache and inflammation, among others.

S. Piotto et al. (Eds.): BIONAM 2019, LNBE, pp. 164–177, 2020.
https://doi.org/10.1007/978-3-030-47705-9_15

NSAIDs inhibit the activity of cyclooxygenase-1 (COX-1) and cyclooxygenase-2 (COX-2) enzymes and, thereby, block the production of prostaglandins, which play a key role in the inflammatory process.

Ketoprofen (KET) is one of the most commonly used NSAIDs with analgesic and antipyretic effects [1], prescribed for rheumatoid arthritis and osteoarthritis, as well as other illness of non-rheumatoid origin. However, like most NSAIDs, KET is a poorly water-soluble drug (KET solubility in water < 0.05 mg/mL); therefore, high and repeated doses are taken to reach the therapeutic drug concentration in the plasma, resulting in many adverse effects, including gastrointestinal irritation in case of oral formulations [2–5].

A NSAID can be administered through various dosage forms, such as granulates, capsules, tablets, topical patches, etc. The use of the polymeric carrier to produce drug delivery systems has various advantages: it protects the active principle against oxidation and deactivation caused by light, oxygen and temperature; it masks the unpleasant taste or odor of the active compound; it can modify the drug dissolution rate.

Oral formulations are preferred when a fast release is desired, for example in the case of a headache. Nevertheless, drawbacks like the NSAIDs low bioavailability and the gastrointestinal irritation have to be solve. A possible way to increase the dissolution rate of a poorly water-soluble drug is its coprecipitation with a hydrophilic polymeric carrier, such as polyvinylpyrrolidone (PVP) [6, 7].

Topical drug delivery can assure a continuous and prolonged release, required for inflammatory diseases like back pain or arthritis. The NSAID penetrates the skin, subcutaneous fatty tissue and muscle without reaching high plasma concentrations and attacking the mucous membrane of the stomach; however, skin irritations may occur [8]. A prolonged-release topical patch can be developed by incorporating the NSAID into a matrix of polymer with hydrophobic behavior, such as calcium alginate.

The drug dissolution rate can be efficiently modified by using the supercritical carbon dioxide ($scCO_2$) assisted processes. The $scCO_2$ guarantees fast mass transfer, high solvent power, high density, near zero surface tension, low viscosity and high diffusivity, that can be modulate by varying the operating pressure and temperature. Among the $scCO_2$ based processes, the supercritical antisolvent (SAS) technique and the supercritical adsorption stand out. Up to now, SAS technique is successfully used to obtain microparticles and nanoparticles of different kinds of materials [6, 8–15]. In this case, the $scCO_2$ acts has an antisolvent: the solute has to be soluble in the organic solvent used but insoluble in the mixture solvent+$scCO_2$, which have to be miscible at the process conditions. Differently, the supercritical adsorption permits to incorporate different kinds of active compounds in various matrices [16–27]. In the case of supercritical adsorption, the solute has to be soluble in $scCO_2$, whereas the polymer matrix chosen as substrate has to possess a proper porosity. Moreover, the $scCO_2$ may have plasticizing and blowing effects on the polymeric matrices.

In this work, the two different $scCO_2$ based processes were used for different applications: SAS process to coprecipitate PVP and KET in form of controlled dimensions microparticles for a KET oral delivery, and the supercritical adsorption to impregnate KET in alginate aerogel for a KET topical delivery.

2 Materials and Methods

2.1 Materials

Carbon dioxide (purity 99%) was purchased from Morlando group (Italy). Ketoprofen (KET, purity $\geq$ 98%), polyvinylpyrrolidone (PVP, average molecular weight 10,000 g/mol), dimethylsulfoxide (DMSO, purity 99.5%), calcium chloride ($CaCl_2$, purity $\geq$ 96%) and sodium alginate were purchased from Sigma-Aldrich (Italy). All materials were used as received. Distilled water was produced using a laboratory water distiller supplied by ISECO S.P.A. (St. Marcel, AO, Italy).

2.2 Supercritical CO_2 Assisted Processes

SAS technique was employed to produce PVP/KET microparticles. The heart of SAS plant (sketched in Fig. 1) is the cylindrical precipitation chamber (PC) with an internal volume of 500 cm^3. A typical SAS experiment starts delivering the CO_2, previously cooled thanks to a refrigerating bath (RB), in the precipitator by means of a high pressure pump (P1). The temperature control (TC) is assured by a proportional integral derivative (PID) controller connected with electrically thin bands, whereas the pressure in the chamber is measured by a test gauge manometer (M) and regulated by a micrometric valve (MV). Once the desired pressure and temperature are reached, the liquid solution, stored in a burette (S1), is co-currently injected into the precipitator by another high pressure pump (P2). The solution passes through a 100 μm internal diameter stainless steel nozzle, that assures the atomization; inside the chamber, a solvent-antisolvent mixture is formed, leading to the precipitation of the solute to be micronized. The precipitated powder is collected on a stainless steel filter (pores size of 0.1 μm), which allows the passage of CO_2 – solvent mixture. Downstream the precipitation chamber, the liquid solvent is recovered in a liquid separator (LS) at a lower pressure (about 2 Mpa) that is regulated by a backpressure valve (BPV). At the exit of the solvent collection vessel, the flow rate and the total quantity of delivered CO_2 are measured by a rotameter (R) and a dry test meter (DM). At the end of the solution injection, only the $scCO_2$ is sent to the precipitation chamber to eliminate the solvent residues. The CO_2 flow is stopped when the washing step is completed and the precipitator is depressurized up to the atmospheric pressure. Finally, the precipitated powder can be collected.

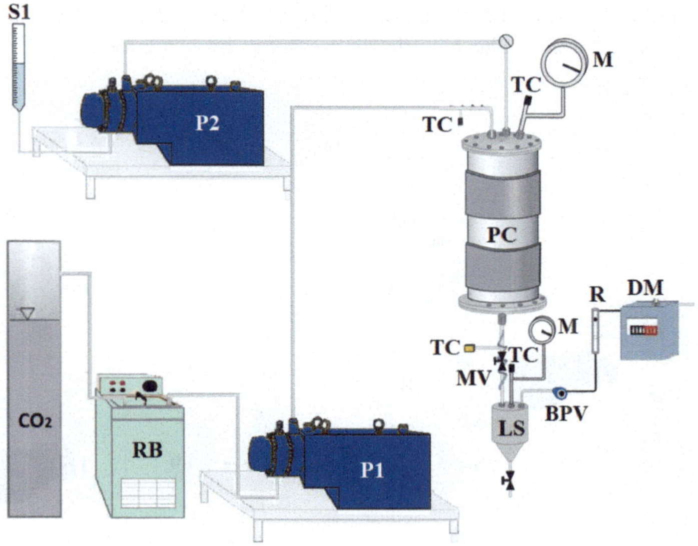

Fig. 1. A schematic representation of SAS plant used to obtain PVP/KET microparticles.

The supercritical adsorption was employed to impregnate KET in calcium alginate aerogels.

The aerogels used as polymeric matrices were prepared by supercritical gel drying (sketched in Fig. 2a), in which the quasi-zero surface tension and high diffusivity of $scCO_2$ are exploited to avoid the pores collapse. The aerogel was produced according to a previously optimized procedure [28]. Briefly, first, alginate hydrogels are prepared as follows: a solution of 5% w/w at sodium alginate in distilled water is poured into cylindrical moulds, then immersed in a coagulation bath at 5% w/w of calcium chloride in distilled water to promote the gelation step and convert the sodium alginate in calcium alginate. Then, the alcogels are obtained by gradually replacing the water in the hydrogels pores with a series of ethanol baths at room temperature, up to have 100% ethanol. Finally, the attainment of the aerogels is achieved by drying the alcogels with $scCO_2$ into a vessel (Fig. 1a) for 5 h at 35 °C and 20 MPa.

The aerogels obtained were used as support for KET adsorption tests, performed in a cylindrical autoclave with an internal volume equal to 100 cm^3 (sketched in Fig. 2a). The vessel was tightly closed at the bottom and at the top with two finger tight clamps. A typical adsorption test starts feeding the CO_2, cooled through a refrigerating bath, to the autoclave with a high pressure pump. Thin band heaters connected with a PID controller assure the thermal heating of the autoclave, whereas the pressure is measured by a digital gauge manometer. The mixing in the vessel is ensured through an impeller, which is located on the top cap and driven by a variable velocity electric motor. When the desired temperature and pressure are reached, CO_2 flow is stopped and the aerogel is left for a certain time in contact with the drug dissolved in the $scCO_2$. Then, the system is depressurized by a micrometric valve to recover the adsorbed aerogel. The CO_2 flow rate at the exit of the cylinder is measured by a rotameter.

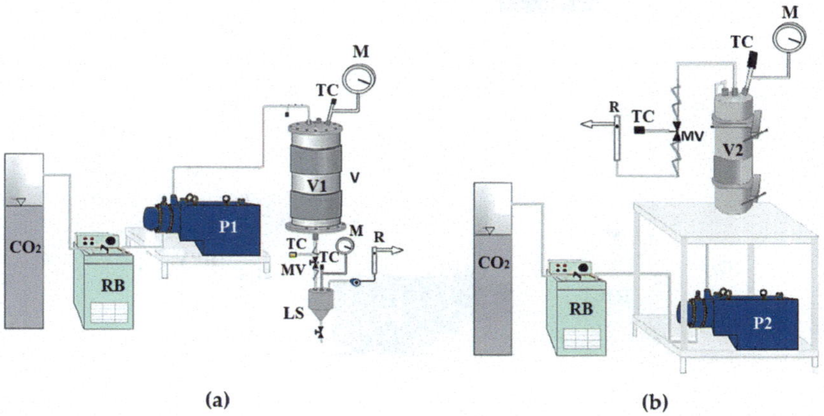

Fig. 2. A schematic representation of the plants used to obtain alginate aerogel loaded with KET: (a) supercritical drying; (b) supercritical adsorption. CO_2: carbon dioxide supply; RB: refrigerating bath; P1, P2: pumps; V1, V2: vessels; TC: thermocouple; M: manometer; MV: micrometric valve; LS: liquid separator; BPV: back pressure valve and R: rotameter.

2.3 Analytical Methods

The samples were observed using a Field Emission Scanning Electron Microscope (FESEM, mod. LEO 1525, Carl Zeiss SMT AG, Oberkochen, Germany). A gold-palladium coating (layer thickness 250 Å) was performed, because the samples has to be conductive to be analyzed by FESEM.

The diameters of 1000 particles produced by SAS process for each sample were estimated by means of FESEM images using the Sigma Scan Pro software (release 5.0, Aspire Software International Ashburn, VA); the obtained data wer analyzed using the Microcal Origin Software (release 8.0, Microcal Software, Inc., Northampton, MA) to determine the particle size distributions (PSDs).

The Fourier transform infrared analysis (FT-IR) was performed by a FT-IR spectrophotometer (IRTracer100, Shimadzu Italia, Milan, Italy), with a resolution of 0.5 cm^{-1} in a range of scan wavenumber 4000–450 cm^{-1} as an average of 16 measurements. Approximately 100 mg of potassium bromide (KBr), which was used as an infrared transparent matrix, were mixed with 1 mg of each sample in a mortar; then, the mixture was compressed through a hydraulic press in form of disks, which were after analyzed.

The KET dissolution studies were performed using UV/vis spectrophotometer (model Cary 50, Varian, Palo Alto, CA).

In the case of SAS PVP/KET samples proposed for oral administration, a HCl 0.1 M solution at pH 2.5 was used as release medium, in order to simulate the gastric acidity. Accurately weighted samples containing an equivalent amount of KET (4 mg) were suspended in 3 mL of HCl solution and placed into a dialysis sack; then, the sack was incubated in 300 mL of the HCl solution at pH 2.5, continuously stirred at 150 rpm and 37 °C.

The dissolution tests of KET adsorbed on alginate aerogels, proposed for topical application, were performed in phosphate buffered saline (PBS) at pH 7.4 to simulate the blood pH. Accurately weighed samples containing an equivalent amount of KET (10 mg) were suspended in 3 mL of PBS in a dialysis sack, which was then incubated in 300 mL of PBS, continuously stirred at 150 rpm and 37 °C. The weight increase of the aerogels measured at the end of the adsorption tests was checked through the drug loadings evaluated considering the absorbance at the end of release; i.e., when all the KET was released from the aerogel to the external PBS. The determined absorbance was converted in NSAID concentration using a calibration curve.

3 Experimental Results

3.1 SAS Coprecipitation of PVP/Ketoprofen

PVP and KET were coprecipitated by SAS process to obtain composite microspheres; i.e., a polymeric matrix in which the drug is homogeneously dispersed that assures an effective coprecipitation [14].

SAS tests were performed by using DMSO as solvent, fixing the temperature at 40 °C and the pressure at 9 MPa, in order to operate above the critical point of the mixture DMSO/scCO_2 (MCP) [29]. The flow rates of CO_2 and the liquid solution were also set respectively at 30 g/min and 1 mL/min, to work with CO_2 molar fractions equal to 0.98; i.e. at the right of the MCP [30]. The total concentration of solutes in DMSO was also fixed at 50 mg/mL to have a massive coprecipitation of the polymer and the drug [15].

Since the polymer/drug ratio mostly influences the drug release kinetics in water [8], the effect of this parameter on morphology and size of precipitated particles was studied. PVP/KET ratio was varied from 10/1 to 3/1, obtaining well-separated microparticles for all the experiments (Fig. 3). Comparing the volumetric cumulative PSDs in Fig. 4, it is observed that the mean diameter increased and the PSDs enlarged by increasing the PVP/KET ratio. Indeed, the particle size ranged from 3.44 ± 1.88 μm to 3.10 ± 1.51 μm by decreasing PVP/KET ratio from 10/1 to 3/1 w/w.

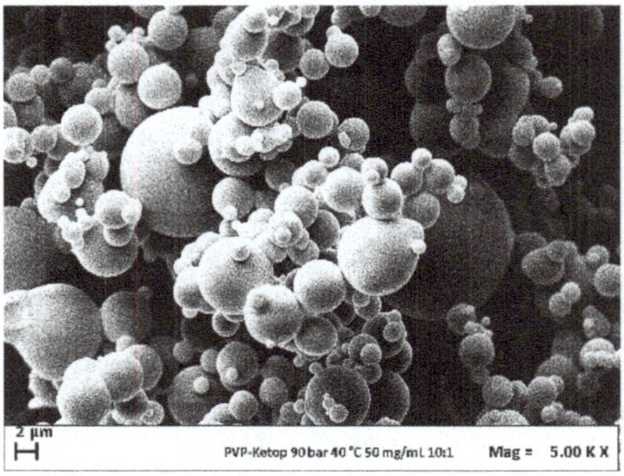

(a)

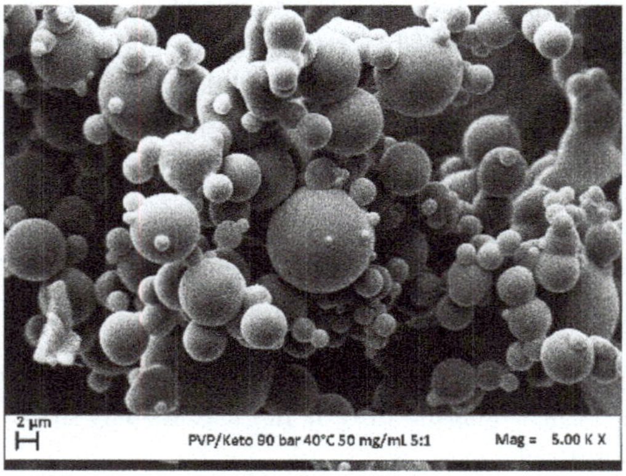

(b)

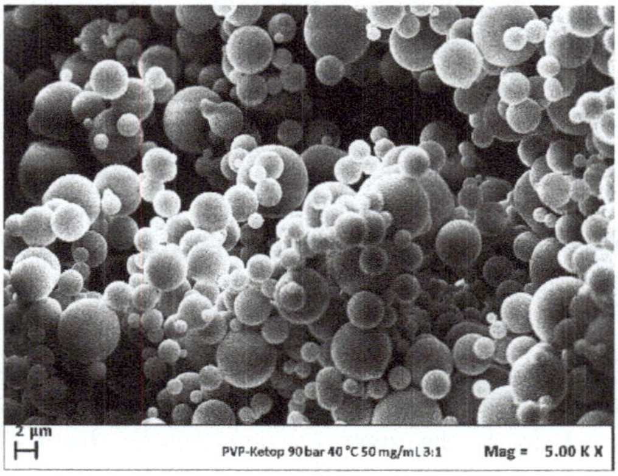

(c)

Fig. 3. FESEM images of PVP/KET microparticles precipitated from DMSO at 9 MPa, 40 °C and 50 mg/mL: (a) 10/1; (b) 5/1; (c) 3/1.

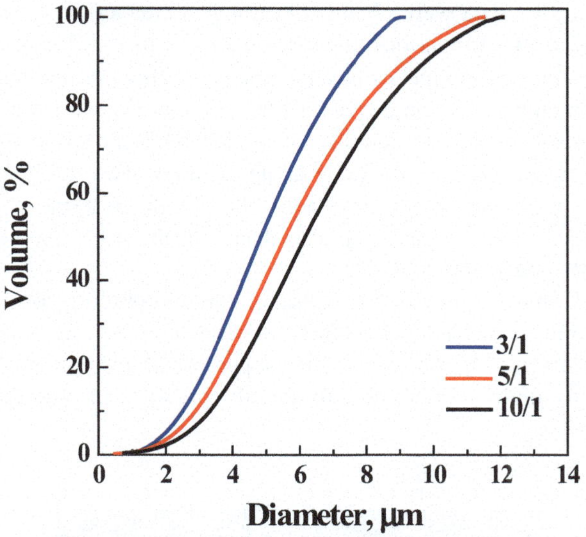

Fig. 4. Volumetric cumulative PSDs of PVP/KET microparticles; effect of polymer/drug ratio.

3.2 Supercritical Adsorption of KET into Alginate Aerogel

The adsorption kinetics were determined at a fixed pressure of 18 MPa to know the time required to reach the maximum amount of KET loaded into the alginate aerogel. Uptake is expressed as q_t; i.e., mmol of KET per gram of aerogel.

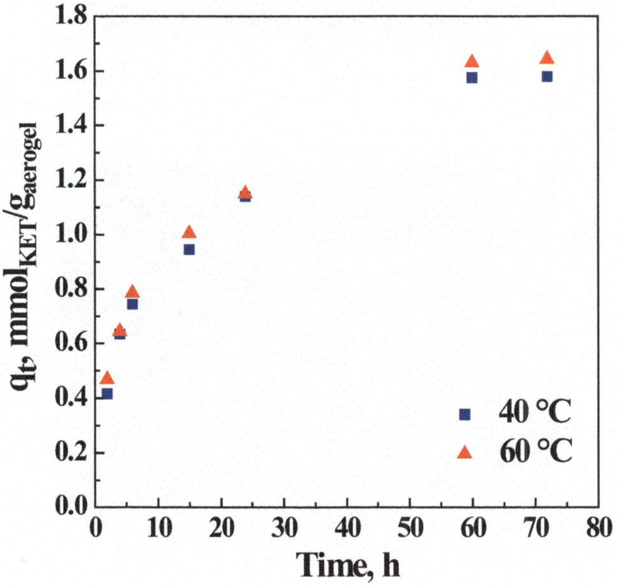

Fig. 5. Kinetic curves at 18 MPa and 40 and 60 °C for the adsorption of KET into alginate aerogels.

Kinetic adsorption data were obtained keeping in contact the alginate aerogel and KET dissolved in $scCO_2$ for various times (from 2 to 72 h) at different temperatures (40 and 60 °C). The adsorption kinetics of KET into aerogels at 40 and 60 °C are reported in Fig. 5. It can be noted that the increase of temperature did not strongly influence the amount of adsorbed KET. The quantity of loaded KET increased by increasing the adsorption time up to a maximum value achieved after about 60 h. Furthermore, the amount of KET adsorbed into calcium aerogels is high, in terms of NSAID/aerogel ratio w/w about 0.42 $mg_{KET}/mg_{aerogel}$, also thanks to the high solubility of KET into $scCO_2$ at 18 MPa and 40/60 °C [31].

The FESEM analyses enabled to observe the morphology of calcium alginate aerogel before and after KET adsorption, as shown in Fig. 6a and b. The porous structure of the aerogel was preserved after supercritical adsorption; moreover, in the case of adsorbed sample, it is possible to note the presence of drug that filled/covered the aerogel pores.

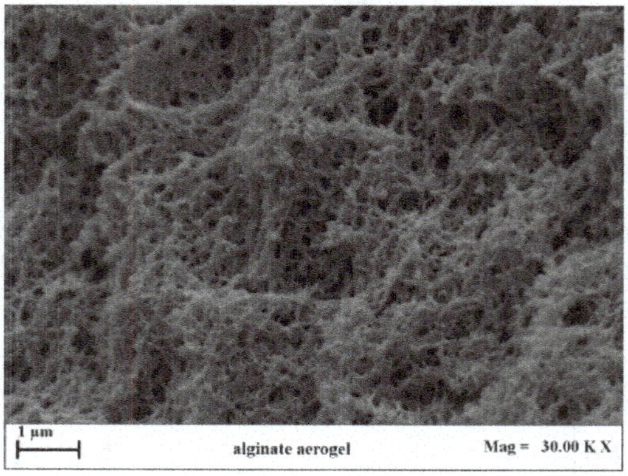

(a)

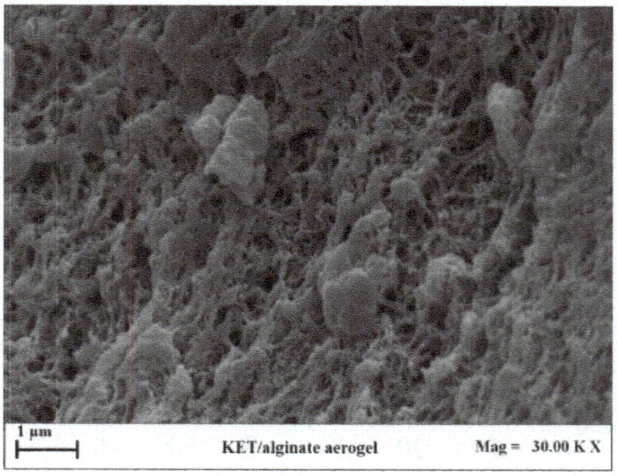

(b)

Fig. 6. FESEM images of alginate aerogel before (a) and after (b) KET adsorption.

3.3 Characterizations of Samples

FT-IR analyses were performed to identify the presence of KET in the composite systems. FT-IR spectra of unprocessed KET, pure PVP, pure calcium alginate aerogel, physical mixtures PVP/KET 3/1 and KET/aeorogel, SAS processed PVP/KET 3/1 and adsorbed sample KET/aerogel are reported in Fig. 7. The FT-IR spectra of PVP/KET physical mixture and SAS coprecipitated powders PVP/KET showed the PVP characteristic peaks and some characteristic bands of KET, namely peaks in the range 1600–1700 cm^{-1} related to the stretching vibrations of carbonyl groups and at 1440 and 1370 cm^{-1} associated to the C-H stretching vibration [32, 33]. The FT-IR spectra of KET/aerogel composites and the corresponding physical mixture exhibited some characteristic peaks of the NSAID; i.e., at 1649 and 1697 cm^{-1} ascribable to the carbonyl group and the band at about 1440 cm^{-1} indicating the stretching vibration of the C-H group [32, 33]. In all studied cases, the spectrum of the polymer prevails in the spectra of the composite systems since the NSAID is hidden in the polymeric structure.

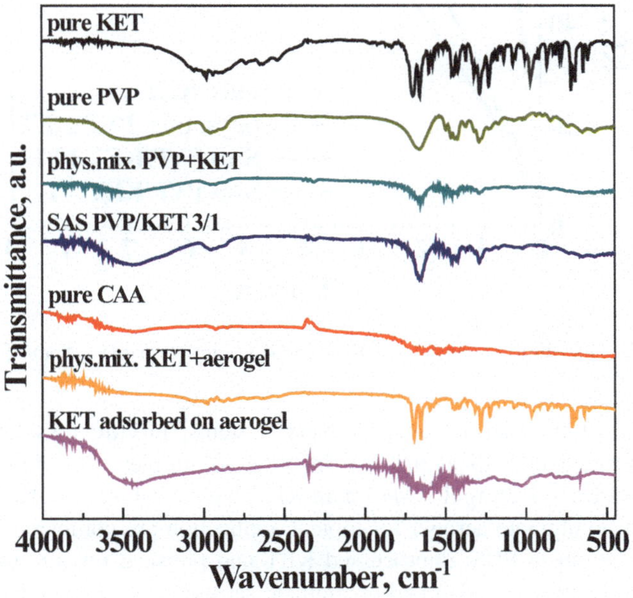

Fig. 7. FT-IR spectra of unprocessed KET, pure PVP and unloaded alginate aerogel, physical mixtures KET+PVP and KET+aerogel, SAS PVP/KET powders and KET adsorbed on aerogel.

When KET is orally administered, it passes through the stomach where KET is largely absorbed. For this reason, the dissolution studies of SAS PVP/KET samples for oral administration were performed in a HCl 0.1 M solution at pH 2.5 and 37°C, to simulate the gastric acidity. From Fig. 8, it can be observed that unprocessed KET completely dissolved in the acid solution in about 8 h, the physical mixture PVP/KET took about 6 h, whereas all SAS coprecipitated PVP/KET powders took about 3 h. Therefore, KET released from PVP-based microparticles is almost 3 times faster than pure KET.

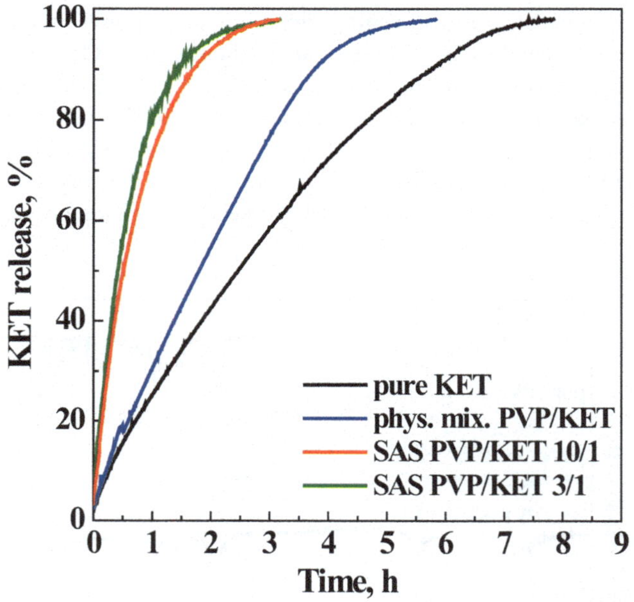

Fig. 8. Dissolution tests of oral KET in HCl solution at pH 2.5 and 37 °C.

In the case of transdermal drug delivery systems, in-vitro dissolution tests are generally performed in PBS at pH 7.4, in order to simulate the blood plasma pH. Dissolution profiles in PBS (pH 7.4) of pure KET, a physical mixture KET/aerogel and KET adsorbed on alginate aerogel for topical application are compared in Fig. 9. For complete dissolution in PBS, unprocessed KET and physical mixture took about 3 h and 4 h, respectively; KET adsorbed on alginate aerogel was completely released after almost 6 h. Therefore, adsorbed samples allowed to prolong KET release about twice.

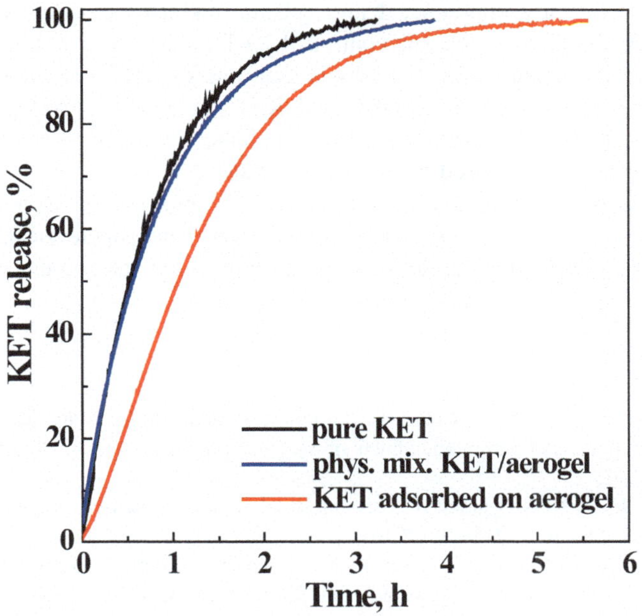

Fig. 9. Dissolution tests of topical KET in PBS at pH 7.4 and 37 °C.

Dissolution tests demonstrated the effectiveness of the supercritical CO_2 based processes to modify the drug release. The important role of the polymeric carriers in the kinetics release was also highlighted: KET dissolution rate can be enhanced by choosing a hydrophilic carrier like PVP to produce microparticles; calcium alginate aerogel is characterized by a hydrophobic behavior, so it promotes a prolonged release of the adsorbed KET.

From a pharmaceutical point of view, SAS PVP/KET microparticles as oral delivery are useful to treat inflammatory diseases that require a rapid therapeutic effect, such as headaches and toothache; whereas, KET-loaded aerogels with a prolonged NSAID release are suitable to reduce the frequency of administration in the case of back ache, rheumatoid arthritis and osteoarthritis.

4 Conclusions

In this work, two different KET delivery systems were successfully proposed by two different $scCO_2$ assisted processes: oral fast and topical controlled KET release were achieved by SAS technique and supercritical adsorption, respectively.

Composite PVP/KET microparticles with mean size ranging from 3.1 ($\pm$1.5) and 3.4 ($\pm$1.9) μm were produced with SAS process, which assured a good control of the particles dimensions varying the operating parameters. An increase in the KET dissolution rate in SAS composite microparticles of about 3 times with respect to the unprocessed NSAID was measured.

KET was efficiently impregnated into calcium alginate aerogels by supercritical adsorption, which allows a possible control of KET loading by varying the operative conditions like the contact time. The dissolution rate of KET adsorbed into alginate aerogel was twice slower than unprocessed KET, suitable to develop topical anti-inflammatory patches that increase patient compliance reducing frequency of administration and side effects caused by NSAID overdoses.

The tuning of the drug dissolution rate can be realized by using supercritical techniques: SAS process for a fast release (oral delivery) and supercritical impregnation for a prolonged release (topical delivery), according to the specific therapy desired.

References

1. Cabré, F., Fernández, M.F., Calvo, L., Ferrer, X., García, M.L., Mauleón, D.: Analgesic, antiinflammatory, and antipyretic effects of S(+)-ketoprofen in vivo. J. Clin. Pharmacol. **38**, 3S–10S (1998)
2. Scheiman, J.M.: NSAID-induced gastrointestinal injury: a focused update for clinicians. J. Clin. Gastroenterol. **50**, 5–10 (2016)
3. Yu, D.-G., Branford-White, C., Shen, X.-X., Zhang, X.-F., Zhu, L.-M.: Solid dispersions of ketoprofen in drug-loaded electrospun nanofibers. J. Dispers. Sci. Technol. **31**, 902–908 (2010)
4. Ivanov, I.T., Tsokeva, Z.: Effect of chirality on PVP/drug interaction within binary physical mixtures of ibuprofen, ketoprofen, and naproxen: a DSC study. Chirality **21**, 719–727 (2009)
5. Tita, B., Fuias, A., Bandur, G., Marian, E., Tita, D.: Compatibility study between ketoprofen and pharmaceutical excipients used in solid dosage forms. J. Pharm. Biomed. Anal. **56**, 221–227 (2011)
6. Prosapio, V., Reverchon, E., De Marco, I.: Formation of PVP/nimesulide microspheres by supercritical antisolvent coprecipitation. J. Supercrit. Fluids **118**, 19–26 (2016)
7. Manna, L., Banchero, M., Sola, D., Ferri, A., Ronchetti, S., Sicardi, S.: Impregnation of PVP microparticles with ketoprofen in the presence of supercritical CO_2. J. Supercrit. Fluids **42**, 378–384 (2007)
8. Arunachalam, A., Karthikeyan, M., Vinay Kumar, D., Prathap, M., Sethuraman, S., Ashutoshkumar, S., Manidipa, S.: Transdermal drug delivery system: a review. Curr. Pharm. Res. **1**, 70–81 (2010)
9. Franco, P., Reverchon, E., De Marco, I.: Zein/diclofenac sodium coprecipitation at micrometric and nanometric range by supercritical antisolvent processing. J. CO2 Util. **27**, 366–373 (2018)
10. Kalogiannis, C.G., Michailof, C.M., Panayiotou, C.G.: Microencapsulation of amoxicillin in poly (l-lactic acid) by supercritical antisolvent precipitation. Ind. Eng. Chem. Res. **45**, 8738–8743 (2006)
11. Franco, P., Martino, M., Palma, V., Scarpellini, A., De Marco, I.: Pt on SAS-CeO_2 nanopowder as catalyst for the CO-WGS reaction. Int. J. Hydrogen Energy **43**, 19965–19975 (2018)
12. Zhong, Q., Jin, M., Davidson, P.M., Zivanovic, S.: Sustained release of lysozyme from zein microcapsules produced by a supercritical anti-solvent process. Food Chem. **115**, 697–700 (2009)
13. Montes, A., Kin, N., Gordillo, M.D., Pereyra, C., Martínez de la Ossa, E.J.: Polymer–naproxen precipitation by supercritical antisolvent (SAS) process. J. Supercrit. Fluids **89**, 58–67 (2014)

14. Montes, A., Wehner, L., Pereyra, C., Martínez de la Ossa, E.J.: Generation of microparticles of ellagic acid by supercritical antisolvent process. J. Supercrit. Fluids **116**, 101–110 (2016)
15. Prosapio, V., De Marco, I., Reverchon, E.: Supercritical antisolvent coprecipitation mechanisms. J. Supercrit. Fluids **138**, 247–258 (2018)
16. Franco, P., Aliakbarian, B., Perego, P., Reverchon, E., De Marco, I.: Supercritical adsorption of quercetin on aerogels for active packaging applications. Ind. Eng. Chem. Res. **57**, 15105–15113 (2018)
17. Smirnova, I., Suttiruengwong, S., Seiler, M., Arlt, W.: Dissolution rate enhancement by adsorption of poorly soluble drugs on hydrophilic silica aerogels. Pharm. Dev. Technol. **9**, 443–452 (2005)
18. Smirnova, I., Mamic, J., Arlt, W.: Adsorption of drugs on silica aerogels. Langmuir **19**, 8521–8525 (2003)
19. García-Casas, I., Crampon, C., Montes, A., Pereyra, C., Martínez de la Ossa, E.J., Badens, E.: Supercritical CO_2 impregnation of silica microparticles with quercetin. J. Supercrit. Fluids **143**, 157–161 (2019)
20. Smirnova, I., Suttiruengwong, S., Arlt, W.: Aerogels: tailor-made carriers for immediate and prolonged drug release. KONA Powder Part. J. **23**, 86–97 (2005)
21. De Marco, I., Reverchon, E.: Starch aerogel loaded with poorly water-soluble vitamins through supercritical CO_2 adsorption. Chem. Eng. Res. Des. **119**, 221–230 (2017)
22. Pantić, M., Knez, Ž., Novak, Z.: Supercritical impregnation as a feasible technique for entrapment of fat-soluble vitamins into alginate aerogels. J. Non Cryst. Solids **432**, 519–526 (2016)
23. Marin, M.A., Mallepally, R.R., McHugh, M.A.: Silk fibroin aerogels for drug delivery applications. J. Supercrit. Fluids **91**, 84–89 (2014)
24. Lovskaya, D.D., Lebedev, A.E., Menshutina, N.V.: Aerogels as drug delivery systems: in vitro and in vivo evaluations. J. Supercrit. Fluids **106**, 115–121 (2015)
25. Salgado, M., Santos, F., Rodríguez-Rojo, S., Reis, R.L., Duarte, A.R.C., Cocero, M.J.: Development of barley and yeast β-glucan aerogels for drug delivery by supercritical fluids. J. CO2 Util. **22**, 262–269 (2017)
26. Bouledjouidja, A., Masmoudi, Y., Sergent, M., Trivedi, V., Meniai, A., Badens, E.: Drug loading of foldable commercial intraocular lenses using supercritical impregnation. Int. J. Pharm. **500**, 85–99 (2016)
27. García-González, C.A., Jin, M., Gerth, J., Alvarez-Lorenzo, C., Smirnova, I.: Polysaccharide-based aerogel microspheres for oral drug delivery. Carbohydr. Polym. **117**, 797–806 (2015)
28. Baldino, L., Concilio, S., Cardea, S., Reverchon, E.: Interpenetration on natural polymer aerogels by supercritical drying. Polymers **8**, 106–117 (2016)
29. Andreatta, A.E., Florusse, L.J., Bottini, S.B., Peters, C.J.: Phase equilibria of dimethyl sulfoxide (DMSO) + carbon dioxide, and DMSO + carbon dioxide + water mixtures. J. Supercrit. Fluids **42**, 60–68 (2007)
30. Reverchon, E., De Marco, I.: Mechanisms controlling supercritical antisolvent precipitate morphology. Chem. Eng. J. **169**, 358–370 (2011)
31. Macnaughton, S.J., Kikic, I., Foster, N.R., Alessi, P., Cortesi, A., Colombo, I.: Solubility of anti-inflammatory drugs in supercritical carbon dioxide. J. Chem. Eng. Data **41**, 1083–1086 (1996)
32. Grimling, B., Górniak, A., Meler, J., Szcześniak, M.: Characterisation and dissolution properties of ketoprofen in binary solid dispersion with chitosan. Prog. Chem. Appl. Chitin Deriv. **19**, 23–31 (2014)
33. Al-Tahami, K.: Preparation, characterization, and in vitro release of ketoprofen loaded alginate microspheres. Int. J. App. Pharm. **6**, 9–12 (2014)

Study on the Use of Biomaterials as Protective Membranes for Certain Functional Foods

Petre Savescu[1(✉)], Fanel Iacobescu[2], and Maria-Magdalena Poenaru[2]

[1] Department 32 TAS Food Control and Expertise, University of Craiova, A.I.Cuza Street, 13, 200585 Craiova, Romania
psavescu@gmail.com

[2] Department 29 Horticulture and Food Sciences, University of Craiova, A.I.Cuza Street, 13, 200585 Craiova, Romania

Abstract. A series of functional and nutraceutical food products from the by-products of the food industry will be designed and produced in a Scientific Partnership Program within the University of Craiova, Romania (2018–2020). These functional foods and nutraceutical products will be used in the personalized nutrition of certain patient groups exhibiting certain medical conditions. In the design and construction of these functional foods, some bioactive compounds will be stabilized in certain organic fruit juices; at the same time, other bioactive compounds will be introduced into gluten-free bakery products. The functional foods which will be produced in the Bioengineering and Biotechnology Laboratory of the INCESA Regional Research Hub of the University of Craiova will be stabilized in bio-resorbable capsules. To preserve the functional properties of nutraceuticals, certain biomaterials (proteins, polysaccharides) will be used- as bio-resorbable membranes. Therefore, the focus will be on the biocompatibility of the membranes used for the best biofilms for consumers, the studies of the redox processes occurring inside the capsules, and the studies on the toxicity of the final products.

1 Introduction

Food supplements are created to bring a surplus of vitamins, natural enzymes, oligo-elements and minerals to the basic nutrition and they address certain varying categories of consumers.

In order to be able to design and produce a good functional food and/or a valuable food supplement, it is very important to use organic food products or parts as a starting point. These products are certified as organic (being in compliance with the requirements of the updated European Regulations 848/2018). Thus, the risk of primary or secondary contamination is excluded. According to the European Bioeconomy Strategy for Sustainable Development, today we are trying to use and reuse a large number of valuable bio-compounds which are separated-concentrated from by-products resulted from some food products' manufacturing chain. This Strategy is developed in compliance with the Environmental Protection Regulations and the European Directive on Food Waste Reduction and Control; this is an opportunity for the EU to improve the efficiency of how we exploit the currently underused resources [1]. Current estimates

S. Piotto et al. (Eds.): BIONAM 2019, LNBE, pp. 178–190, 2020.
https://doi.org/10.1007/978-3-030-47705-9_16

indicate that around one-third of worldwide foodstuffs produced for human consumption are wasted or lost, leading to significant economic and environmental costs [2].

In many cases, when designing and constructing nanomolecular systems specific to functional foods (or to similar systems of dietary supplements), certain oxidation-reduction processes are continuously monitored – for example, the activity of several enzymes such as oxidoreductases. All these specific elements (the *p*H, the redox potential- E*h*, the hydrogen ion pressure logarithm – the r_H) are extremely important for both the construction and the maintenance of these functional foods and for the consumers' health. The activity of certain oxidoreductases such as those NAD-dependent, NADP-dependent, FAD-dependent, and FMN-dependent, the ratios of the oxidized and reduced forms' concentrations, and their functional kinetics have been some of the objectives the present scientific research.

The special effects of natural ionized water and partially deionized water on the activity of certain oxidoreductases and on the ion kinetics in the chemical composition of some natural extracts have been proven, based on results obtained during a scientific research cycle. Using mixtures of natural ionized water in very precise concentrations to remove lipoxygenases (or other enzymes that induce certain undesirable tastes in the final food or certain changes in the shelf life due to strong oxidations) presents a number of considerable advantages [3, 4]. Testing certain natural membranes which are kept for a certain period under light and gravitational magnetic fields and moistened with nanostructured and natural ionized water represents a technological advantage both for increasing the extraction capability of valuable bio-compounds and for transferring valuable cations in the final environment of the foods.

The paper presents an innovative application of biomaterials in the field of bioeconomy. The main purpose of the present scientific work is to obtain valuable food supplements by using hemp (*Cannabis Sativa* hemp pomaces left over after pressing and extracting the oil) and Jerusalem Artichoke sediments (*Helianthus Tuberosus* - a perennial plant which easily grows on various soil types and whose food potential is not sufficiently exploited at present).

The *novelty of the paper* is underlined by several factors:

- Presenting an innovative technology for extracting valuable bio-compounds with the help of supercritical fluids (Supercritical Fluid Extraction - *hereinafter referred to as SFE*). Within this technology, the extraction efficiency is influenced by the using protective cellulose membranes which have been previously ionized by plasma (the extraction efficiency of valuable bio-compounds thus increases by over 22%). This is a novelty worldwide and will be patented.
- A modified *cold plasma* type of technology was used in the separation-concentration of valuable bio-compounds from these plants. In conventional research, Cold Plasma results from using a Tesla coil string. Cold plasma is produced along with static electricity, which can also release hot plasma (which can affect less thermally stable bio-compounds). Therefore, we used a new Tesla T coil array system ("T coil in T coil" - the inner coil is directly fed, the outer coil has an inverted power supply circuit and a nanocoated CO_2 gas circulates between them). This innovative system removes any static electricity and any form of heating from the mixture. This results in separating the bio-compounds from certain hydro-alcoholic extracts in a stable

plasma field - as can be seen in the Fig. 4. This is also new worldwide and could revolutionize the crystallization and easy separation of undesirable remnant compounds in certain products (certain preservatives, carbohydrates, pesticides - which could increase the toxicity concentration of the final product).

- The innovative product thus obtained is packaged under inert conditions in LICAPS (Liquid Filled Hard Capsules) vegetable capsules which have certain characteristics (solvent free, preservative free, gluten free, sugar free, GMO free) which make the product a "premium" one, unique in the food supplements market. This type of bioresorbable capsules allows antioxidants to be stored in the final product and released when they reach the consumer's digestive system.

Using special Mild Food Processing technologies, certain extraction techniques using supercritical fluids (in this case, carbon dioxide at about 32 °C and high pressures up to 650 bar), and non-invasive treatments with the help of starter extracts and natural ionized vegetal membranes are extremely important in the preservation and volumetric transfer of valuable bio-compounds from organic raw materials. Certain special separation - concentration technologies should be used for producing innovative high nutritional density supplements from hemp sediments (pomaces from the oil extraction process), and *Helianthus Tuberosus* and Sorghum leaves and roots, as well as achieving highly efficient levels in bio-compound separation. As part of the project *"Valuing the Food Potential of Selected Agricultural Crops (artichoke tubers, sorghum seeds, hemp pomace - resulting from cold pressed hemp oil)"* we used (for the first time in the world and on these food residues) plasma technologies such as improved cold plasma and improved Supercritical Fluid Extraction with the help of nanostructured gases. Some of the results are presented in this paper. The use of a portable plasma generator induces a weaker field, with action on the surface of the materials to be treated [5].

Choosing the constructive characteristics of the equipment and the technological flow in any separation-concentration unit is primarily influenced by the properties of the raw materials. The present paper presents the results of experimental research on characterizing the raw materials and the Supercritical Fluid Extraction technology (SFE) used in producing innovative foods.

The applied technologies are not very cheap, but the system can develop food and agricultural products with no artificial toxicity and at the same time the environment is also protected. Specialists in the food and pharmaceutical industry have developed many food supplements from organic farm products [6]. The choice of raw materials and the use of the GPS techniques in culture technologies were also very important [7]. These allowed the best orientation of the crops in the field, the increasing the insulation period and the production valuable compounds enriched crops (especially minerals: Ca, Mg, K).

In order to develop special dietary supplements, the local Romanian hemp varieties were used - made in the Secuieni Agricultural Development Research Institute, Romania, and *Helianthus tuberosus* leaves from the Dabuleni Agricultural Development Research Institute, Romania. The cultivation technology of these plants (used for the development of valuable food supplements) has been free of pesticide, heavy metals or growth hormone treatments. The field of processing these transformation or even organic products has been very poorly developed. Among the species used so far, we could mention tubers and roots such as potatoes, carrots, Jerusalem artichokes. Of these, the Jerusalem artichoke (*Helianthus tuberosus*) was less used to produce valuable dietary supplements or to separate the inulin concentrate (to replace sweeteners forbidden for people with diabetes or cardiovascular diseases) [8].

The hemp seeds contain: 36% oil, 28% protein, 14–27% non-extractive extract, 17.8–26.3% cellulose and 2.5–6.8% ash [9].

An increasing trend at the EU level also refers to organic crops of technical plants. This trend also includes organic hemp crops (*Cannabis sativa*) which are of particular interest to both growers and producers. In developing special dietary supplements in Romania, the following hemp seed species have been used: Secuieni 1, Zenit, Diana, Dacia-Secuieni, Secuieni-Jubileu, which were developed and cultivated in the Secuieni Agricultural Development Center, Romania [10, 11]. The Dacia-Secuieni and Jubileu Secuieni varieties are very important in the development of really valuable non-narcotic and non-hallucinogenic food supplements (as they have 31–33% fibre and 0.013% THC, and 20–24% fibre and 0.002% THC, respectively). From a nutritional point of view, hemp seeds are richer in protein, essential amino acids, and Omega 3 and 6 [12]. The hemp made in Secuieni, Romania has a low THC content (delta-9-tetrahydrocannabinol) well below 0.2%, the level required by the EU for industrial crops.

2 Experimental Procedure

In order to design and produce an innovative food supplement, it is very important to characterize the raw materials and the technological operations.

The plants (*Helianthus tuberosus*) are subject to a series of sorting, calibrating, drying, conditioning, packaging, storage, and preservation operations and physic-chemical and microbiological tests, before undergoing the extraction process.

The simplest form of preparing a food supplement from plants is in the form of powder (from a single plant or a mixture of powders from several plants). In this manner, the *Helianthus tuberoses* leaves (or roots) are organically dried by airflow in the shade or by exposure to the sun (under a certain light intensity and a limited time). All of these have the advantages of precise dosing, the accurately measuring the nutritional density, a rigorous analytical management, ensuring traceability, establishing a longer shelf life and easy storage in safe packaging [13].

According to several papers in the field, the area of super critical extraction is a nascent one in international research; this important technology has a number of advantages (high dissolution power of fluids in supercritical states, high diffusion power, organic extraction processes, low solvent removal costs, low solvent costs, valuable bio compound extraction at low temperatures - which preserve their nutritional character, reduced surface tension and reduced viscosity) [14, 15]. Eventually, extraction products need to be subjected to further analysis to eliminate toxic compounds.

For the SFE tests performed on the Jerusalem artichoke (*Helianthus tuberosus*), several important experimental variants were developed from the Dabuleni Development Research Station and University of Craiova, Romania:

- V_1 - An experimental variant made from Jerusalem artichoke leaves, the Dacian variety
- V_2 - An experimental variant made from Jerusalem artichoke leaves, the Rustic variety
- V_3 - An experimental variant made from Jerusalem artichoke leaves, the Olimp variety
- V_4 - An experimental variant made from Jerusalem artichoke leaves, the local variety from Dabuleni
- V_5 - An experimental variant made from Jerusalem artichoke leaves, the organic variety from Dabuleni

The experimental variants of hemp were:

- V_6 - Experimental variant in which peeled hemp seeds were used
- V_7 - Experimental variant in which peeled and pressed hemp seeds were used
- V_8 - Experimental variant in which dried and ground hemp seeds were used
- V_9 - Experimental variant in which protein hemp extract was used
- V_{10} - Experimental variant – hemp extract oil

These experimental variants were carefully prepared according to standardized working methods. Atomic Absorption Spectroscopy (AAS), NIR, FT-IR, UV-Vis Molecular Absorption Spectrometry, the Single Addition Method and GC-MS Chromatography Techniques were used in the physic-chemical tests to quantify the main bio compounds of these products. The extraction with fluids in a supercritical state (supercritical carbon dioxide at 32 °C and variable pressure 150–650 bar) and ion charged bio-membranes stabilized in magnetic and gravitational fields were used in these case.

2.1 The Microwave Digestion Method

In order to correctly determine the traces and influence of certain compounds in the redox processes that take place when obtaining food supplements, a series of chemical tests have been performed to determine the heavy metals and minerals in several varieties of Jerusalem artichoke and hemp (with no toxicity). Therefore, several edible parts of the Jerusalem artichoke and hemp seed pomaces (which can be processed and consumed) were checked (before the experiment) while the concentrations of certain chemical elements (heavy metals and impurities) were tested by A.A.S. Tech to avoid using parts of the plant that contain traces of pollution.

Conditions: In the first stage the samples were prepared for atomic absorption spectrometry; the samples were mineralized (using the microwave digestion). Acid digestion decomposes the matrix of the main sample to concentrate the elements to be analysed and to prepare the final test. Compared to the conventional synthesis technologies, microwave heating has a number of advantages which include: shorter reaction times, better reproducibility, improved purity, a higher efficiency and improved management control. The microwave accelerated reaction system is designed to digest, dissolve, and hydrolyse a wide variety of materials in laboratory conditions. The main purpose of using such a technique is to prepare samples for atomic absorption (AA) testing. For this purpose, a CEM Mars, 1200 W microwave mineralization system was used. The temperature is internally controlled by a fibre optic in one reference vessel for each series of vessels.

The method: In brief, 10 g of dry substance (d.s.) were weighed, with analytical accuracy. For mineralization, to each digestion cartridge we added 10.0000 g of the product (Jerusalem artichoke leaves separated by variety for the first digestion variant and in another stage, in each bowl, hemp seeds from the Secuieni Research Center, separated by variety), 6 mL of concentrated nitric acid and 3 mL of 30% oxygenated water. For the BLANK cartridge a single digestion was used in cartridges containing only the reagents, without the products to be tested. The method is summarized in Table 1.

Table 1. The microwave digestion steps.

Power, W	Time, min	Agitation	Comment
300	5	Yes	For protect of cartridges
0	2	No	For helping the sedimentation process in bottom of cartridge
400	3	Yes	
600	3	Yes	
200	1	No	
800	3	Yes	
1000	5	Yes	
0	10	No	Cooling for open the cartridges

3 Results and Discussion

The analysis of the molecular absorption spectre graphs shows that, depending on the wavelength, in the UV field (200–400 nm), the widest variations were recorded in variant V_9 (when extracting proteins from hemp seeds: R^2 was 0.872), whereas the smallest variations of these spectral curves were recorded in the pressed hemp seeds variant (R^2 was 0.9545).

By analysing the concentrations of the main indicators in the flour obtained from hemp pomaces (after oil extraction), one can be notice that there is a lot of raw protein (28.64% dry matter), total fibres (47.71% dry matter) and very valuable minerals: K (1300 mg/100 g sample), Mg (612.40 mg/100 g sample), Ca (286.42 mg/100 g sample). At the same time, the hemp sediments (pomaces - which remain after the hemp seeds are pressed and the oil is extracted) contain various valuable fatty acids (especially polyunsaturated fatty acids - about 77%). That is why it is extremely important to recover these valuable bio-compounds, to concentrate them and use them in valuable nutritional density food supplements.

In addition, the tests of the NAD oxidized forms show that the same phenomenon occurs. The highest concentrations are found in the peeled seeds variant (2.053 abs. units. in V_6) and, respectively, in the peeled and pressed seeds (2.039 abs. units in V_7). The concentration of the $NADH_2$ reduced forms has the highest values when the seeds are pressed (1.883 in V_7) and the lowest ones when the hemp cakes are dried and ground after oil extraction (0.431 abs. units in V_8) (Fig. 1). The ratio of [NAD] / [NADH + H +] in experimental variants is shown in Fig. 1.

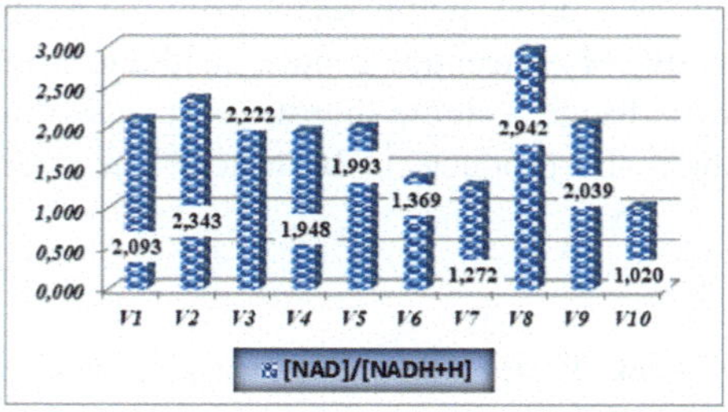

Fig. 1. The ratio of [NAD]/[NADH+H+] in experimental variants.

When analysing the FMN reduced and oxidized forms, the same surface oxidation phenomenon occurs and therefore the highest values are found in the case of seed pressing (V_7 versus V_6) and protein extraction from the seeds (V_9). The lowest concentrations of FMN and FMNH + H oxidized and reduced forms were found in V_8 variants - as a result of the drying and grinding processes. A as result of cold processing, removing hemp oil creates a reduced environment in pomaces.

Mild food processing of these intermediate or even organic products, has been very poorly developed. Light processing hemp seeds can be improved in the near future, which can produce valuable nutraceuticals and dietary supplements (Fig. 2). The ratio of [FMN] / [FMNH + H +] in experimental variants is shown in Fig. 2.

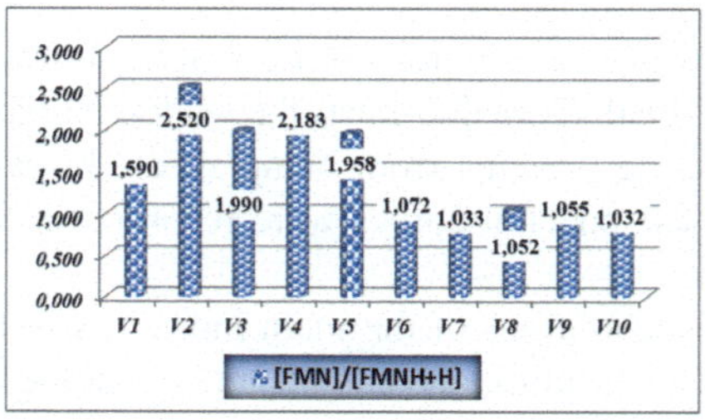

Fig. 2. The ratio of [FMN]/[FMNH+H+] in experimental variants.

The Jerusalem artichoke leaf samples recorded important nitrogen values (mg/100 g): 1,263 in V_5 - in the organic farming system, and 1.940 in V_4 - the local variety in Dabuleni. Following the atomic absorption spectrometry (AAS) testing, the Jerusalem artichoke also traced values of *phosphorus* concentrations (2.95 mg/100 g in

V_5 – the variety from the organic area and 3.68 mg/100 g in the Olimp variety - V_3), *potassium* (24.33 mg/100 g in the V_4 variety from the organic area and 31.34 mg/100 g from the local variety in Dabuleni), *calcium* (between 1.45 mg/100 g in the variety from the organic area and 5.47 mg/100 g from the local variety in Dabuleni), *magnesium* (0.95 mg/100 g in the variety from the organic area and 2.36 mg/100 g from the local variety in Dabuleni). Other elements were recorded as p. p.m. (parts per million): *Copper* (5.37 ppm in the variety from the organic area and 7.24 ppm in the Dabuleni variety), *Zinc* (19.86 ppm in the variety from the organic area and 28.16 ppm in the Dabuleni variety), *Iron* (95.47 ppm in the variety from the organic area and 147.42 ppm in the Dabuleni variety), *Manganese* (25.44 ppm in the variety from the organic area and 33.26 ppm in the Dabuleni variety), *Molybdenum* (3.86 ppm in the variety from the organic area and 4.25 ppm in the Dabuleni variety). There were no traces of Lead, Cadmium, Mercury and Arsenic.

At the same time, the Jerusalem artichoke leaves exhibited a number of complex substances: proteins (24.2–26.8% Dry Substance), simple carbohydrates (0.6–1.8% Dry Substance), cellulose (5.8–6.4% Dry Substance). As regards the silicon content, as these plants were not treated with pesticides, they registered an average of 4.4 mg per 100 g of Fresh Substance.

By using protective membranes (which were previously charged in the plasma field by immersing them in nano-nanostructured gas atmosphere), special results were obtained in increasing the concentrations of the NADH+H^+ and FMNH+H^+ reduced forms and in the extraction of the Helianthus Tuberosus active biocompounds. Thus the extracts' antioxidant character increased and a high nutrition density extract was obtained (Fig. 3).

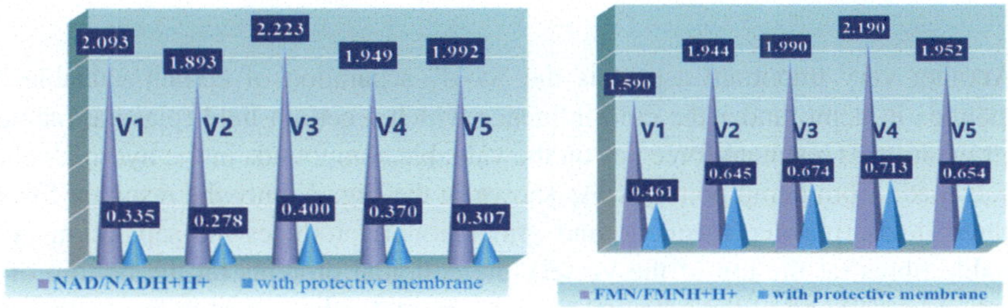

Fig. 3. The ratio [NAD/NADH+H+] and [FMN]/[FMNH+H+] of experimental variants.

The best option for processing in this case was the Dabuleni Local Variety Variant and, close behind, the Organic Variant.

An extraction technique using supercritical fluids (carbon dioxide SFE as extraction agent) was used to extract the main valuable bio compounds from pressed hemp seed cakes, dried and ground hemp cakes and Jerusalem artichoke leaves.

In these extractions, we used Helix Natural Product SFE equipment delivered to the Research Infrastructure of the Bioengineering and Biotechnology Laboratories, the

Research HUB INCESA, University of Craiova, Romania by the world leader in the field, Applied Separations.

Prior to the SFE extraction, certain interfering substances were separated, as these could be drawn in the solvent or in the final (aqueous) extracts. To this end, we used a combination of calcification and carbonation (defecation and carbonation) purifying techniques, as well as filtration separation techniques including special membranes. Thus, a series of soluble pectin, lignin, celluloses, and hemi-celluloses were separated. The natural protective membranes are obtained from vegetal cellulose which were pressed and charged in ionic, magnetic, and gravitational fields. Thus, active nano-structures are produced, which directly influence the activity of the co-enzymes and greatly increase the efficiency of separating the active bio compounds. Such a separation produced by the cumulative effect of the fields can be seen in Fig. 4, which shows a pectin gel at the bottom, a crystal structure in the middle, and lignin, cellulose and hemicellulose at the top.

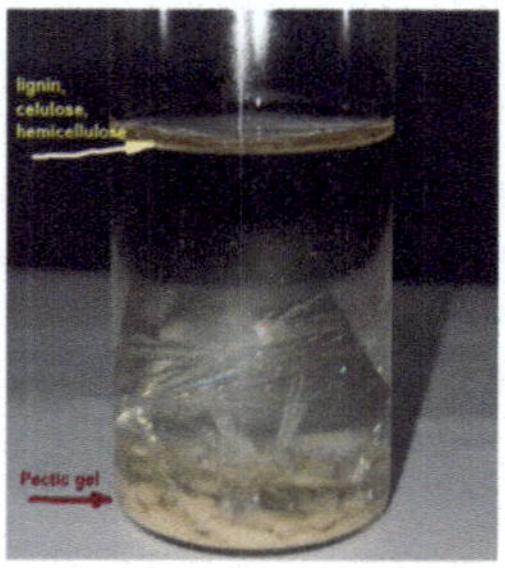

Fig. 4. The separation process influenced by the field of natural nanocomposite membranes

Another very important aspect is the "cold" separation of certain valuable bio-compounds in hemp under the cold influence which a certain field (plasma, magnetic and gravitational) resultant force has on the valuable compounds in the hydro alcoholic extracts made from hemp pomaces. As shown in the Fig. 4, only the resultant force of the three fields (plasma, magnetic and gravitational) produces the separation of the vegetable fibres (at the top of the vessel), the colloidal proteins (at the bottom of the extraction vessel) and of the potential pesticide crystals (the crystal in the centre). This separation process is innovative and can be the basis of a new cold separation technology, as a novelty in separation techniques worldwide. The separations thus achieved are stable (even in a dynamic stirring regime) and can be completed by filtration through inert protective membranes.

The Tesla coil system (direct and inverted) which also contains nanostructured gas is attach such as Fig. 5b and is the subject of a European patent application.

After a brief ionic blast in which weak bases (CaO solutions) and carbon anhydride were used, rich precipitates formed. These precipitates were removed by specific separation techniques (filtration, sedimentation-filtration).

By analysing the free -SH groups in the Jerusalem artichoke experimental (by using UV Molecular Absorption Spectrometry, Pure Analysis Substances and DNA modification analysis software) can see the consequences of the microorganisms' infection in variant V_2 (Rustic). In the Dabuleni variant (V_4) and the Organic one (V_5), the peaks of molecular absorption spectra for free-SH groups (252 nm) are identical, without DNA modifications. The organic variant V_5 and the Dabuleni Variety V_4 have the lowest concentrations of NAD, $NADH_2$, FMN, $FMNH_2$, oxidized and reduced hem proteins, riboflavin, flavoprotein and other forms.

In order to improve the supercritical extraction efficiency, new technological ways have been tested. The relevant tests which produced good extraction results required the use of special air filters, which were placed on the compressor's discharge path (Fig. 5a) and inside the central extraction unit (Fig. 5b). Inside the central extraction unit, the increase in the efficiency of the new technologies required the optimizing certain routes and minimizing the heat transfer.

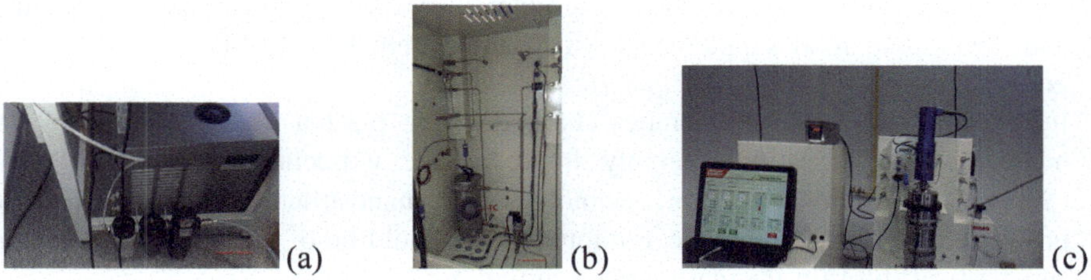

Fig. 5. Ensure of superior air filtration (a) and change of some technological ways with Tesla Coil Circuit -TC (b) - essential conditions for obtaining well extraction efficiencies. The best extraction are on the regime of dynamic extraction (c)

New natural osmotic membranes were used both for extraction and filtration; these membranes were charged in a magnetic-gravitational field and naturally ionized (by soaking them in balneal-therapeutic water and electrolyzing them with Nano coated electrodes), which facilitated extraction efficiency and selectivity. These new membranes as well as the modifications made to SFE technology will be the object of a Patent Application in June 2020. Preparing the extraction system also required ensuring the special environmental condition. In order to check the extraction and concentration degree of important bioactive compounds, we did calibration tests and set reference lines for the optical analysis equipment (UV-VIS spectrophotometer, NIR analyser and Fourier – FT IR transform spectrophotometer). The UV-UV spectrophotometer was a *T92 plus* controlled by special software that can record the DNA changes. For the IR spectral domain, we used *Tango NIR* and *Vertex V70* Analyser made by Bruker Optics. For calibrations and references, we used known concentration lines and Pure Analysis solutions.

Very good results were obtained by using the *regime of dynamic extraction* during which the different stirring speeds had a special contribution (between 120 and 200 rotations/minute, depending on the consistency of the raw material). The most valuable results were recorded at lower pressures (200–450 bar) and temperatures of 30–32 °C.

4 Conclusions and Recommendations

As a result of the studies and the technological level we achieved during the scientific research, the following conclusions can be drawn:

- Measuring the levels of NAD, FMN, FAD concentrations can be a valuable indicator of the leaves' contamination with pathogenic microorganisms; these leaves can be used as raw materials in the creation of food supplements.
- The Redox processes in these technologies are extremely important and coordinated by a series of oxidoreductases which have NAD-dependent, FMN-dependent and FAD-dependent active co-enzymes.
- The activity of these co-enzymes is strongly influenced by the ion exchange produced by the ion charged natural membranes during the filtering and supercritical extraction processes.
- The organic farming system is an important source of raw materials for designing and developing food supplements which are extremely valuable for certain categories of people.
- Introducing protective membranes charged in the plasma field generated by the nanostructured gases considerably improves the extraction efficiency of SFE (Supercritical Fluid Extraction) technology. The nanostructured gas in which the protective membranes have been impregnated should be of the same nature as the supercritical fluid (in this case carbon dioxide).
- Packing LICAPS resorbable capsules in a controlled atmosphere keeps the antioxidant concentrations in the food supplement unchanged and may prolong their shelf life.
- Hemp residues resulting from seed pressing and specific oil separation can be successfully used when designing valuable food supplements.
- The rheological characteristics of the mixture containing the supercritical fluid and the bio compounds extracted from the material are extremely important in the efficiency formulas and in choosing the technology.
- Physical, chemical, and microbiological tests specific to raw materials, sub-products obtained from various technological operations, and finished products must comply with all the rules of standardized and accredited methods.
- By using special innovative techniques in "mild food processing" technologies (which use the effects of the plasma, magnetic and gravitational fields' resultant forces), high quality extracts can be obtained in a shorter time, while having an improved extraction efficiency.
- The tests continue in several dynamic SFE systems in which the extraction efficiency of the supercritical fluid is improved by a continuous stirring process and by the influence of the carbon dioxide plasma field (used as a supercritical fluid) combined with the natural membranes, which are charged with ion exchangers.
- The rheological characteristics of the mixture containing the supercritical fluid and the bio compounds extracted from the material are extremely important in the efficiency formulas and in choosing the technology.
- The most valuable results of the technological development stages were recorded at lower pressures (200–450 bar) and temperatures of 30–32 °C.

Acknowledgments. This work was supported by a grant of the Romanian National Authority for Scientific Research and Innovation, CNCS/CCCDI – UEFISCDI, project number PN-III-P1-1.2-PCCDI-2017- 0566 (contr. nr. 9 PCCDI/2018/P4 - ***Complex system of full capitalization of some agricultural species with energy and food potential***). *"Valuing the Food Potential of Selected Agricultural Crops (artichoke tubers, sorghum seeds, hemp pomace - resulting from cold pressed hemp oil)".*

References

1. http://www.europarl.europa.eu/sides/getDoc.do?pubRef=-//EP//TEXT+REPORT+A7-2011-0430+0+DOC+XML+V0//EN
2. www.eca.europa.eu/ECADocuments/SR16_34/SR_FOOD_WASTE_RO.pdf
3. Savescu, P., Preda, M.: Research on the variation of the redox potential in to soy milk obtained from deionized water. Revista de Chimie (Rev. Chim.) **56**(8), 853–857 (2005). ISSN 0034-752
4. Savescu, P., Preda, M.: Study on the variation of redox potential of soy milk obtained from ionic modified waters. Revista de Chimie (Rev. Chim.) 56(9), 895–900 (2005). ISSN 0034-752
5. Tudoran, C.D.: High frequency portable plasma generator unit for surface treatment experiments. Rom. J. Phys. **56**, 103, (2011)
6. Savescu, P.: Habilitation Thesis. The use of new innocuous resources, designing and development of new functional food by improving redox specific food industry, The University USAMVBT of Timisoara, Romania (2018)
7. Badescu, G., Badescu, R., Stefan, O., et al.: Using GPS-GNSS global position systems in agriculture. In: International Conference on Agricultural and Biosystems Engineering. Advances in Biomedical Engineering, Hong Kong, pp. 449–452 (2011)
8. Apostol, L., Belc, N., Gaceu, L., Vladut, V., Oprea, O.B.: Chemical composition and rheological parameters of helianthus tuberosus flour used as a sources of bioactive compounds in bakery. Revista de Chimie (Rev. Chim.) **70**(6), 2048–2053 (2019)
9. Savescu, P.: Recovery the food potential of products, by-products and residues of hemp seeds and artichoke - a source of valuable food supplements. In: The 9th Central and East European Congress on Food Annals, Sibiu, Romania, p. 22 (2018). ISBN 978-606-12-1546
10. Popa, L.D., Buburuz, A.A., Isticioaia, S.F., Leonte, A., Naie, M., Teliban, G.C.: The impact of "Secuieni method" on the seed, stems and fiber yields of two new genotypes of monoecious hemp, under A.R.D.S. Secuieni conditions, *AN. I.N.C.D.A. Fundulea,* vol. LXXXV, pg. 153–162 (2017)
11. Popa L.D., Buburuz A.A., Isticioaia S.F.: The behavior of a monoecious hemp hybrid from the second generation (HUF2), in different technological conditions. Homage volume at 55 Research years at S.C.D.A. Secuieni. Ed. Ion Ionescu de la Brad - Iaşi, pg. 195 – 208, (2017). ISBN 978-973-147-253-9
12. Savescu, P.: Study on the influence of oil extraction from Cannabis Sativa seeds on the activity of natural hemp bio components. In: 19th International Multidisciplinary Scientific GeoConference SGEM 2019, Section 25: Section Advances in Biotechnology, SGEM Proceedings, no. 6.1, pp. 933–938 (2019). https://doi.org/10.5593/sgem2019/6.1/s25.120. ISSN 1314-2704

13. Savescu, P., Badescu, G., Milut, M., Ciobanu, A., Apostol, L., Vlădut, V.: Healthy food – through innovative technologies. In: Volume of Symposium ISB-INMA Tech, Bucuresti, pp. 509–514 (2019). ISSN 2344 – 4118
14. Ackerley, N., Sertkaya, A., Lange, R.: Food transportation safety: characterizing risks and controls by use of expert. Food Prot. Trends **30**, 212–222 (2010)
15. Angela, M., Meireles, A.: Extracting Bioactive Compounds for Food Products-Theory and Applications. By Technology and Engineering (2008)

Bioartificial Sponges for Auricular Cartilage Engineering

Marta Feula[1], Mario Milazzo[2], Giulia Giannone[3], Bahareh Azimi[4], Luisa Trombi[4], Ludovica Cacopardo[1], Stefania Moscato[5], Andrea Lazzeri[3], Arti Ahluwalia[1], Stefano Berrettini[6], Carlos Mota[7], and Serena Danti[1,2,3(✉)]

[1] Research Center "E. Piaggio", University of Pisa, 56122 Pisa, Italy
serena.danti@unipi.it

[2] The BioRobotics Institute, Scuola Superiore Sant'Anna, 56025 Pontedera, PI, Italy

[3] Department of Civil and Industrial Engineering (DICI), University of Pisa, 56122 Pisa, Italy

[4] Research Unit of DICI-Pisa, Interuniversity Consortium for Materials Science and Technology (INSTM), 50121 Florence, Italy

[5] Department of Clinical and Experimental Medicine, University of Pisa, 56126 Pisa, Italy

[6] Department of Surgical, Medical, Molecular Pathology and Emergency Medicine, University of Pisa, 56126 Pisa, Italy

[7] Institute for Technology Inspired Regenerative Medicine (MERLN), Complex Tissue Regeneration Department, Maastricht University, 6229ER Maastricht, The Netherlands

Abstract. Auricle reconstruction due to congenital, post-infective or post-traumatic defects represents a challenging procedure in the field of aesthetic and reconstructive surgery due to the highly complex three-dimensional anatomy of the outer ear. Tissue engineering aims to provide alternatives to overcome the shortcomings of standard surgical reconstructive procedure. In the present study, poly(vinyl alcohol)/gelatin (PVA/G) sponges at different weight ratios were produced via emulsion and freeze-drying, and crosslinked by exposure to glutaraldehyde vapors. PVA/G sponges gave rise to highly porous, water stable and hydrophilic scaffolds. Characterization of PVA/G sponges showed round-shaped interconnected pores, high swelling capacity (>200%) and viscoelastic mechanical behavior. The PVA/G 70/30 (w/w) scaffold was selected for *in vitro* biological studies. Bone marrow derived human mesenchymal stromal cells (hMSCs) were used and differentiated towards chondrogenic lineage under different culture conditions: *1)* commercial versus handmade differentiation medium; *2)* undifferentiated versus pre-differentiated hMSC seeding; and *3)* static versus dynamic culture [i.e. ultrasound (US) or bioreactor stimulation]. Histological results highlighted intense glycosaminoglycan, glycoprotein and collagen syntheses after three weeks, mostly using the commercial medium, whereas round morphology was observed in pre-differentiated cells. In static culture, immunohistochemistry for chondrogenic markers revealed an early differentiation stage, characterized by the expression of Sox-9 and collagen type I fibers. The application of US on cell/scaffold constructs increased extracellular matrix deposition and resulted in

S. Piotto et al. (Eds.): BIONAM 2019, LNBE, pp. 191–209, 2020.
https://doi.org/10.1007/978-3-030-47705-9_17

30% higher collagen type II expression at the gene level. Bioreactor culture induced collagen type II, aggrecan and elastin formation. This study demonstrated that 70/30 PVA/G sponge is a suitable candidate for auricle reconstruction.

Keywords: Auricle · Tissue engineering · Mesenchymal Stromal Cells · Poly (vinyl alcohol) (PVA) · Emulsion

1 Introduction

Auricle malformation or absence due to congenital or acquired defects represents a relevant clinical problem, since it also negatively affects patient's psychological well-being. Surgical auricle reconstruction represents one of the most challenging procedure in the field of aesthetic and reconstructive surgery due to the highly complex three-dimensional anatomy of the external ear [1, 2]. Current surgical treatments include the use of rib cartilage autograft and alloplastic biomaterials for auricle reconstruction [1, 3, 4]. The application of a prosthetic auricle has been investigated as an alternative to surgical reconstruction, since it provides good aesthetic results minimizing risks associated with the surgical procedure [5]. The main shortcomings associated to the aforementioned treatments are related to surgical complications and foreign body reaction; therefore, scientists are investigating tissue engineering techniques to regenerate auricle cartilage in vitro [3, 4, 6]. In 1997, Cao *et al.* generated the first ear-shaped cartilage construct in an animal model [7], thus disclosing tissue engineering potential in auricle reconstruction. Since then, many researchers have focused their attention on the possibility of engineering the human auricle. The first successful results were obtained by employing a multistep approach: in the first step, chondrocytes were seeded on a polymeric ear-shaped scaffold; then, cell/scaffold construct was implanted in an animal at subcutaneous level; successively, the regenerated cartilage was harvested, manipulated and re-implanted in patients [1, 8]. Recently one of the most investigated option is the combination of computer tomography (CT) scan and additive manufacturing technologies to fabricate patient-specific ear-shaped scaffolds in order to regenerate cartilage constructs *in vitro* [3, 9].

The auricle, the external auditory meatus and the Eustachian tube consist of elastic cartilage [6, 10]. The properties of auricular cartilage depend on characteristics and spatial distribution of its main structural components: collagen fibers, elastic fibers, glycosaminoglycans (GAGs), chondrocytes and water [11–13]. Chondrocytes are round-shaped and can be observed as single cells or isogenic groups composed by two or three cells surrounded by abundant extracellular matrix (ECM) [10, 14]. In the ear, collagen fibers mainly consist of collagen type II, which are responsible for tensile strength, while elasticity and flexibility of the tissue are due to the presence of elastic fibers, peculiar to elastic cartilage [15, 16]. On the other hand, GAGs possess negatively charged sulfate groups able to attract water molecules and bind proteoglycans to collagen fibrils, thus providing unique biomechanical properties and a viscoelastic component [17]. Cartilage predominant proteoglycan is aggrecan, which provides cartilage elasticity and viscosity retention [18]. In addition, the elastic cartilage ECM contains smaller proteoglycans such as biglycan, decorin and fibromodulin, glycoproteins as fibronectin and other non-collagenous proteins [13].

Auricular cartilage, as any other biological tissue, shows viscoelastic behavior. As such, elastin biomechanically behaves as a linear elastic solid and can be elastically deformed up to 160% strain values [19]. The stress-strain curve obtained by mechanical testing of elastin samples is essentially linear, even though there is a slight difference between loading and unloading curve due to a phenomenon of energy dissipation within the material [19]. Elastin elasticity is a direct consequence of the entropic recoil of elastin molecules [20]. Collagen is a basilar structural protein able to provide mechanical integrity and strength to soft tissues, thus playing a key role in determining tissue biomechanical properties [19, 21].

Data dealing with auricular cartilage mechanical properties available in literature are limited. Griffin *et al.* tested samples of human auricular cartilage harvested from cadavers (average age 69 $\pm$ 10 years) in order to explore the biomechanical properties of this tissue [4]. Compressive test reported an overall elastic modulus of 1.66 $\pm$ 0.63 MPa without significant differences among the different anatomical regions in which human auricle can be divided [4]. Elastic cartilage does not undergo calcification; however, its biomechanical properties change as a consequence of ageing [10, 14]. Individual's age advancing is related to significant changes in matrix thickness and composition: the amount of elastin decreases, elastic fibers are thinner and more fragmented, chondrocytes number and size drop [22–24]. These changes progressively compromise cartilage mechanical integrity [23, 25].

Studies conducted in the nineties have demonstrated that both bovine chondrocytes and human chondrocytes are able to replicate and generate cartilage matrix when seeded on synthetic scaffolds [25, 26]. However chondrocytes isolated from articular, costal or septal cartilage are not able to synthesize elastin, thus regenerated tissues show fibrocartilaginous nature and present lower biochemical and mechanical properties than native elastic cartilage [3, 27]. Nevertheless, only a small amount of tissue can be harvested from the patient, thus an extensive *in vitro* expansion of chondrocytes is required in order to obtain a sufficient number of cells to produce cartilage engineered constructs of clinically relevant size [27]. Furthermore, chondrocyte proliferative potential is physiologically low and their capability to produce cartilage matrix decline over time [27].

Therefore, to allow the regeneration of such specific ECM, it is fundamental to provide a suitable scaffold in which chondrogenesis and elastic cartilage formation can take places. It is well known that once chondrocytes are cultured on tissue culture plastic, namely, in a bidimensional (2D) setting, they lose their morphotype, from round to spindle-like, and finally de-differentiate into fibroblasts, which in turn produce different ECM molecules, including a prevalence of collagen type I [27]. Differently, a tridimensional (3D) setting is considered the best option to obtain a cartilaginous tissue *in vitro*.

Previous studies have already proved that bone marrow derived mesenchymal stromal cells (MSCs) are able to differentiate into chondrocytes and express chondrogenic markers, such as aggrecan and collagen type II when adequately stimulated [28, 29]. The employment of MSCs for *in vitro* cartilage engineering presents several benefits as a low number of cells is initially required and their isolation from bone marrow is relatively easy and already adopted in clinical procedures [6]. However, prolonged *in vitro* cultivation is necessary in order to obtain noticeable chondrogenic differentiation. MSCs are not involved in the physiological development of auricular cartilage and so far there is no evidence that elastic fibers can be obtained [27, 30]. To induce the regeneration of auricular cartilage, suitable chemical and physical signals should be provided. Chemical signals can be introduced in the scaffold by blending synthetic with biological polymers [31–35], and MSCs can be activated towards chondrogenic cascade by using proper supplements in the culture medium [3, 6, 29, 35, 36]. Dynamic stimuli, such as low intensity ultrasound (US) stimulation, have been reported to promote chondrogenic marker expression and can be applied to the cell/scaffold constructs during the culture [18, 35, 37–39].

The aim of this study is the fabrication and characterization of a bioartificial scaffold for auricular cartilage engineering. Specifically, poly(vinyl alcohol) (PVA), a biocompatible synthetic polymer was blended with gelatin (G) and additionally combined Alginate (Alg) to produce hydrophilic sponges able to emulate native ECM and facilitate cell adhesion. Human MSCs (hMSCs) from the bone marrow were used as a cell source since they can be easily isolated from bone marrow and are able to extensively replicate *in vitro* and differentiate in chondrocytes. As chondrogenic differentiation can be enhanced by the application of dynamic stimulation, low frequency US was applied, and a pilot study in a bioreactor was also performed. Specific objectives of the study are the identification of the optimal scaffold composition and the assessment of the optimal hMSC differentiation protocol.

2 Materials and Methods

2.1 Sponge Fabrication

PVA-based scaffolds were produced via emulsion and freeze-drying, as detailly described by Ricci & Danti [40]. A 11.7% aqueous solution of PVA (Mw = 89000–98000 g/mol, >99% hydrolyzed from Sigma-Aldrich, St. Louis, MO, USA) in Milli-Q® deionized water (to avoid Ca^{2+} ions contamination, thus unwanted sodium Alginate crosslinking during emulsification) was autoclaved 1 h at 120 °C and then was cooled down to 50 °C inside a thermostatic bath under stirring at 1000 rpm. Gelatin (gelatin from bovine skin, type B, from Sigma-Aldrich) and Alginic acid sodium salt (Alg; Fluka BioChemika) were added in order to obtain different polymer/biomolecule composition: PVA/G 90/10, 80/20, 70/30 and PVA/G/Alg 80/10/10 and 90/5/5 (w/w)%. Then, 0.18 g of sodium dodecyl sulfate (SDS; from Sigma-Aldrich) were added to obtain a dense foam. After 10 min under stirring, the foam was poured into a six-well plate, quenched in liquid nitrogen and lyophilized. Sponges were crosslinked by exposure to glutaraldehyde (GTA; grade II, from Sigma-Aldrich) vapors for 72 h at 37 °C in a sealed cabinet and then flushed under the chemical hood for 72 h.

2.2 Scaffold Characterization

Scanning Electron Microscopy (SEM) Analysis
PVA/G samples 90/10, 80/20 and 70/30 (w/w)% were mounted on aluminum stumps, sputter-coated with gold (Sputter Coater Emitech K550, Quorum Technologies Ltd,

West Sussex, United Kingdom) and examined on a scanning electron microscope (JEOL JSM-5200, JEOL Ltd, Tokyo, Japan). Approximative pore size distribution was evaluated using ImageJ software (version 1.51i, NIH, USA).

Swelling Analysis
Swelling analysis was performed by measuring sample volume before and after swelling in distilled water at different time points. Volume swelling ratio (Q, Eq. 1) was calculated according to the following equation:

$$Q = \frac{\text{Volume of hydrated sample}}{\text{Volume of dry sample}} \times 100 \quad (1)$$

Differential Scanning Calorimetry (DSC) Analysis
Phase transition properties of the samples were determined using a DSC Q200 Differential Scanning Calorimeter controlled by a TA module (TA Instruments, New Castle, USA). 7–8 mg of dry samples were placed in hermetically sealed aluminum pans and heated from −35 °C to +250 °C at a heating rate of 10 °C/min in an inert atmosphere. An empty pan was used as the reference cell. Transition temperatures were calculated using the Universal Analysis 2000 software (TA Instruments).

Fourier Transformed Infrared Spectroscopy (FTIR) Analysis
FTIR spectra were recorded using Nicolet 380 FT-IR Spectrometer (ThermoFisher Scientific, USA) equipped with the Thermo Scientific™ Smart™ iTX accessory. The spectra were recorded between 4000 cm^{-1} and 550 cm^{-1} with a 8 cm^{-1} spectral resolution. For each spectrum 256 scans were co-added. The FTIR spectrum was taken in a transmittance mode. Data were analyzed using EZ OMNIC Software (ThermoFisher Scientific, USA).

Gelatin Release Evaluation
Gelatin release from crosslinked and non-crosslinked PVA/G sponges was evaluated by means of a spectrophotometric method. Briefly, 5 mg of PVA/G sponges at different (w/w)% composition were immersed in 1 ml of phosphate buffer saline (PBS) and incubated at 37 °C for 96 h. Then, released solution was collected from each sample, diluted in PBS and 200 μL of Bradford reagent (Bio-Rad, USA) were added. The absorbance of the solution at 595 nm was measured using a photometer (BioPhotometer plus, Eppendorf, Germany). Gelatin concentration in the release solution was determined through comparison with a calibration curve.

Mechanical Testing
Viscoelastic properties of PVA/G sponges at different (w/w)% ratios were investigated by using epsilon dot method [41, 42]. Mechanical tests were performed in triplicate on hydrated samples. Short compressive tests were performed at different strain rates (0.005, 0.001, 0.0005 s^{-1}) using the twin column ProLine Z005 testing machine (Zwick Roell) equipped with a 10 N load cell (Zwick KAP-TC). Force and displacement data were recorded over time (sampling rate 1 kHz). Firstly, data were processed to obtain stress-time series. Apparent elastic moduli were calculated as the slope of stress-strain curves and linear viscoelastic region (LVR) was identified. A standard linear solid (SLS) model was chosen to derive PVA/G viscoelastic parameters.

Viscoelastic parameters, i.e. the instantaneous (E_{inst}) and equilibrium elastic moduli (E_{eq}) and the characteristic relaxation time (τ), were estimated by globally fitting stress-time series in a combined parameter space using OriginLab (Northampton) fitting toolbox [41, 42].

2.3 Biological Study Design

PVA/G 70/30 (w/w)% sponges were cut in cylindrical scaffolds (5 mm in diameter and 1.5 mm in thickness) using a biopsy puncher and a microtome blade. Scaffolds were sterilized in absolute ethanol (Bio-Optica, Milan, Italy) for 24 h and then treated with 2% glycine (Sigma-Aldrich) for 1 h in order to block unreacted sites of GTA. Therefore, scaffolds were washed three times in PBS supplemented with antibiotics.

Study Design

Human mesenchymal stromal cells (hMSCs) (Merk, Germany) were defrosted and expanded according to the manufacturer's recommendations. HMSCs at passage 2 were trypsinized, seeded on PVA/G 70/30 scaffolds (puncher cut into 5 mm in diameter and 1.5 mm in thickness) at a density of 500,000 cell/scaffold and differentiated for 21 days in standard culture conditions, namely, 95% air, 5% CO_2 in humidified incubator at 37 °C. At the endpoint samples were processed for metabolic activity tests, SEM and histological analyses. If not otherwise specified, all reagents were purchased from Sigma–Aldrich.

The following culture conditions were applied: *1)* commercially available chondrogenic differentiative medium *versus* homemade culture medium; *2)* seeding undifferentiated *versus* pre-differentiated hMSC; *3)* static *versus* dynamic cultures. Specifically:

1) Cells were seeded on PVA/G 70/30. After 1 day the culture medium was replaced with chondrogenic differentiative media. Two chondrogenic differentiation media were used:
1a) StemMACS ChondroDiff medium was purchased from Miltenyi Biotec (Bergisch Gladbach, Germany).
1b) Homemade chondrogenic medium was obtained as reported by Barachini *et al.* [43]. Briefly, homemade chondrogenic medium consisted of DMEM/F12, 1.25 μg/ml bovine serum albumin, 5.35 μg/ml linoleic acid, 50 μg/ml ascorbic acid, 100 μg/ml sodium pyruvate, insulin–transferrin–selenium (ITS premix), 10^{-7} M dexamethasone and 10 ng/ml transforming growth factor beta 1 (TGF-β1; PeproTech, Rocky Hill, NJ).
2) Two pre-differentiation strategies were considered:
2a) Pre-differentiation in pellet conditions as described by Barachini *et al.* [43]. Briefly, 250,000 hMSCs/tube, were centrifuged at 1200 rpm for 7 min to obtain chondrogenic pellets, which were washed in sterile saline and cultured in chondrogenic medium StemMACS Chondrodiff medium (Miltenyi). Pellets were cultured 1 week in Milteniy ChondroDiff medium, then trypsinized, seeded on PVA/G 70/30 scaffolds and differentiated for more 14 days.
2b) Pre-differentiation in 2D conditions on tissue culture plastic flasks for 4 days, as reported by Bajpai *et al.*, by adding StemMACS ChondroDiff (Miltenyi) medium

[44]. After the pre-differentiation in 2D conditions, the cells were trypsinized, seeded on PVA/G 70/30 scaffolds and differentiated for ~~more~~ further 17 days.

3) Dynamic stimulations:

3a) Cell/scaffold constructs were daily treated 3 times for 5 s with US at 40 kHz frequency and 20 W power in a sonicator bath (Bransonic 2510; Bransonic, Danbury, USA) as reported by Barachini *et al.* [43]. After 1 day the culture medium was replaced with chondrogenic differentiative media: Milteniy ChondroDiff medium was added to a set of samples (n = 3) as in *1a)* and the home-made chondrogenic medium was added to the other set (n = 3) as in *1b)*. Cells were differentiated for 21 days.
In addition, hMSCs, pre-differentiated in 2D conditions as in *2b)*, were seeded on PVA/G 70/30 scaffolds (n = 8) and differentiated for 21 days. A set of samples (n = 4) was daily treated with US three times for 10 s each time, while the other set (n = 4) served as negative control.

3b) In a pilot study, pre-differentiasted hMSCs as in *2b)*, were trypsinized and seeded on a PVA/G 70/30 (w/w)% scaffold (25 mm in diameter and 1 mm in thickness to fit the bioreactor holder) at a density of 800,000 cells/scaffold and then cultured in StemMACS Chondrodiff medium (Miltenyi) into a patented bioreactor (WO/2015/040554) which moved the construct at 0.23 Hz for 15 min every 30 min rest intervals for 14 days.

2.4 Biological Analysis

Cell metabolic activity
AlamarBlue® assay was performed once a week in order to assess cell viability. Cell/scaffold constructs to be tested were placed in 15 ml Falcon tubes and alamarBlue® reagent (Life Technologies, USA) was added, according to manufacturer's prescriptions. AlamarBlue solution without cells served as negative controls. After 3 h of incubation, 100 μl samples were collected in triplicate and analyzed with a plate reader (Victor3, PerkinElmer, Waltham, USA). Absorbance values at 570 nm and 600 nm were recorded and used to calculate dye reduction percentage. At the end of the test, fresh culture medium was added.

Histological Analysis
Cell/scaffold constructs were fixed in 4% w/v neutral buffered formalin diluted in 1× PBS (0.1 M, pH 7.2) (Bio-Optica) overnight at 4 °C. Thereafter samples were washed in PBS, dehydrated with a graded series of ethanol, clarified in xylene and finally paraffin-embedded. Finally, 8 μm-thick sections were obtained by standard microtomy and mounted onto glass slides and stained with hematoxylin-eosin (H&E), Alcian Blue at pH 1 and pH 2.5, Toluidine Blue, orcein-Van Gieson; periodic acid-Schiff (PAS) reaction was also performed.

Immunohistochemical analysis was carried out on cell/scaffold constructs, by following a standard protocol described in details by Barachini *et al.* [43]. The following antibodies with specified concentrations were used: mouse monoclonal anti-human elastin 1:50 (AB9519, AbCam, Cambridge, United Kingdom); rabbit polyclonal anti-human type I collagen 1:1000 (AB34710, AbCam); mouse monoclonal anti-human aggrecan 1:50 (SC33695, Santa Cruz Biotechnology, USA); rabbit polyclonal anti-human Sox9 1:100 (SC20095, Santa Cruz Biotechnology); mouse monoclonal anti-human type II collagen 1:50 (SC52658, Santa Cruz Biotechnology). All the histological analysis were observed with a Nikon Eclipse Ci microscope (Nikon Instruments, Amsterdam, The Netherlands) and the images were acquired by a digital camera equipped on the microscope.

Gene Expression
Real time polymerase chain reaction (RT-PCR) was performed in order to investigate type II collagen gene expression on US-irradiated and non-US irradiated cell/scaffold constructs. Total RNA was isolated with Tri reagent® (Sigma-Aldrich) and 500 ng of this were reverse-transcribed into complementary DNA (cDNA) using M-MLV-Reverse transcriptase (Promega), at 37 °C for 45 min, according to the manufacturer's instructions. RT-PCR for quantification of expression levels human Collagen-2 was carried out with the LC Fast Start DNA Master SYBR Green kit (Roche) using 2 μl of cDNA, corresponding to 500 pg of total RNA in a 20 μl final volume, 3 mM $MgCl_2$ and 0.5 μM sense primer (5'-CAACACTGCCAACGTCCAGAT-3') and antisense primer (5'- CTGCTTCGTCCAGATAGGCAA-3'), using β-actin as housekeeping gene (5'-GACGACGACAAGATAGCCTAGCAGCTATGAGGATC-3' and 5'- GAG GAGAAGCCCGGTTAACTTCCGCAGCATTTTGCGCCA3'). Melting curves were generated after each run to confirm the amplification of specific transcripts. Reactions were carried out in duplicate, and the relative expression of a specific mRNA was determined by calculating the fold change relative to the β-actin control. The fold change of the tested gene mRNA was obtained with a LightCycler® software, by using the amplification efficiency of each primer, as calculated by the dilution curve. The specificity of the amplification products was verified by subjecting the amplification products to electrophoresis on 1.5% agarose gel and visualization by ethidium bromide staining (Sigma Aldrich).

2.5 Statistical Analysis

One-way analysis of variance with post-hoc Tukey's test was performed using OriginLab (Northampton). Statistical significance was assumed at a value of $p < 0.05$. Data are reported as mean ± standard deviation.

3 Results and Discussion

3.1 Sponge Characterization

Morphological Characterization
The mixture of a biocompatible and hydrophilic synthetic polymer, such as PVA, and natural polymers, such as protein and polysaccharide components (i.e., bioartificial material) was considered in order to mimic a preliminary ECM. The PVA/G/Alg foam collapsed after freeze-drying. Similar results in terms of structure collapse and shrinkage were observed by Gurikov *et al.* [45]. Vice versa, PVA/G sponges gave rise

to highly porous, water stable and hydrophilic scaffolds, which were selected for further characterization (Fig. 1) [33, 40, 43].

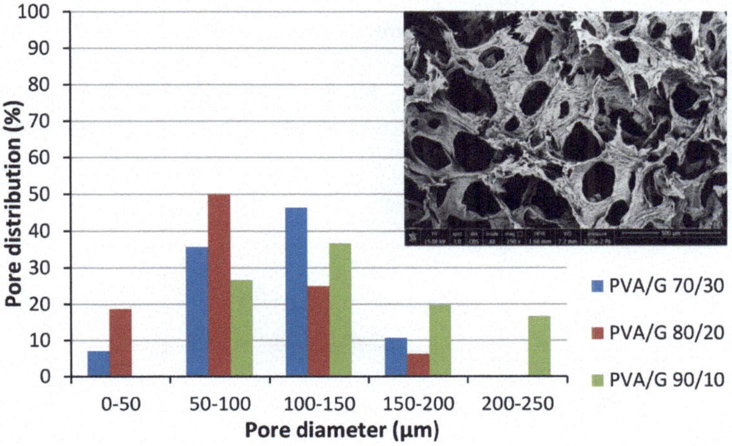

Fig. 1. Pore diameter distribution of PVA/G sponges at different (w/w)%. Lens shows a typical section of PVA/G sponge with round and interconnected pores.

The morphology of the PVA/G sponges was analyzed via SEM to assess pore interconnectivity. SEM analysis confirmed the porous nature of the sponges, highlighting round shaped interconnected pores with diameters ranging in 50–250 µm (Fig. 1), suitable to cell infiltration. Pore size distribution results, acquired via ImageJ, were in line with those observed by De la Ossa *et al.*, evaluated with mercury intrusion porosimetry [46].

Physico-Chemical Characterization

DSC thermograms showed the presence of two endothermal events. The first endothermal event, between 83 °C and 102 °C, was due to water evaporation. As reported by Chiellini *et al.*, the highly energetic water evaporation process may interfere with the detection of G endothermic relaxation, which typically occurs at about 91 °C in dried gelatin samples [47, 48]. Differences between crosslinked and non-crosslinked samples at these temperatures demonstrated that the latter sponge had a greater capability to interact with water. The second endothermal event was represented by a sharp peak at 226–230 °C, associated to the melting of PVA crystals. PVA crystals melting temperature observed in these blends was significantly higher than the melting temperature of about 191 °C found by Chiellini *et al.* and Alves *et al.* for pure PVA films [47, 49]. The increasing of melting temperature in PVA/G blends can be explained as a consequence of the intense interaction between G molecules and PVA, thus proving a good compatibility between the two components at different (w/w)% ratios [50]. The FTIR spectra obtained presented the typical peaks of PVA/G blends reported in previous works on the characterization of polymeric films for biomedical applications [49–51]. In general, the complexity of the spectrum within material fingerprint region increased by increasing G content. A characteristic absorption peak at 1450 cm^{-1} demonstrated the formation of CH = N groups due to the reaction of GTA with $-NH_2$ groups of lysine residues [46]. The difference between crosslinked and non-crosslinked spectra became more pronounced by increasing G content, as previously reported by De la Ossa *et al.*, on PVA/G sponges [46]. There was no appreciable G release from all the crosslinked samples, thus proving sponge water stability at physiological temperature after chemical cross-linking with GTA. Swelling analysis is reported in Fig. 2.

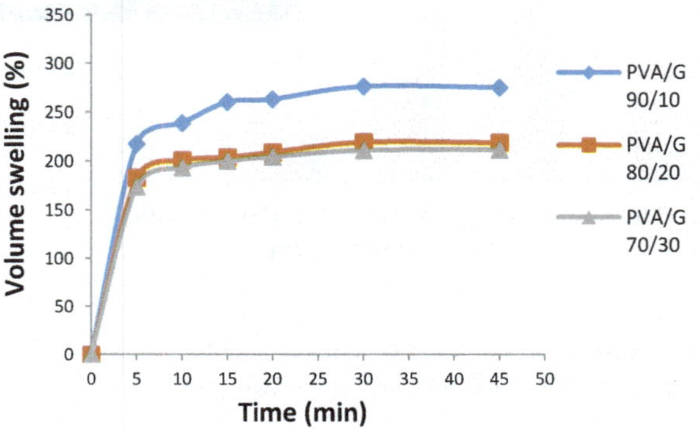

Fig. 2. Volume swelling ratio of PVA/G sponges at different G concentrations (90/10, 80/20, 70/30 w/w%).

It demonstrated the good capability of crosslinked sponges to interact with water, by acquiring a highly hydrated configuration. Samples quickly reached swelling equilibrium, presenting volume swelling ratio higher than 200%, thanks to the hydrophilic nature of PVA and G (Fig. 2). Volume swelling ratio decreased with increasing G content as a consequence of the dense network of crosslinks established after chemical crosslinking with GTA.

Mechanical Characterization

Experimental stress-time series within the LVR (2% strain) were used to estimate viscoelastic parameters, according to epsilon dot method. The sponges became less stiff (E_{eq} decreased) and more elastic and less liquid-like (τ increased) as the G content decreased (Table 1).

Table 1. PVA/G viscoelastic parameters derived from fitting of experimental data.

	PVA/G 70/30	PVA/G 80/20	PVA/G 90/10
E_{inst} [kPa]	6.85 ± 1.52	8.86 ± 1.42	2.12 ± 0.33
E_{eq} [kPa]	5.96 ± 1.53	6.79 ± 1.02	1.61 ± 0.50
τ [s]	1.34 ± 0.32	3.06 ± 3.70	9.29 ± 5.95
R^2	>0.99	>0.99	>0.99

The apparent elastic moduli of PVA/G samples, calculated as the slopes of stress-strain curves, are reported in Table 2. Slight variations with strain rate revealed that hydrated PVA/G sponges exhibited viscoelastic behavior. Moreover, G content $\geq$ 20% resulted in a significant increase in material stiffness. This phenomenon is likely to be a consequence of crosslinking. Indeed, the reaction of GTA aldehyde groups with free lysine or hydroxylysine residues of G polypeptide chains limits polymeric chain mobility by creating a network of crosslinks [48]. Statistical analysis underlined that PVA/G 90/10 presented a significantly lower elastic modulus, thus demonstrating that an increasing in G content and consequently, increasing of crosslinking sites, affected the material stiffness.

Table 2. Apparent compressive elastic modulus [kPa] of PVA/G sponges at different strain rates [s^{-1}].

Strain rate	PVA/G 70/30	PVA/G 80/20	PVA/G 90/10
0.005	6.61 ± 1.57	7.27 ± 0.23	2.03 ± 0.35
0.001	6.11 ± 1.60	7.25 ± 0.07	1.81 ± 0.49
0.0005	5.95 ± 1.46	7.09 ± 0.08	1.73 ± 0.53

Although the elastic moduli of PVA/G sponges was about three orders of magnitude lower than compressive elastic modulus of human auricular cartilage reported in literature [4, 23], an increase in construct stiffness is expected as a consequence of ECM deposition and in particular collagen synthesis within scaffold pores. The mechanical properties of cell/scaffold constructs after differentiation time were not tested because the small dimensions of the engineered constructs was not suitable for mechanical testing with the equipment available. To address this issue, further studies may include nanoindentation tests using the nano-epsilon dot method [52], which allows the investigation of viscoelastic parameters before and after cell culture. In addition, new devices which enable monitoring of the mechanical properties of tissue constructs during cell culture such as the Mechano Culture Testing System [53] could also be considered.

3.2 Characterization of Cell/Scaffold Constructs

The aforementioned characterization highlighted significant differences between PVA/G sponges in terms of pore size distribution, swelling behavior, physico-chemical and mechanical properties. PVA/G 70/30 and PVA/G 80/20 (w/w)% had more similar characteristics, namely, pore size distribution suitable for cell colonization, good swelling behavior and higher stiffness than PVA/G 90/10 (w/w)%. Finally, PVA/G 70/30 was selected as the best scaffold candidate, on the basis of stiffness, highest G content, which facilitates cell adhesion, and suitable poral features for cell colonization, as previously studied by De la Ossa *et al.* [46].

Chondrogenic Differentiation of hMSC/Scaffold Constructs
AlamarBlue assay, performed at different time points during chondrogenic differentiation, demonstrated that hMSCs cultured on PVA/G 70/30 scaffolds were viable up to three weeks in all tested samples (Fig. 3).

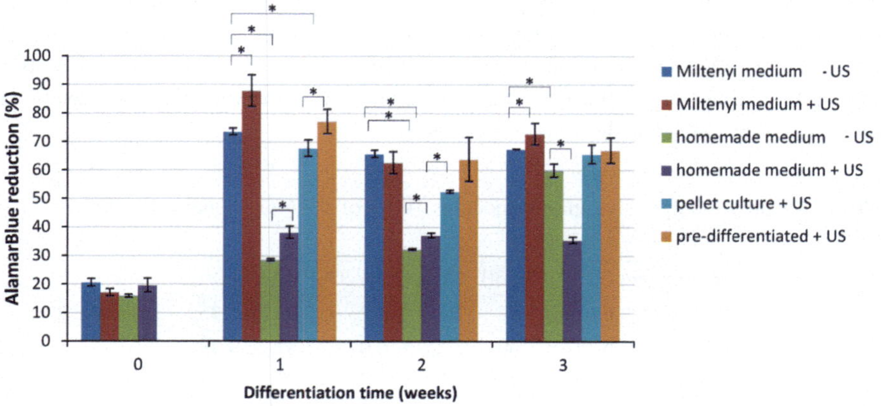

Fig. 3. Viability of chondroinduced hMSCs over differentiation time (0 week refers 12 h after seeding) in different culture conditions.

Miltenyi ChondroDiff medium was the best-performing differentiative medium. Indeed, significant lower cell metabolic activity was observed in constructs cultured in the homemade chondrogenic medium. Moreover, statistically significant differences between US-stimulated samples and negative controls demonstrated that US stimulation treatment affected cell metabolism. These results agree with the outcomes reported by Barachini *et al.* in their work on chondrogenic differentiation of human dental pulp stem cells on PVA/G 80/20 [43].

Histochemistry and Immunohistochemistry
In agreement with the metabolic activity results (Fig. 3), the scaffolds hosted varying amounts of cells, as observed via H&E staining. High positivity for Sox9 chondrogenic marker, observed in all tested samples, indicated an early differentiation stage of all the constructs (data not shown). A meaningful representation of the results achieved in the different conditions in terms of ECM maturity is reported in Fig. 4. Cell/scaffold constructs cultured in homemade chondrogenic medium displayed low glycoprotein, GAG (Fig. 4 C, D) and collagen (Fig. 4 I, J) expression and a non-uniform cell colonization. Conversely, a good and uniform cell spreading was detected in all samples cultured in Miltenyi StemMACS ChondroDiff medium (Fig. 4 A, B, E, F, G, H, K, L). Intense GAG deposition and roundish morphology was observed in hMSCs pre-differentiated in pellet conditions; however, cell colonization was restricted to a small region of the scaffold since the pelletized cells remained locally confined, thus providing no physiologically relevant results. HMSCs cultured in Miltenyi StemMACS ChondroDiff medium showed very good amount of collagen (Fig. 4 A, B, E, F) and GAGs (Fig. 4 G, H, K, L), as well as glycoproteins (data not shown).

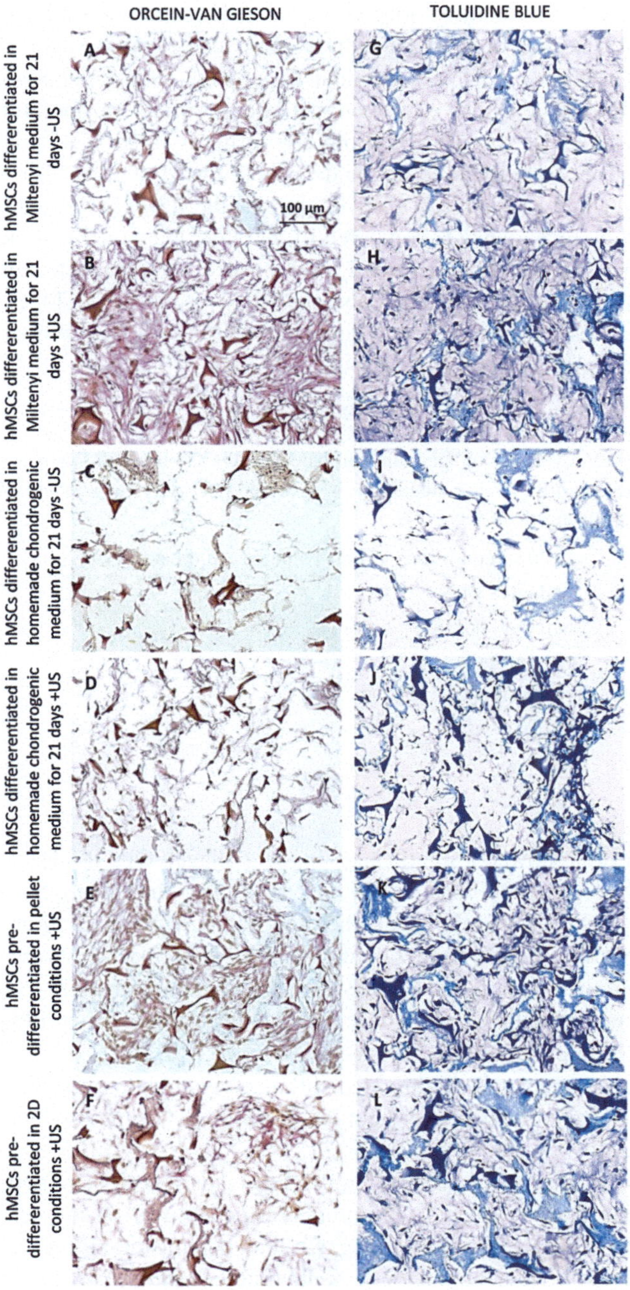

Fig. 4. Chondroinduced hMSCs cultured on PVA/G 70/30 scaffolds. Left column: Orcein-van Gieson staining: Orcein highlights the presence of elastic fibers in dark violet (absent in our samples). Van Gieson stains collagen fibers in pink-red. Right column: Toluidine Blue staining showing sulfated GAGs in purple. Original magnification 200×; scale bar: 100 µm.

In particular, a more intense collagen (Fig. 4 B, D) expression was observed in US-stimulated hMSCs compared to the non-US stimulated counterparts (Fig. 4 A, C); aggrecan was also detected in US-stimulated samples (data not shown), overall suggesting powerful US beneficial effects on matrix deposition, corroborating the findings of previous studies [18, 35, 37, 38]. Both elastic fibers (Fig. 4 A-F) and elastin (immunoistochemical reaction) could not be detected in these samples. The observation of larger and round shaped cells in 2D pre-differentiated hMSCs suggested the occurrence of some morphological changes compared to undifferentiated cells. A downregulation of type I collagen expression indicated an enhanced chondrogenic differentiation in US-stimulated samples, as previously observed by Barachini *et al.* [43]. The positivity levels of the expression of collagen and sulphated GAGs, as also shown in Fig. 4, are summarized in Table 3.

Table 3. Positivity levels of collagen and sulphated GAGs as observed via histochemistry under different culture conditions. − = negative; ∓ = weak positivity; + = positivity; ++ = good positivity; +++ = strong positivity.

Culture conditions	Total collagen	Sulphated GAGs
Miltenyi - US	∓	+
Miltenyi + US	+++	++
Home made − US	−	−
Home made + US	∓	−
Pellet + US	+	∓
2D + US	+	−

RT-PCR results showed a significantly higher collagen type II mRNA expression in US-stimulated constructs (130% ± 7% fold difference) than non-US stimulated constructs (100% ± 5% fold difference), therefore a more powerful chondrogenesis, in line with the results reported in literature [18, 37]. All in all, the constructs showed an intense cell colonization and ECM deposition, with a commitment to the chondrogenic phenotype, which was more pronounced using StemMACS ChondroDiff (Miltenyi) medium and US stimulation for 21 days.

However, these outcomes highlighted still an early chondrogenic phase, with weak production of collagen type II at protein level, good GAG synthesis but low aggrecan expression were observed.

Differently, synthesis of aggrecan, reduced collagen type I, increased collagen type II and even elastin was detected via immunohistochemistry on the cell/scaffold construct cultured within the bioreactor (Fig. 5), thus demonstrating that dynamic stimulation allows an improvement in chondrogenic differentiation of hMSCs with respect to static culture conditions. This bioreactor provides a cyclic membranal-like stress/deformation field that stretched the scaffolds and the cells cultured within. This outcome is suggestive that chondro-differentiated hMSCs can produce elastic fibers in presence of proper stimulation.

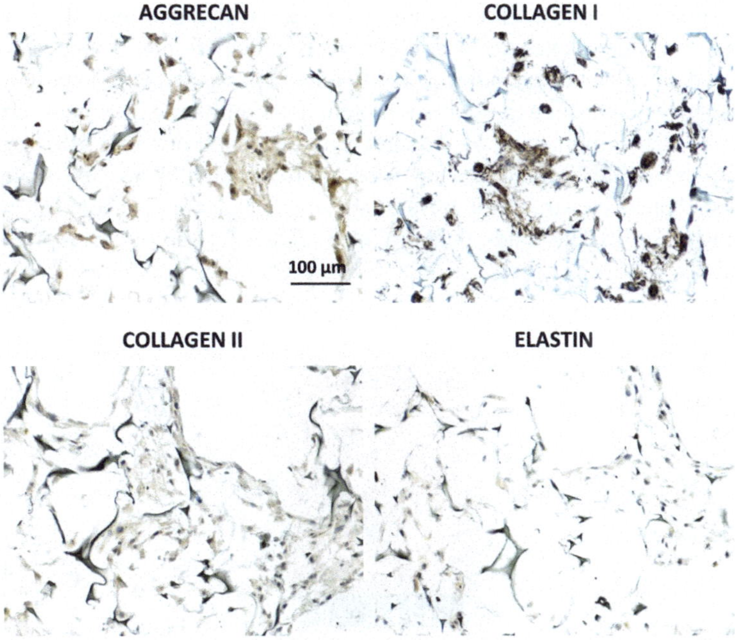

Fig. 5. Immunohistochemistry on chondroinduced hMSCs cultured on a PVA/G 70/30 (w/w)% scaffold in dynamic culture conditions using a bioreactor (WO/2015/040554): aggrecan, collagen type I, collagen type II, elastin. Original magnification: 200×; scale bar: 100 µm.

4 Conclusions

The aim of this work was the fabrication and characterization of bioartificial spongy scaffolds for the regeneration of auricular cartilage *in vitro* using bone marrow hMSCs. Morphological, physico-chemical and biomechanical characterization of PVA/G sponges demonstrated round-shaped interconnected pores, high swelling capacity, water stability and viscoelastic behavior. In particular, PVA/G 70/30 (w/w)% was selected for hMSC culture and chondrogenic lineage commitment, as this scaffold formulation presented higher G content, but similar morphological, physico-chemical and mechanical properties when compared to PVA/G 80/20 (w/w)%. *In vitro* studies were performed to identify the best chondrogenic differentiation conditions. Chondroinduced hMSCs were all over viable up to three weeks, thus proving scaffold cytocompatibility. StemMACS ChondroDiff commercially available medium positively affected cell metabolic activity and ECM deposition, thus appearing as the best differentiative medium. Histochemistry performed on cell/scaffold constructs proved intense sulphated GAG (but low aggrecan), glycoprotein and collagen synthesis after three weeks of differentiation. Immunohistochemistry for chondrogenic markers revealed an early differentiation stage, characterized by collagen type I and Sox-9. Since collagen type II was not detected at protein level, RT-PCR was used to assess its expression at gene level, revealing that mRNA was expressed both in non-US stimulated and US-stimulated constructs, with a 30% increase in the latter. US showed beneficial effects on hMSC chondrogenesis in terms of enhancement of metabolic activity and type II collagen gene expression. Dynamic culture within a bioreactor finally proved an enhanced chondrogenic differentiation, by showing intense aggrecan, collagen type II and elastin expression at protein level. Further experiments are needed to investigate the combination of dynamic culture conditions and US to induce a mature cartilage ECM. Moreover, the used approach should also be scaled up to generate ear-shaped constructs of physiologically relevant dimensions by means of mold casting (Fig. 6). This study thus provided a basis for future development in this area.

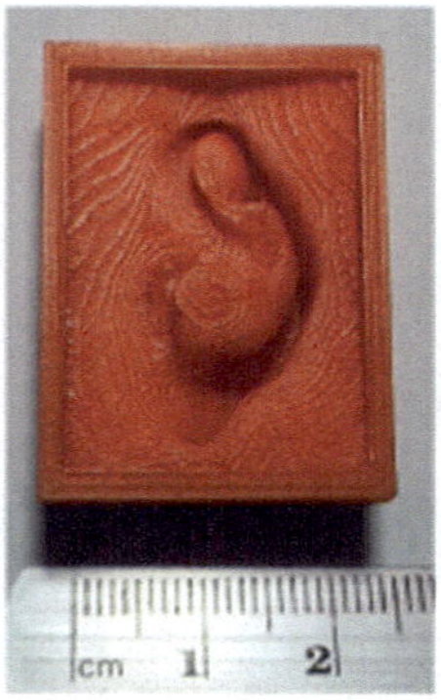

Fig. 6. Example of a 1:2 size left auricle mold (2 cm × 3 cm) as obtained using reverse engineering and a Zortrax M200 3D printer.

Acknowledgements. AURICULAE Project, funded by Stem Cells & Life Foundation, Padova, Italy is greatly acknowledged. Dr. Delfo D'Alessandro (University of Pisa, Pisa, Italy), as well as Dr. Alessandra Fusco and Dr. Giovanna Donnarumma (University of Campania "Luigi Vanvitelli", Naples, Italy) are thanked for their fundamental technical support to this work.

References

1. Kamil, S.H., Vacanti, M.P., Vacanti, C.A., Eavey, R.D.: Microtia chondrocytes as a donor source for tissue-engineered cartilage. Laryngoscope **114**, 2187–2190 (2004)
2. Han, S.E., Lim, S.Y., Pyon, J.K., Bang, S.I., Mun, G.H., Oh, K.S.: Aesthetic auricular reconstruction with autologous rib cartilage grafts in adult microtia patients. J. Plast. Reconstr. Aesthetic Surg. **68**, 1085–1094 (2015)
3. Jessop, Z.M., Javed, M., Otto, I.A., Combellack, E.J., Morgan, S., Breugem, C.C., Archer, C.W., Khan, I.M., Lineaweaver, W.C., Kon, M., et al.: Combining regenerative medicine strategies to provide durable reconstructive options: auricular cartilage tissue engineering. Stem Cell Res. Ther. **7**, 19 (2016)
4. Griffin, M.F., Premakumar, Y., Seifalian, A.M., Szarko, M., Butler, P.E.M.: Biomechanical characterisation of the human auricular cartilages; implications for tissue engineering. Ann. Biomed. Eng. **44**, 3460–3467 (2016)

5. Thorne, C.H., Brecht, L.E., Bradley, J.P., Levine, J.P., Hammerschlag, P., Longaker, M.T.: Auricular reconstruction: Indications for autogenous and prosthetic techniques. Plastic Reconstr. Surg. **107**, 1241–1252 (2001)
6. Ciorba, A., Martini, A.: Tissue engineering and cartilage regeneration for auricular reconstruction. Int. J. Pediatr. Otorhinolaryngol. **70**, 1507–1515 (2006)
7. Cao, Y., Vacanti, J.P., Paige, K.T., Upton, J., Vacanti, C.A.: Transplantation of chondrocytes utilizing a polymer-cell construct to produce tissue-engineered cartilage in the shape of a human ear. Plastic Reconstr. Surg. **100**, 297–302 (1997)
8. Shieh, S.J., Terada, S., Vacanti, J.P.: Tissue engineering auricular reconstruction: in vitro and in vivo studies. Biomaterials **25**, 1545–1557 (2004)
9. Zhou, G., Jiang, H., Yin, Z., Liu, Y., Zhang, Q., Zhang, C., Pan, B., Zhou, J., Zhou, X., Sun, H., et al.: In vitro regeneration of patient-specific ear-shaped cartilage and its first clinical application for auricular reconstruction. EBioMedicine **28**, 287–302 (2018)
10. Schultz, T.W., Geneser, F.: Textbook of Histology. Transactions of the American Microscopical Society (1987)
11. Lai, C.H., Chen, S.C., Chiu, L.H., Yang, C.B., Tsai, Y.H., Zuo, C.S., Chang, W.H.S., Lai, W.F.: Effects of low-intensity pulsed ultrasound, dexamethasone/TGF-β1 and/or BMP-2 on the transcriptional expression of genes in human mesenchymal stem cells: chondrogenic vs. osteogenic differentiation. Ultrasound in Med. Biol. **36**, 1022–1033 (2010)
12. Murakami, W.T., Wong, L.W., Davidson, T.M.: Applications of the biomechanical behavior of cartilage to nasal septoplastic surgery. Laryngoscope **92**, 300–309 (1982)
13. Van Osch, G.J.V.M., Van Den Berg, W.B., Hunziker, E.B., Häuselmann, H.J.: Differential effects of IGF-1 and TGFβ-2 on the assembly of proteoglycans in pericellular and territorial matrix by cultured bovine articular chondrocytes. Osteoarthritis Cartilage **6**, 187–195 (1998)
14. Ackert, J.E., Maximow, A.A., Bloom, W.: A Textbook of Histology. Transactions of the American Microscopical Society (1942)
15. Gosline, J., Lillie, M., Carrington, E., Guerette, P., Ortlepp, C., Savage, K.: Elastic proteins: biological roles and mechanical properties. Philos. Trans. R. Soc. B: Biol. Sci. **357**, 121–132 (2002)
16. Lotz, M., Loeser, R.F.: Effects of aging on articular cartilage homeostasis. Bone **51**, 241–248 (2012)
17. Ross, M.H.P., Pawlina, W.: Histology a Text and Atlas with Correlated Cell and Molecular Biology (2014). ISBN 9780874216561
18. Xia, P., Wang, X., Qu, Y., Lin, Q., Cheng, K., Gao, M., Ren, S., Zhang, T., Li, X.: TGF-β1-induced chondrogenesis of bone marrow mesenchymal stem cells is promoted by low-intensity pulsed ultrasound through the integrin-mTOR signaling pathway. Stem Cell Res. Ther. **8**, 281–292 (2017)
19. Fung, Y.C., Skalak, R.: Biomechanics: Mechanical Properties of Living Tissues. Journal of Applied Mechanics (1982)
20. Urry, D.W., Hugel, T., Seitz, M., Gaub, H.E., Sheiba, L., Dea, J., Xu, J., Parker, T.: Elastin: a representative ideal protein elastomer. Philos. Trans. R. Soc. B: Biol. Sci. **357**, 169–184 (2002)
21. Milazzo, M., Jung, G.S., Danti, S., Buehler, M.J.: Wave propagation and energy dissipation in collagen molecules. ACS Biomater. Sci. Eng. **6**, 1367–1374 (2020)
22. Sherratt, M.J.: Tissue elasticity and the ageing elastic fibre. Age **31**, 305–325 (2009)
23. Nimeskern, L., Utomo, L., Lehtoviita, I., Fessel, G., Snedeker, J.G., van Osch, G.J.V.M., Müller, R., Stok, K.S.: Tissue composition regulates distinct viscoelastic responses in auricular and articular cartilage. J. Biomech. **49**, 344–352 (2016)

24. Riedler, K.L., Shokrani, A., Markarian, A., Fisher, L.M., Pepper, J.P.: Age-related histologic and biochemical changes in auricular and septal cartilage. Laryngoscope **127**, 399–407 (2017)
25. Vacanti, C.A., Vacanti, J.P.: Bone and cartilage reconstruction with tissue engineering approaches. Otolaryngol. Clin. North Am. **27**, 263–276 (1994)
26. Rodriguez, A., Cao, Y.L., Ibarra, C., Pap, S., Vacanti, M., Eavey, R.D., Vacanti, C.A.: Characteristics of cartilage engineered from human pediatric auricular cartilage. Plastic Reconstr. Surg. **103**, 1111–1119 (1999)
27. Otto, I.A., Levato, R., Webb, W.R., Khan, I.M., Breugem, C.C., Malda, J.: Progenitor cells in auricular cartilage demonstrate cartilage-forming capacity in 3D hydrogel culture. Eur. Cells Mater **35**, 132–150 (2018)
28. Ciuffreda, M.C., Malpasso, G., Musarò, P., Turco, V., Gnecchi, M.: Protocols for in vitro differentiation of human mesenchymal stem cells into osteogenic, chondrogenic and adipogenic lineages. In: Mesenchymal Stem Cells, pp. 149–158. Springer (2016)
29. Pittenger, M.F., Mackay, A.M., Beck, S.C., Jaiswal, R.K., Douglas, R., Mosca, J.D., Moorman, M.A., Simonetti, D.W., Craig, S., Marshak, D.R.: Multilineage potential of adult human mesenchymal stem cells. Science **284**, 143–147 (1999)
30. Kusuhara, H., Isogai, N., Enjo, M., Otani, H., Ikada, Y., Jacquet, R., Lowder, E., Landis, W.J.: Tissue engineering a model for the human ear: assessment of size, shape, morphology, and gene expression following seeding of different chondrocytes. Wound Repair Regener. **17**, 136–146 (2009)
31. Milazzo, M., Contessi Negrini, N., Scialla, S., Marelli, B., Farè, S., Danti, S., Buehler, M.J.: Additive manufacturing approaches for hydroxyapatite-reinforced composites. Adv. Funct. Mater. **29**, 1903055 (2019)
32. Cascone, M.G., Lazzeri, L., Sparvoli, E., Scatena, M., Serino, L.P., Danti, S.: Morphological evaluation of bioartificial hydrogels as potential tissue engineering scaffolds. J. Mater. Sci. Mater. Med. **15**, 1309–1313 (2004)
33. Moscato, S., Mattii, L., D'Alessandro, D., Cascone, M.G., Lazzeri, L., Serino, L.P., Dolfi, A., Bernardini, N.: Interaction of human gingival fibroblasts with PVA/gelatine sponges. Micron **39**, 569–579 (2008)
34. Kamoun, E.A., Chen, X., Mohy Eldin, M.S., Kenawy, E.R.S.: Crosslinked poly(vinyl alcohol) hydrogels for wound dressing applications: a review of remarkably blended polymers. Arab. J. Chem. **8**, 1–14 (2015)
35. Lee, H.J., Choi, B.H., Min, B.H., Son, Y.S., Park, S.R.: Low-intensity ultrasound stimulation enhances chondrogenic differentiation in alginate culture of mesenchymal stem cells. Artif. Organs **30**, 707–715 (2006)
36. Bernardo, M.E., Fibbe, W.E.: Mesenchymal stromal cells and hematopoietic stem cell transplantation. Immunol. Lett. **168**, 215–221 (2015)
37. Parvizi, J., Wu, C.C., Lewallen, D.G., Greenleaf, J.F., Bolander, M.E.: Low-intensity ultrasound stimulates proteoglycan synthesis in rat chondrocytes by increasing aggrecan gene expression. J. Orthop. Res. **17**, 488–494 (1999)
38. Jonnalagadda, U.S., Hill, M., Messaoudi, W., Cook, R.B., Oreffo, R.O.C., Glynne-Jones, P., Tare, R.S.: Acoustically modulated biomechanical stimulation for human cartilage tissue engineering. Lab Chip **18**, 473–485 (2018)
39. Aliabouzar, M., Lee, S.J., Zhou, X., Zhang, G.L., Sarkar, K.: Effects of scaffold microstructure and low intensity pulsed ultrasound on chondrogenic differentiation of human mesenchymal stem cells. Biotechnol. Bioeng. **115**, 495–506 (2018)
40. Ricci, C., Danti, S.: 3D models of pancreatic ductal adenocarcinoma via tissue engineering. Methods Mol. Biol. **1882**, 81–95 (2019)

41. Mattei, G., Tirella, A., Gallone, G., Ahluwalia, A.: Viscoelastic characterisation of pig liver in unconfined compression. J. Biomech. **47**, 2641–2646 (2014)
42. Tirella, A., Mattei, G., Ahluwalia, A.: Strain rate viscoelastic analysis of soft and highly hydrated biomaterials. J. Biomed. Mater. Res. - Part A **102**, 3352–3360 (2014)
43. Barachini, S., Danti, S., Pacini, S., D'Alessandro, D., Carnicelli, V., Trombi, L., Moscato, S., Mannari, C., Cei, S., Petrini, M.: Plasticity of human dental pulp stromal cells with bioengineering platforms: a versatile tool for regenerative medicine. Micron **67**, 155–168 (2014)
44. Bajpai, V.K., Mistriotis, P., Andreadis, S.T.: Clonal multipotency and effect of long-term in vitro expansion on differentiation potential of human hair follicle derived mesenchymal stem cells. Stem Cell Res. **8**, 74–84 (2012)
45. Gurikov, P., Smirnova, I.: Non-conventional methods for gelation of alginate. Gels **4**, 14 (2018)
46. De la Ossa, J.G., Trombi, L., D'Alessandro, D., Coltelli, M.B., Serino, L.P., Pini, R., Lazzeri, A., Petrini, M., Danti, S.: Pore size distribution and blend composition affect in vitro prevascularized bone matrix formation on poly(vinyl alcohol)/gelatin sponges. Macromol. Mater. Eng. **302**, 1700300 (2017)
47. Chiellini, E., Cinelli, P., Fernandes, E.G., Kenawy, E.R.S., Lazzeri, A.: Gelatin-based blends and composites. Morphological and thermal mechanical characterization. Biomacromol **2**, 806–811 (2001)
48. Bigi, A., Cojazzi, G., Panzavolta, S., Rubini, K., Roveri, N.: Mechanical and thermal properties of gelatin films at different degrees of glutaraldehyde crosslinking. Biomaterials **22**, 763–768 (2001)
49. Alves, P.M.A., Carvalho, R.A., Moraes, I.C.F., Luciano, C.G., Bittante, A.M.Q.B., Sobral, P.J.A.: Development of films based on blends of gelatin and poly(vinyl alcohol) cross linked with glutaraldehyde. Food Hydrocolloids **25**, 1751–1757 (2011)
50. Gao, X., Tang, K., Liu, J., Zheng, X., Zhang, Y.: Compatibility and properties of biodegradable blend films with gelatin and poly (vinyl alcohol). J. Wuhan Univ. Technol.-Mater. Sci. Ed. **29**, 351–356 (2014)
51. Pawde, S.M., Deshmukh, K.: Characterization of polyvinyl alcohol/gelatin blend hydrogel films for biomedical applications. J. Appl. Polym. Sci. **109**, 3431–3437 (2008)
52. Mattei, G., Gruca, G., Rijnveld, N., Ahluwalia, A.: The nano-epsilon dot method for strain rate viscoelastic characterisation of soft biomaterials by spherical nano-indentation. J. Mech. Behav. Biomed. Mater. **50**, 150–159 (2015)
53. Cacopardo, L., Mattei, G., Ahluwalia, A.: A new load-controlled testing method for viscoelastic characterisation through stress-rate measurements. Materialia **9**, 100552 (2020)

Following the Growth and Division of Lipid Boundaries by Using Glass Microsphere-Supported Protocells

Augustin Lopez, Carolina Chieffo, and Michele Fiore(✉)

Université de Lyon, Institut de Chimie et Biochimie Moléculaires et Supramoléculaires (ICBMS, UMR 5246) Claude Bernard Lyon 1, Bâtiment Lederer, 1 Rue Victor Grignard, 69622 Villeurbanne Cedex, France
michele.fiore@univ-lyon1.fr

Abstract. Protocells are compartmented molecular networks which can be designed to study the origins of life. Glass microsphere-supported giant vesicles (MSGVs) are model protocells for which monodispersed glass beads are coated with a lipid bilayer thanks to avidin and biotinylated phospholipids. These supramolecular assemblies have proved to be extremely effective to understand certain phenomena related to the self-reproduction of protocells thanks to a series of intriguing experiments. First, the growth and division (G&D) of these giant vesicles was observed by epifluorescence and confocal microscopy when they were fed with fatty acids solutions at different feeding rates. Second, chemical analyses performed by a combination of GC-MS, UPLC-HRMS and phospholipid-specific assay, allowed to independently study the composition of the vesicles obtained after G&D.

Keywords: Phospholipids · Fatty acids · Membranes · Vesicles · Protocells · Self-reproduction · Autopoietic systems

In memory of Océane Fiore.

1 Introduction

Protocells are compartmented molecular networks that, when prepared with prebiotically plausible amphiphiles, are convenient to address the transition from inanimate to animate in the emergence of life. In this purpose, these compartments are made of a bilayer of lipids bearing an inner aqueous volume and encapsulate simple molecules such as peptides, nucleotides, sugars, etc. (Stano 2018; Toparlak and Mansy 2019). Thus, protocells are used to study and to reproduce the basic properties attributed to functional cells such as an operational metabolism or evolvability (Lopez and Fiore 2019). Among others, the possibility to form an autopoietic system (Varela et al. 1974) with the self-reproduction of vesicles was deeply investigated during the last thirty years.

S. Piotto et al. (Eds.): BIONAM 2019, LNBE, pp. 210–222, 2020.
https://doi.org/10.1007/978-3-030-47705-9_18

The growth and division (G&D) is a process in which a growing vesicle – the "mother" – first transforms its spherical shape into a dumbbell shape, and then splits into two spherical "daughter" vesicles (Božič and Svetina 2004). This process is depicted in Fig. 1. It represents a key point for the preparation of cellular autopoietic systems able to self-reproduce (Stano and Luisi 2010; Szostak 2017).

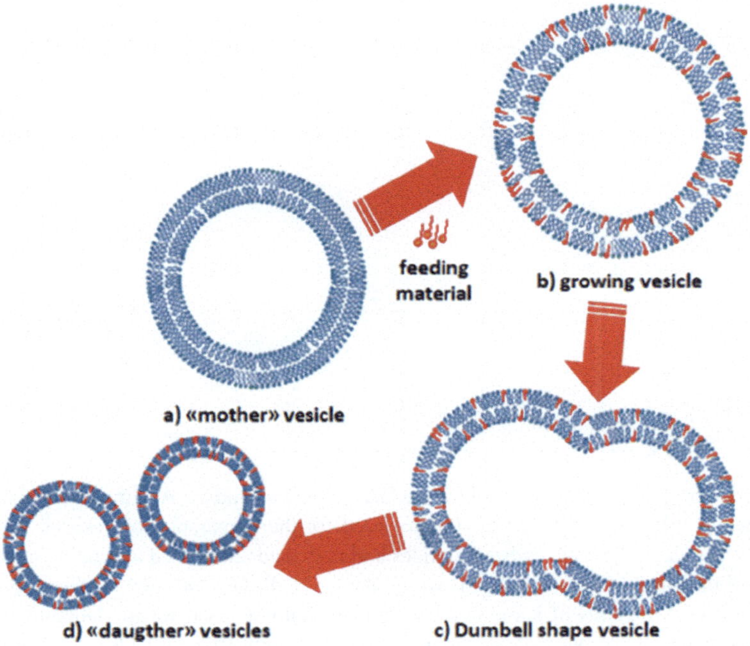

Fig. 1. A representative scheme of growth and division as described by Božič and Svetina (2004). Vesicle self-reproduction is a process in which a growing vesicle (a→b) first transforms its spherical shape into a dumbell shape (b→c), and then splits into two spherical daughter vesicles (c→d); "Mother" vesicles refers to the vesicles that were submitted to G&D and "daughter" vesicles to those that generate from this process.

In a very general way, the preparation of vesicles (Vs) is achieved with the hydration of a film of phospholipids, fatty acids or other amphiphilic compounds producing vesicles of various types and size. Depending on the studies, vesicles of variable dimensions are used even though extruded vesicles (*ext*-Vs) with monodispersed size are the most common. Giant vesicles (GVs), characterized by a micrometric size, are also widely favored since they are easily observable and have dimensions comparable to actual cells. The lamellarity as well can be controlled in order to obtain either giant unilamellar vesicles (GUVs) or giant mutlilamellar vesicles (GMVs) (Walde et al. 2010). Many of the fatty acids are considered as prebiotic molecules and most of them can be used for the formation of stable lipid boundaries in vesicles serving as protocells models (Hargreaves and Deamer 1978; Monnard and Deamer

2003). In addition, mono-alkyl phosphates are also considered as plausible membrane constituents of protocells in synthetic biology (Fiore 2018) and prebiotic system chemistry (Albertsen et al. 2014a; Fiore et al. 2017; Fayolle et al. 2017).

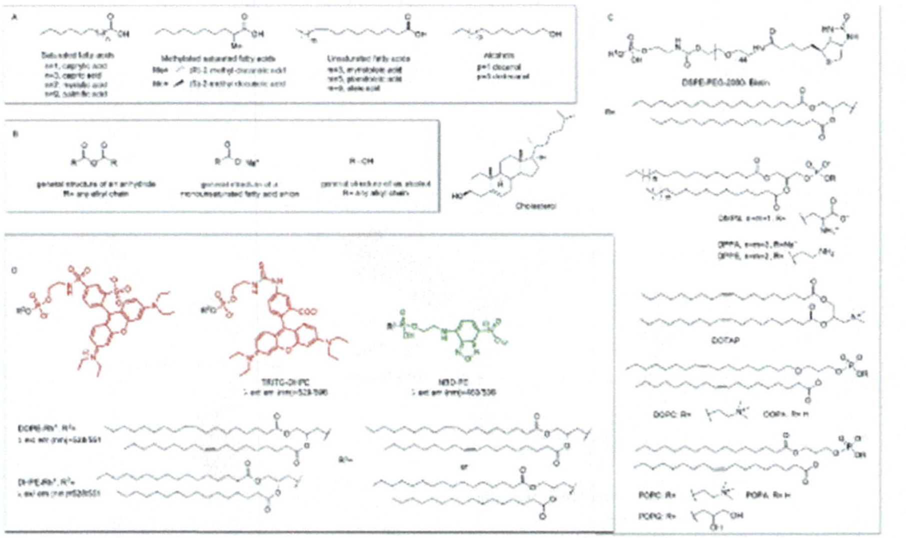

Fig. 2. Structures of some lipids used for the G&D division studies. A) Structures of some fatty acids, alcohols, phospholipids and cholesterol used for the preparation of vesicles; B) General structure of the alcohols, fatty acids anhydrides and fatty acid anion used as feeding materials; C) Structure of the fatty acid, phospholipids and of the DSPE-PEG-2000 used for the preparation of the MSGVs; D) Structures of some fluorescent phospholipids used for the labelling of vesicle membranes. POPC: 1-palmitoyl-2-oleoyl-*sn*-glycero-3 phosphocholine; DOPC, (1,2-dioleoyl-*sn*-glycero-3-phosphocholine; POPA, (1-palmitoyl-2-oleoyl-*sn*-glycero-3-phosphate); DOPA (1,2-dioleoyl-sn-glycero-3-phosphate); POPG: 1-palmitoyl-2-oleoyl-*sn*-glycero-3; DMPS (1-phosphoglycerol,2-Dimyristoyl-*sn*-glycero-3-phospho-L-serine; DSPE-PEG2000-Biotin (1,2-distearoyl-sn-glycero-3-phosphoethanolamine-N [biotinyl(polyethylene glycol)-2000]); DOPE-Rh^+ (1,2-dioleoyl-*sn*-glycero-3-phosphoethanolamine-N-(lissamine rhodamine B sufonyl ammonium salt); DHPE-Rh^+ (1,2-Dihexadecanoyl-sn-Glycero-3-Phosphoethanolamine-N-(lissamine rhodamine B sulfonyl), Triethylammonium salt); DOPE-NBD (1,2-Dioleoyl-sn-glycero-3-phosphoethanolamine-N-(7-nitro-2-1,3-benzoxadiazol-4-yl); DHPE-NBD (1,2-Dihexadecanoyl-sn-Glycero-3-Phosphoethanolamine-N-(7-nitro-2-1,3-benzoxadiazol-4-yl); TRITC-DHPE (N-(6tetramethylrhodaminethiocarbamoyl)-1,2-dihexadecanoyl-sn-glycero-3- phosphoethanolamine, triethylammonium salt).

In order to study the cellular self-reproduction, vesicles or various sizes and lipid composition have been used. Regarding lipid composition, fatty acids, mixed fatty acids, phospholipids or ternary mixture of fatty acids, phospholipids and cholesterol have been reported (Fig. 2A). The G&D of these vesicles was notably performed by providing them amphiphiles which could be monomers (Fig. 2B), delivered in the medium and/or synthesized within the vesicles, or aggregates such as micelles and vesicles (Stano and Luisi 2010; Hardy et al. 2015; Kurihara et al. 2015; Szostak 2017). Despite the major progress achieved on the conditions enabling the G&D or the characterization of the different steps, the clear distinction and then characterization of the two different populations (mothers and daughters) is still problematic. Different devices have been reported in order to sort out the populations of vesicles after a G&D process.

In pioneer studies, the size of the vesicles obtained after G&D was measured through electron microscopy (Walde et al. 1994) (Table 1, Entry 1) or light microscopy (Wick et al. 1995) (Table 1, Entry 2). In both cases, after a feeding period with oleic anhydrides the studied oleic acid vesicles showed an enhanced average size. Associated to the higher number of vesicles detected and the microscopic observations of the vesicular self-reproduction, these measures assessed that G&D was occurring. Even though the vesicles could be classified by their size it was not possible to discriminate with certainty which were the mothers and which the daughters formed after G&D (Fig. 3A). Then, detection cargos have been used since the first investigations performed with ferritin, a protein containing dense cores of iron which can be visualized by cryo-TEM. This protein was entrapped into the mother vesicles (made of POPC or oleic acid) during their preparation. After a feeding period (with oleate micelles or oleic an-hydrides, respectively) both vesicles filled with ferritin and empty ones have been identified. The decreased concentration of ferritin entrapped into the filled vesicles (and the increased number of vesicles containing ferritin) after a feeding period assessed that a G&D happened and that the protein coming from the mother vesicles was distributed into the daughter vesicles. The empty vesicles were also de-scribed as daughter vesicles formed without partitioning of the cargos (Berclaz et al. 2001) (Table 1, Entry 3). Other revelation cargos such as fluorescent dyes (Hanczyc et al. 2003) (Table 1, Entry 4) or fluorescent RNAs (Zhu and Szostak 2009) (Table 1, Entry 5) have been used. In all these studies it was possible to differentiate empty and filled vesicles, however, among the cargos-containing vesicles, it is not possible to distinguish the mothers from the daughters (Fig. 3B). Lately, a separation of the vesicles generated by a G&D process based on their lipid composition was performed by free-flow electropho-resis fractionation. POPC vesicles, electrically not charged, were fed with oleate micelles negatively charged due to the carboxylate. After G&D, it was possible to separate the vesicles obtained depending on their charges. This setup allowed to show that two populations of vesicles were formed, an oleate-rich one and an oleate-poor one (Pereira de Souza et al. 2015) (Table 1, Entry 6). Despite the physical sorting which was possible in that case, the absolute differentiation between the mother and the daughters was still unreachable (Fig. 3C).

Thus, despite their ingenuity to classify the vesicles formed after a G&D none of these studies allowed a physical separation of each population of vesicles (mothers vs. daughters). Therefore a method is urgently needed to independently characterize the size, the lipid composition or even the content of both categories of vesicles. Such a method would be of great help to unambiguously establish the transmission of physicochemical characteristics from mother to daughter vesicles and then to deepen our knowledge about the G&D process.

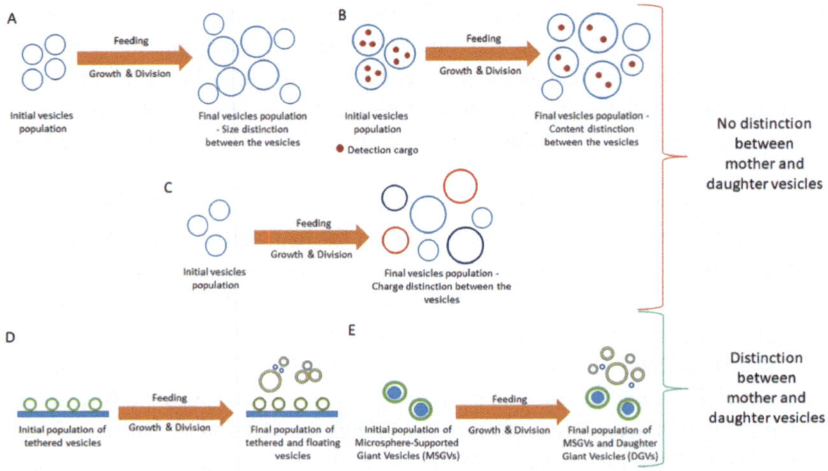

Fig. 3. Some examples of experimental setups used to characterize the vesicles formed by a self-reproduction process. After G&D, A) the size, B) the content or C) the charge of the vesicles is characterized. These first three examples don't show any distinction between mother and daughter vesicles. The use of D) tethered membranes or E) supported giant vesicles allow distinction and separation between the two populations of vesicles.

2 Discussion

2.1 Surface-Mediated Vesicle Replication

Even if the use of surface-mediated vesicle replication allows the distinction of mother from daughter vesicles after a G&D, only a few examples are reported in the literature. Vesicles can be anchored to a surface by either a specific integral membrane-bound linkage or through adsorption (Rebaud et al. 2014). A pertinent example is that vesicles may be tethered to an avidin-coated surface via biotinylated phospholipids (Pignataro et al. 2000). Surface-attached vesicles would grow through the uptake of additional membrane components such as fatty acids in the form of free molecules or as micelles or through the fusion of phospholipid vesicles. Vesicles adsorbed to a surface have been shown to fuse with additional vesicles introduced in fluid flowing over the surface regions of a glass surface coated by hydrocarbons (Johnson et al. 2002). In the latter case, microfluidic devices in which fluid flow can be precisely controlled and vesicles adsorbed to a surface fuse with vesicles introduced in a flow field. (Morigaki and Walde 2002). This study may form the basis for surface-mediated vesicle G&D since the grown membrane bilayer would eventually become unstable leading to the budding off of daughter vesicles (Fig. 3D).

Table 1. A short overview of the different systems used to characterize the vesicles obtained after a G&D process.

Entry	"Mother vesicles types"[a]	Vesicles lipid boundaries[b]	Feeding materials[c]	Devices used	Observed processes	Types of distinction	Physical separation between the different populations	Reference
1	ext-Vs of 100 and 400 nm	Caprylic and oleic acids	Caprylic and oleic anhydrides	Cryo-TEM	Increase of the size of the vesicles (Growth) and of their number (Division)	Size	No	(Walde et al. 1994)
2	GMVs (>0.5 μm)	Oleic acid	Oleic anhydride	Light Microscopy	Increase of the size of the vesicles (Growth) and of their number (Division); G&D transformations observed	Size	No	(Wick et al. 1995)
3	ext-Vs of 100 nm	POPC and oleic acid	Oleate micelles and oleic anhydrides	Cryo-TEM; Ferritin encapsulation	Increase of the size of the vesicles (Growth) and of their number (Division); The obtained vesicles contain less ferritin than the mother vesicles	Size; Ferritin content	No	(Berclaz et al. 2001;
4	ext-Vs of 100 nm	Myristoleic acid	Myristoleate micelles	Dynamic Light Scattering; Flow-field fractionation coupled with size analysis by multiangle laser-light scattering (FFF-MALLS); FRET assay; Calcein encapsulation (fluorescent dye)	Increase of the size of the vesicles (Growth); The obtained vesicles contain an amount of calcein which is about the one expected from a division process	Size; Calcein content	No	(Hanczyc et al. 2003
5	ext-GMVs of 5 μm	Oleic acid, myristoleic acid, decanoic acid, decanoic acid:decanol (2:1), POPC	Oleate, myristoleate, decanote micelles and decanoate micelles with decanol emulsion	HPTS (fluorescent dye) encapsulation; DY547-A_{10} (fluorescent RNA) encapsulation	Increase of the size of the vesicles (Growth) and of their number (Division); G&D transformations observed; The obtained vesicles contain less fluorescent RNA (oleic acid and decanoic acid:decanoate (2:1) vesicles) or dye (myristoleic and decanoic acid vesicles) than the mother vesicles	Size; HPTS and DY547-A_{10} content	No	(Zhu and Szostak 200
6	ext-Vs of 100 nm	POPC	Oleate micelles	Cryo-TEM; Ferritin; Free-Flow Electrophoresis	Formation of two populations of vesicles: large oleate-rich vesicles and small oleate-poor vesicles	Size; Ferritin content; Charge	Yes, in the case of the charge analysis	(Pereira de Souza et al. 2

[a] the vesicles were obtained by different methods; [b] the amphiphiles used are listed in Figure 2A; [c] the amphiphiles used are listed in Figure 2B

2.2 Preparation of Microsphere-Supported Giant Vesicles (MSGVs)

A valid alternative is the use of microsphere-supported giant vesicles (MSGV), lipid bilayers supported by monodispersed glass beads. The size of the supported vesicles (>5 μm) classifies them as giant vesicles that could in principle possess the same complex lipid composition of a minimal cell (Walde et al. 2010). They represent an effective tool to separate by simple centrifugation of the mother vesicles (denser due to the inner microsphere) from the daughters generated by the G&D (Fig. 3E). The use of MSGVs permits *i*) the direct observation of membrane growth and division on the anchored bilayers supplied with amphiphiles; *ii*) the microscopic distinction of the mother from daughter vesicles; *iii*) the sedimentation, by centrifugation, of the heavier mother vesicles hence the possibility to exchange host solutions and to recover the daughter vesicles as floating lipid material; *iv*) the independent analysis of the composition of two vesicle populations after the feeding procedure; as well as *v*) investigating the influence of multiple parameters such as the feeding rate on the G&D process (Albertsen et al. 2014a). The first example of MSGVs was reported by the group of Bruce Lennox a few years ago (Gopalakrishnan et al. 2009). They called their supramolecular constructs spherical supported bilayer membranes (SS-BLMs) and the construction of these supported vesicles is very similar to the one used for the lipid membranes tethering. To form the population of SS-BLMs, monodispersed glass microspheres (5.02:0.01 mm in diameter) were first coated with avidin (Fig. 4A, step 1). Then, they were incubated with small unilamellar vesicles (SUVs) made of a low proportion of DSPE-PEG2000-Biotin and of various phospholipids (Table 2, Entry 1). The strong bio-affinity between avidin and biotinylated lipids (DSPE-PEG200-Biotin, Fig. 2C) enabled the transfer of the lipid bilayer on the beads (Fig. 4A, step 2A) (Gopalakrishnan et al. 2009). The obtained supported vesicles were fully characterized by confocal microscopy. In that case, fluorescent probes such as TRITC-DHPE and NBD-DOPE (Fig. 2D) were added to the lipid composition in low percentage. The SS-BLMs were also characterized by cryo-TEM analysis using a well-known technique (Dubochet et al. 1985). Fiore and co-workers prepared similar supported vesicles, called MSGVs, with a very similar procedure to that used by Monnard and co-workers to coat decanoic acid on the surface of glass microspheres (Table 2, Entry 2) (Albertsen et al. 2014b). First, glass microspheres were incubated with avidin (Fig. 4A, step 1) and then DSPE-PEG2000-Biotin was added in order to build up a solid DSPE-PEG2000-Biotin/avidin architecture (Fig. 4B, step 2b).

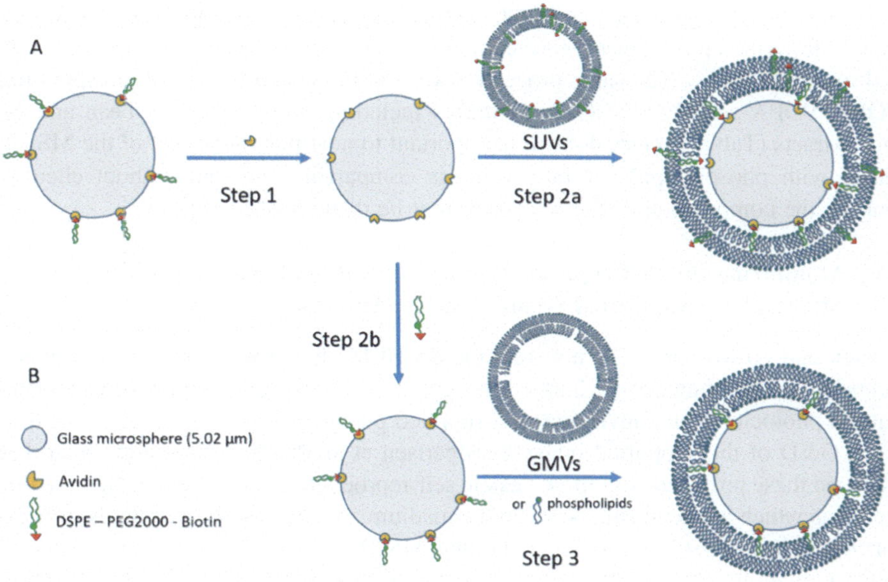

Fig. 4. The two different methodologies used to prepare MSGVs. A) Lennox and co-workers procedure B) Fiore and co-workers method corresponding to Monnard and co-workers protocol with some modifications.

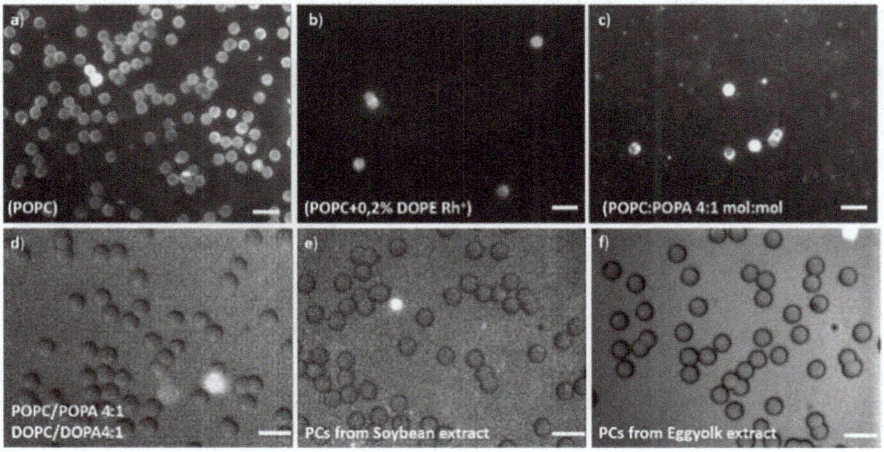

Fig. 5. Micrographs of different MSGVs (cf. Table 2) produced by the method described by Fiore and co-workers. Micrographs a, c (using a 50x lens) and d–f (using a 100x oil immersion lens) were obtained by staining with 0.5 mM Nile Red®; Micrograph b was obtained without staining with Nile red while the sample contains 0.2% of DOPE-Rh (Fig. 2D). Scale bar is always 10 µm.

Then, phospholipids such as DOPC were added as free compounds or as GMVs in order to form the lipid bilayer over the microspheres (Fig. 4B, step 3 and Table 2, entry 3) (Fiore et al. 2018). The same procedure was used to coat a variety of phospholipids (POPC, POPA and DOPC/DOPA mixtures) including non-purified soybean and egg yolk extracts (Table 2, entry 4–8). It is important to note that the aspect of the MSGVs coated with phospholipids or fatty acids is comparable and that without chemical analysis the composition of the bilayer cannot be distinguished (Fig. 5).

2.3 Monitoring the Self-reproduction of Lipid Boundaries Using Microsphere-Supported Giant Vesicles (MSGVs)

Lennox and co-workers have not used the SS-BLMs for growth and division experiments (Gopalakrishnan et al. 2009). The group of Pierre-Alain Monnard established that the photochemical conversion of a supplied precursor lipid into decanoic acid led to the G&D of the supported bilayers (Albertsen et al. 2014b). Fiore and co-workers also used these protocells to follow vesicle self-reproduction and to understand the lipid exchange which are occurring between the medium and the vesicles during the process. For that, a hosting solution (HS) containing MSGVs (1.8×10^4 mL^{-1}) was supplied with an ethanolic solution of fatty acid. Such solvent evaporates during the "feeding" period and allows to deliver the lipids as monomers without aggregates (Walde et al. 2010). The MSGVs used as starting material were named as M_0SGVs: these MSGVs were not fed and their lipid boundary did not change in composition yet. In order to distinguish each population of vesicles, we called M_1SGVs the supported giant vesicles obtained after one feeding period. With D_1GVs – where D stands for "daughter" – we designated the vesicles generated after a feeding period by the G&D process.

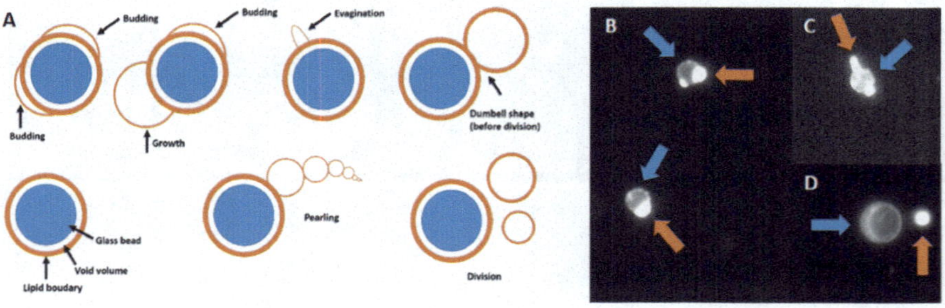

Fig. 6. A) Graphical representation and B) C) D) epifluorescence microscopy observations of the main phenomena observable during G&D processes. These observations were monitored on MSGVs made of DOPC and 0.2% of DOPE-Rh (Fig. 2D) fed with oleic acid (Fiore et al 2018). A) The blue circle corresponds the glass microsphere and B) C) D) the blue arrows designate the bead. A) The orange object represents the lipid boundary irrespectively from its composition and the composition of the feeding material used. Among the different phenomena represented, B) budding, C) evagination and D) division are highlighted with orange arrows on the microscopic pictures.

The mother vesicles (M1SGVs) containing the microspheres were isolated by cycles of centrifugation and washings from the empty daughter vesicles (D1GVs) floating in the hosting solution (HS) containing free lipids as well.

Firstly, the feeding method was optimized using type C M0SGVs (Table 2, Entry 3) trying two different methods (Fiore et al. 2018). The slow feeding method (SFM) consists of the gradual addition of an ethanolic solution with a 0.5 mM concentration in oleic acid (OA) during four hours at a feeding ratio of 33 μl h^{-1}. The fast feeding method (FFM) consists of a single injection of the OA solution at once. In both cases, the amounts of OA added to the host solution were the same and a final concentration of 0.05 mM was reached (CVC of OA = 47.0 mm) (Maurer et al. 2009). During the slow feeding period, the investigated samples contained M_1SGVs showing budding and evagination. Some M_1SGVs also showed thickening of the amphiphilic microsphere coating, accompanied by longer tubular evaginations growing out of the microsphere bilayer coating and floating in the hosting solution (Fig. 6A, B and C). As the feeding experiment progressed, an increased frequency of these growth-induced structural features, as well as the presence of freely floating filamentous vesicles, division products, were unambiguously detected (Fig. 6A and D). Thus, all these phenomena previously described on free vesicles were observed on MSGVs.

Table 2. Overview of the MSGVs produced and the associates G&D.

Entry	MSGVs[a]	Vesicle product	Phenomena			Reference
1	A	No G&D performed				(Gopalakrishnan et al. 2009)
Entry	MSGVs[a]	PL_{found}/ $PL_{initial}$	OA_{found}/ $OA_{delivered}$	PL_{found}/ $PL_{initial}$	**Phenomena**[b]	**Reference**
2	B	**D$_1$GVs** **M$_1$SGVs**	Vesicles replenished after G&D		Budding, Growth and Division	(Albertsen et al. 2014b)
3	C	**D$_1$GVs**	0.65 ± 0.01	0.85	Budding, Growth, Pearling and Division	(Fiore et al. 2018)
		M$_1$SGVs	0.35 ± 0.01	0.15		
4	D	**D$_1$GVs** **M$_1$SGVs**	Not calculated		Budding, Growth, Pearling and Division	Present research
5	E	**D$_1$GVs** **M$_1$SGVs**	Not calculated		Budding, Growth, Pearling and Division	Present research
6	F	**D$_1$GVs** **M$_1$SGVs**	Not calculated		Budding, Growth, Pearling and Division	Present research
7	G	**D$_1$GVs** **M$_1$SGVs**	Not calculated		Budding, Growth, Pearling and Division	Present research
8	H	**D$_1$GVs** **M$_1$SGVs**	Not calculated		Budding, Growth, Pearling and Division	Present research

[a] lipid composition of A = DOPC or DOPC: DMPS 3:2; or DOPC: DOTAP 3:1; or DOPC: DOTAP: DPPE 1:1:2; or DOPC: DPPA 3:2; B = Decanoic acid; C = DOPC; D = POPC; E = POPC: POPA 4: 1; F = POPC: POPA: DOPC: DOPA 4:1:4:1; G = Soybean exracts; H = Eggyolk extracts; [b] Observed by epifluorescence microscopy

After separation by centrifugations and washings, the lipid composition of both types of populations (M_1SGVs and D_1GVs) was characterized by a combination of UPLC-HRMS, GC-MS and Stewart assay (Stewart 1980).

In both cases, (SFM and FFM) the two types of population showed blended membranes illustrating that a lipid exchange is occurring (as an example, the values obtained with the FFM are given in Table 2, Entry 3) (Fiore et al. 2018).

With the same method (Fig. 4B), other types of supported vesicles were prepared including protocells made of mixtures of naturally occurring phospholipids (Table 2, Entry 4–8). In all the cases, feeding the MSGVs with oleic acid led to their division. Moreover, all the best known phenomena associated to G&D such as budding, growth, pearling and division (Fig. 6A) (Stano and Luisi 2010; Szostak 2017) were observed by epifluorescence microscopy (unpublished data). These processes associated to G&D were initially observed on floating vesicles.

The possibility to reproduce them with a wide variety of MSGVs is leading us to the assumption that they are well-suited systems to study the G&D as it occurs with free vesicles.

3 Conclusions

Cell-like protocells are useful to understand the self-reproduction of prebiotic compartments. The G&D phenomenon and the formation of a cellular autopoietic system were extensively studied during the last thirty years. Even though the process was described in numerous cases, its full comprehension is still out of reach. This is notably since separating the different population of vesicles generated by the G&D causes a problem. Lately, it has been shown that the study of the growth and the division could be improved using microsphere-supported giant vesicles (MSGVs). These supramolecular constructs were firstly designed as cellular models and then used to assess the transmission of a catalytic function after a division. However, they were never exploited to investigate the lipid exchange which is occurring between the vesicles and the medium during the G&D phenomenon. Indeed, the convenient separation of the MSGVs from the floating lipid material (including the DGVs) allows to independently analyze their lipid compositions. Moreover, the observation of the self-reproduction patterns with the MSGVs could assess that similar lipid exchange is occurring when free vesicles grow and divide. Otherwise, the possibility to prepare MSGVs with complex membranes opens strong perspectives for the study of the G&D process but also for the study of protocells made of mixtures of amphiphiles obtained in so-called prebiotic conditions in prebiotic systems chemistry (Ruiz-Mirazo et al. 2014; Fiore et al. 2017; Fayolle et al. 2017; Fiore 2018). Finally, we wish to highlight that the MSGVs may be used as cellular models in various biology studies. To give an example, monitoring the lipid trafficking by simulating the lipid bilayer of a cell by including the membrane proteins in the supported bilayer could be achieved with these giant vesicles.

Acknowledgements. MF thanks Prof. Stefano Piotto and Prof. Federico Rossi that gave him the opportunity to present preliminary results on this topic at the 3rd International Conference on Bio and Nanomaterials (BIONAM) – September 29–October 3, 2019.

Conflict of Interest. The authors declare no conflict of interest.

References

Albertsen, A.N., Duffy, C.D., Sutherland, J.D., Monnard, P.-A.: Self-assembly of phosphate amphiphiles in mixtures of prebiotically plausible surfactants. Astrobiology **14**, 462–472 (2014a). https://doi.org/10.1089/ast.2013.1111

Albertsen, A.N., Maurer, S.E., Nielsen, K.A., Monnard, P.-A.: Transmission of photo-catalytic function in a self-replicating chemical system: in situ amphiphile production over two protocell generations. Chem. Commun. **50**, 8989–8992 (2014b). https://doi.org/10.1039/C4CC01543F

Berclaz, N., Müller, M., Walde, P., Luisi, P.L.: Growth and transformation of vesicles studied by ferritin labeling and cryotransmission electron microscopy. J. Phys. Chem. B **105**, 1056–1064 (2001). https://doi.org/10.1021/jp001298i

Božič, B., Svetina, S.: A relationship between membrane properties forms the basis of a selectivity mechanism for vesicle self-reproduction. Eur. Biophys. J. **33**, 565–571 (2004). https://doi.org/10.1007/s00249-004-0404-5

Dubochet, J., Adrian, M., Chang, J.-J., Homo, J.-C., Lepault, J., McDowall, A.W., Schultz, P.: Cryo-electron microscopy of vitrified specimens. Q. Rev. Biophys. **21**, 129–228 (1985). https://doi.org/10.1017/S0033583500004297

Fayolle, D., Altamura, E., D'Onofrio, A., Madanamothoo, W., Fenet, B., Mavelli, F., Buchet, R., Stano, P., Fiore, M., Strazewski, P.: Crude phosphorylation mixtures containing racemic lipid amphiphiles self-assemble to give stable primitive compartments. Sci. Rep. **7**, 18106–18114 (2017). https://doi.org/10.1038/s41598-017-18053-y

Fiore, M.: The synthesis of mono-alkyl phosphates and their derivatives: an overview of their nature, preparation and use, including synthesis under plausible prebiotic conditions. Org. Biomol. Chem. **16**, 3068–3086 (2018). https://doi.org/10.1039/C8OB00469B

Fiore, M., Madanamoothoo, W., Berlioz-Barbier, A., Maniti, O., Girard-Egrot, A., Buchet, R., Strazewski, P.: Giant vesicles from rehydrated crude mixtures containing unexpected mixtures of amphiphiles formed under plausibly prebiotic conditions. Org. Biomol. Chem. **15**, 4231–4240 (2017). https://doi.org/10.1039/C7OB00708F

Fiore, M., Maniti, O., Girard-Egrot, A., Monnard, P.-A., Strazewski, P.: Glass microsphere-supported giant vesicles for the observation of self-reproduction of lipid boundaries. Angew. Chem. Int. Ed. **57**, 282–286 (2018). https://doi.org/10.1002/anie.201710708

Gopalakrishnan, G., Rouiller, I., Colman, D.R., Lennox, R.B.: Supported bilayers formed from different phospholipids on spherical silica substrates. Langmuir **25**, 5455–5458 (2009). https://doi.org/10.1021/la9006982

Hanczyc, M.M., Fujikawa, S.M., Szostak, J.W.: Experimental models of primitive cellular compartments: encapsulation, growth, and division. Science **302**, 618–622 (2003). https://doi.org/10.1126/science.1089904

Hardy, M.D., Yang, J., Selimkhanov, J., Cole, C.M., Tsimring, L.S., Devaraj, N.K.: Self-reproducing catalyst drives repeated phospholipid synthesis and membrane growth. Proc. Natl. Acad. Sci. U.S.A. **112**, 8187–8192 (2015). https://doi.org/10.1073/pnas.1506704112

Hargreaves, W.R., Deamer, D.W.: Liposomes from ionic, single-chain amphiphiles. Biochemistry **17**, 3759–3768 (1978). https://doi.org/10.1021/bi00611a014

Johnson, J.M., Ha, T., Chu, S., Boxer, S.G.: Early steps of supported bilayer formation probed by single vesicle fluorescence assays. Biophys. J. **83**, 3371–3379 (2002). https://doi.org/10.1016/S0006-3495(02)75337-X

Kurihara, K., Okura, Y., Matsuo, M., Toyota, T., Suzuki, K., Sugawara, T.: A recursive vesicle-based model protocell with a primitive model cell cycle. Nat. Commun. **6**, 8352–8359 (2015). https://doi.org/10.1038/ncomms9352

Lopez, A., Fiore, M.: Investigating prebiotic protocells for a comprehensive understanding of the origins of life: a prebiotic systems chemistry perspective. Life **9**, 49–70 (2019). https://doi.org/10.3390/life9020049

Maurer, S.E., Deamer, D.W., Boncella, J.M., Monnard, P.-A.: Chemical evolution of amphiphiles: glycerol monoacyl derivatives stabilize plausible prebiotic membranes. Astrobiology **9**, 979–987 (2009). https://doi.org/10.1089/ast.2009.0384

Monnard, P.-A., Deamer, D.W.: Preparation of vesicles from nonphospholipid amphiphiles. In: Methods in Enzymology, pp. 133–151. Elsevier (2003). https://doi.org/10.1016/S0076-6879(03)72008-4

Morigaki, K., Walde, P.: Giant vesicle formation from oleic acid/sodium oleate on glass surfaces induced by adsorbed hydrocarbon molecules. Langmuir **18**, 10509–10511 (2002). https://doi.org/10.1021/la026579r

Pereira de Souza, T., Holzer, M., Stano, P., Steiniger, F., May, S., Schubert, R., Fahr, A., Luisi, P.L.: New insights into the growth and transformation of vesicles: a free-flow electrophoresis study. J. Phys. Chem. B **119**, 12212–12223 (2015). https://doi.org/10.1021/acs.jpcb.5b05057

Pignataro, B., Steinem, C., Galla, H.-J., Fuchs, H., Janshoff, A.: Specific adhesion of vesicles monitored by scanning force microscopy and quartz crystal microbalance. Biophys. J. **78**, 487–498 (2000). https://doi.org/10.1016/S0006-3495(00)76611-2

Rebaud, S., Maniti, O., Girard-Egrot, A.P.: Tethered bilayer lipid membranes (tBLMs): Interest and applications for biological membrane investigations. Biochimie **107**, 135–142 (2014). https://doi.org/10.1016/j.biochi.2014.06.021

Ruiz-Mirazo, K., Briones, C., de la Escosura, A.: Prebiotic systems chemistry: new perspectives for the origins of life. Chem. Rev. **114**, 285–366 (2014). https://doi.org/10.1021/cr2004844

Stano, P.: Is research on "Synthetic Cells" moving to the next level? Life **9**, 3–32 (2018). https://doi.org/10.3390/life9010003

Stano, P., Luisi, P.L.: Achievements and open questions in the self-reproduction of vesicles and synthetic minimal cells. Chem. Commun. **46**, 3639 (2010). https://doi.org/10.1039/b913997d

Szostak, J.W.: The narrow road to the deep past. In search of the chemistry of the origin of life. Angew. Chem. Int. Ed. **56**, 11037–11043 (2017). https://doi.org/10.1002/anie.201704048

Toparlak, O.D., Mansy, S.S.: Progress in synthesizing protocells. Exp. Biol. Med. (Maywood) **244**, 304–313 (2019). https://doi.org/10.1177/1535370218816657

Varela, F.G., Maturana, H.R., Engel, H., Uribe, R.: Autopoiesis: the organization of living systems, its characterization and a model. Biosystems **5**, 187–196 (1974). https://doi.org/10.1016/0303-2647(74)90031-8

Walde, P., Cosentino, K., Engel, H., Stano, P.: Giant vesicles: preparations and applications. Chem. Eur. J. Chem. Bio. **11**, 848–865 (2010). https://doi.org/10.1002/cbic.201000010

Walde, P., Wick, R., Fresta, M., Mangone, A., Luisi, P.L.: Autopoietic self-reproduction of fatty acid vesicles. J. Am. Chem. Soc. **116**, 11649–11654 (1994). https://doi.org/10.1021/ja00105a004

Wick, R., Walde, P., Luisi, P.L.: Light microscopic investigations of the autocatalytic self-reproduction of giant vesicles. J. Am. Chem. Soc. **117**, 1435–1436 (1995). https://doi.org/10.1021/ja00109a031

Zhu, T.F., Szostak, J.W.: Coupled growth and division of model protocell membranes. J. Am. Chem. Soc. **131**, 5705–5713 (2009). https://doi.org/10.1021/ja900919c

Single Compartment Approach for Assembling Photosynthetic Protocells

Emiliano Altamura[1], Paola Albanese[1], Francesco Milano[2], Massimo Trotta[3], Pasquale Stano[4], and Fabio Mavelli[1,5(✉)]

[1] Department of Chemistry, University of Bari Aldo Moro, Bari, Italy
fabio.mavelli@uniba.it
[2] CNR-ISPA, Institute of Sciences of Food Production, Lecce Unit, Lecce, Italy
[3] CNR-IPCF, Institute for Physical and Chemical Processes, Bari, Italy
[4] Department of Biological and Environmental Sciences and Technologies, University of Salento, Lecce, Italy
[5] Institute of Nanotechnology, CNR NANOTEC, Bari, Italy

Abstract. In this work, we show the extraction, purification and reconstitution of the bacterial protein complex bc1 in giant lipid vesicles used as versatile micro-sized enzymatic reactors. Firstly, it is demonstrated that the bc1 shows an enzymatic activity also as micellar suspension when the proper substrates are added. Moreover, it has been possible to reconstitute, with the droplet transfer method, the bc1 complex in artificial giant lipid vesicles. The reconstitution has been proved by confocal microscopy analysis demonstrating a change of internal pH, thanks to a pH-sensitive fluorescent dye, upon the addition of a trigger substrate.

1 Introduction

The construction from scratch of synthetic cells capable of displaying characteristics and behavior of living cells is the most ambitious goal in bottom-up synthetic biology [1–6]. In this framework, for long time, lipid vesicles have been studied since they are especially suitable for investigating the physico-chemical properties of lipid membranes. From the early 1990s, these compartmentalized systems have been considered, by pioneers in the field of origin-of-life, as cell mimicking structures in which several complex reactions can occur [7–11]. After the synthesis of the first functional protein within the liposome lumen in 2001 [12], several important results have been achieved, thanks to the integration of liposome technology and cell-free systems, and occasionally microfluidics (reviewed in P. Stano, 2011 [13]). This new and promising research field has a strong theoretical grounding in the autopoiesis theory [14, 15], and it is often supported by rigorous stochastic numerical modelling [16–23]. In these artificial structures both lipid boundaries and hydrophilic lumen have crucial roles. Giant lipid vesicles can be used as versatile micro-sized enzymatic reactors with catalytic proteins encapsulated in their lumen or embedded in their membrane [24]. Indeed, from the origin-of-life point of view, it is very interesting to understand what could be the crudest lipid mixture that can produce the artificial membrane [25] and, then, what is its role in the origin of the metabolism [26]. On the other hand, from the biotechnological point of

S. Piotto et al. (Eds.): BIONAM 2019, LNBE, pp. 223–232, 2020.
https://doi.org/10.1007/978-3-030-47705-9_19

view, the autonomous generation of chemical energy by the synthetic cells is one of the key functions not yet developed. One strategy could be the extraction, purification and reconstitution of biological organelles from living organisms, and encapsulate them in an artificial compartment (Multi Compartment Approach) [27].

Moreover, we focus our research and scientific efforts on a second strategy: obtaining artificial photosynthetic protocells by the reconstitution of the overall bacterial photosynthetic machinery in the lipid membrane of giant vesicles. We define this synthetic procedure as "Single Compartment Approach" (SCA), since the photosynthetic enzymes are all extracted, purified and reconstituted in the protocell membrane and not encapsulated in form of independent organelles in the internal aqueous lumen. The photosynthetic machinery is extracted from *Rhodobacter sphaeroides* and consists of three protein complexes: the photosynthetic reaction center (RC), cytochrome bc1 complex (bc1), and ATP synthase (ATPase) (Fig. 1). The reactions catalyzed by these proteins are the following: RC converts – under infrared irradiation – the quinone Q to quinol QH_2, extracting two protons from the cytoplasm-like side and oxidizing two cytochrome c_2 ($2cyt^{2+} \rightarrow 2cyt^{3+}$). In contrary, bc1 catalyzes the opposite reaction and further contributes to the proton gradient by decreasing the periplasm pH. Eventually, ATP synthase exploits the proton gradient to catalyze the formation of ATP (Fig. 1).

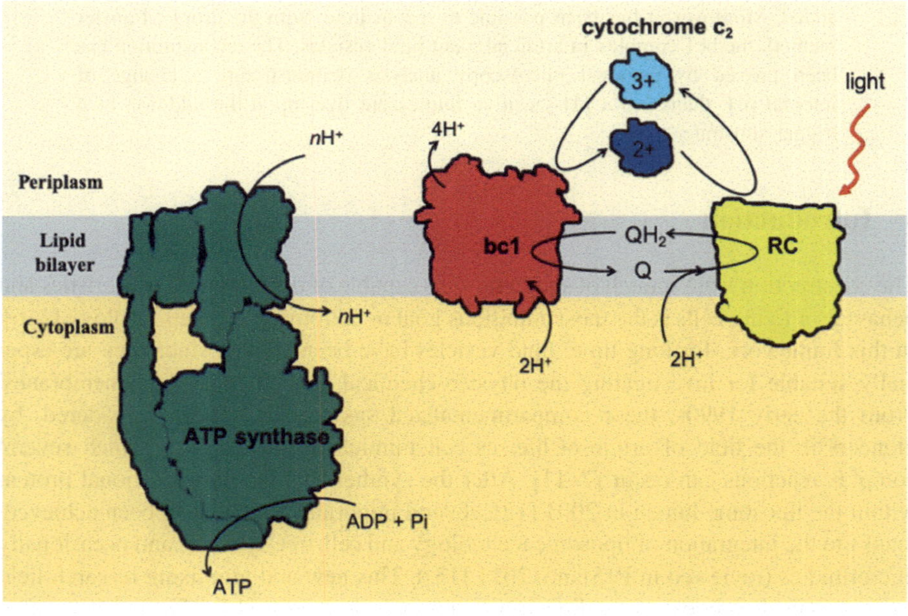

Fig. 1. Schematic representation of the light-transducing system in vivo. It consists of three protein complexes: the photosynthetic reaction centre (RC in yellow), cytochrome bc1 complex (bc1 in red), and ATP synthase (in green). The reactions catalyzed by these proteins are shown, based on current biochemical knowledge. RC converts – under irradiation – the quinone Q to quinol QH_2, abstracting two protons from the cytoplasm-like side and oxidizing two cytochrome c_2 ($2cyt^{2+} \rightarrow 2cyt^{3+}$). In contrary, bc1 catalyzes the opposite reactions and further contributes to the proton gradient, which is acidic in the periplasm. Eventually, ATP synthase exploits the proton gradient to catalyze the ADP phosphorylation.

Recently, we reported the reconstitution of the bacterial photosynthetic reaction center (RC) in the membrane of GVs in a one-step procedure by keeping a high degree of the RC physiological orientation [29]. In the GVs embedding RC on their lipid membrane, a transmembrane proton gradient is generated (alkaline pH inside) upon irradiation of infrared light. This is the first step for realizing energetically autonomous protocells, i.e. synthetic cells transducing light energy in chemical energy, able of sustaining internal metabolic processes. Namely, RC transduces light into chemical energy in form of a transmembrane pH gradient that can sustain the ATP production catalyzed by the ATP synthase. Thus, following this approach, the construction of photo-autotrophic synthetic cells can be implemented by the incorporation of the photosynthetic set of membrane proteins in the vesicle lipid bilayer, leading to ATP synthesis from photon energy by means of the creation and dissipation of a proton gradient. It is worthwhile to remark that this mechanism is made possible by the lipid membrane capability of separating efficiently the inner aqueous lumen of GVs from the external solution. However, the reconstitution of RC in GVs is just the initial step toward the realization of a sustainable photo-autotrophic synthetic cell. The second protein complex of the photosynthetic apparatus is the *R. sphaeroides* bc1, a trans-membrane protein formed by four subunits, three of which exhibit a redox activity (cytochrome b, cytochrome c1, and the Rieske iron-sulfur protein). The structure of bc1 [29] is asymmetric with respect to the membrane, as it has a larger polar end exposed to the periplasm, while the other, the smaller one, is exposed to the cytoplasm (Fig. 1). This physiological orientation is somehow "opposite" to the alignment of RC which has the large polar side (i.e., the H subunit) exposed to the cytoplasm. Based on these considerations, we have started an experimental program having the construction and functionalization of GVs with RC and bc1 (RC-bc1@GVs) as long-term goal.

Following the synthetic biology paradigm, we have proceeded firstly with the investigation and characterization of the single proteins and of their alone and coupled activity. Secondly, we focused on the reconstitution of the two proteins in the GV lipid membrane and the characterization of the protocell: RC-bc1@GVs. In fact, recently, we have already coupled RC and bc1 in form of simple micellar suspensions to study the enzymatic activity. Oscillations of the concentrations of the redox pair cyt^{2+}/cyt^{3+} have been induced by light irradiation and rationalized by means of a minimal kinetic model that reproduces quite well the experimental time courses [30]. Based on these results, we move a step forward by showing the reconstitution and the activity of bc1 in the GV membrane.

Thus, in this paper we will recall the extraction and purification procedures of bc1 complex from living bacteria then, we will test the protein functionality in bulk as micellar suspension and finally, we will show preliminary results related to enzyme reconstitution in the GV lipid membrane and the bc1 activity as a proton pump.

2 Methods

Bacterial bc1 Extraction and Purification

His-tagged cytochrome bc1 complex (bc1) was extracted and purified as described by Guergova-Kuras [31] with minor modifications. One volume of chromatophores was suspended at OD_{860} = 50 and then diluted with 0.3 volumes of buffer B (50 mM MOPS (pH 7.8), 100 mM NaCl, 1 mM $MgSO_4$, 20% (w/v) glycerol, and 5 mM histidine) and 0.7 volumes of Ni-NTA resin equilibrated in buffer B. Dodecyl maltoside (DM) (Sigma) was added dropwise from a stock solution 10% (w/v) to a final concentration of 0.6% (w/v). The mixture was incubated for 40 min at 4 °C, then transferred in a chromatographic column and washed with about 20 column volumes of buffer C (50 mM MOPS (pH 7.8), 100 mM NaCl, 1 mM $MgSO_4$, 0.01% DM), until the eluent was clear. A further washing with 2 column volumes of buffer C supplemented with 5 mM histidine was also performed. The purified bc1 was eluted with buffer containing 200 mM histidine and 15 μg/mL POPC (Avanti Polar-Lipids, Inc., Alabaster, AL). The histidine was removed by overnight dialysis against 50 mM MOPS (pH 7.0), 100 mM NaCl, 1 mM $MgSO_4$, 20% (w/v) glycerol, 0.01% DM, and 15 μg/mL POPC. Purified bc1 was aliquoted and frozen in liquid nitrogen and stored at −20 °C until used. The concentrations of cytochrome b and cytochrome c1 in the final preparation were measured from dithionite-reduced minus ferrocyanide-oxidized difference spectra as previously described [32].

bc1 Reconstitution in Giant Vesicle

For giant vesicle preparation, the droplet transfer method was applied [33], with a protocol optimized in the recent past years [28, 34]. In a 1.5 mL Eppendorf tube (tube A), 300 μL of organic phase consisting in 0.5 mM 1-palmitoil-2-oleoyl-*sn*-glycero-3-phosphocholine (POPC) (Lipoid GmbH, Steinhausen, Switzerland) dispersion in mineral oil (Sigma-Aldrich, #M5904), were gently overlaid over 500 μL of aqueous O-solution (200 mM glucose and sodium phosphate buffer 10 μM pH 7.4), in order to create a lipid-covered interfacial region (Fig. 2a). In another 1.5 mL Eppendorf tube (tube B), 20 μL of I-solution (200 mM sucrose, bc1 micelles and the same buffer of the O-solution, and other solutes when required) were added to 600 μL of 0.5 mM POPC in mineral oil. A water-in-oil (w/o) emulsion was obtained by pipetting repeatedly up and down the mixture for 30 s (Fig. 2b). Next, the w/o emulsion (tube B) was gently poured on top of the organic phase in tube A. Tube A was centrifuged at 2,500 g for 10 min at room temperature letting the droplets to cross the interface region (Fig. 2c). After the centrifugation, the mineral oil appeared clear and it was removed. The resulting protocells, forming a pellet on the bottom of the tube, were gently collected by direct aspiration with a polypropylene micropipette tip (50 μL). To obtain an oil-free protocells suspension, care was taken to wipe off, with a tissue paper, the small amount of oil possibly collected on the outside of the pipette tip. Protocells were washed twice by centrifugation, supernatant removal, and re-suspension in fresh O-solution to remove non-entrapped substances.

When a bc1 micellar suspension is added to the I-solution, as soon as the bilayer is formed the protein can be reconstituted in the membrane exposing, preferentially, the hydrophilic domain towards the water core of the vesicle (Fig. 2d).

I-solution and O-solution, although isotonic, are prepared at two different densities (sucrose in I-solution, glucose in O-solution) to facilitate the crossing of the interface region and sediment at the bottom of the tube. (d) Schematic representation of the behavior of a single bc1 surrounded by a toroid of detergent molecules when added in the I-solution. As soon as the bilayer is formed the protein can be reconstituted in the membrane exposing the hydrophilic domain towards the water core of the vesicle.

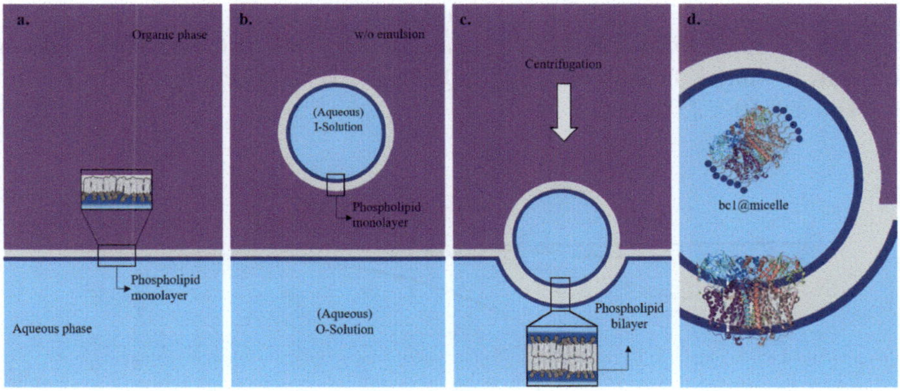

Fig. 2. Preparation of GVs by the droplet transfer method. (a) A lipid-in-oil solution is gently stratified over an "outer" aqueous solution (O-solution) which will constitute the external phase of GVs. The lipid molecules will organize in form of a monolayer at the water/oil interface (w/o interface) as schematically shown in the zoom on the left. (b) In another vial, a small amount of "inner" aqueous solution (I-solution) is emulsified in a lipid-containing oil solution by pipetting. The resulting emulsion is gently poured over the system prepared in (a). For the sake of simplicity, only one w/o droplet is shown. The w/o droplet are coated (stabilized) by a lipid monolayer oppositely oriented respect the one at the interface region in figure (a). (c) The system is then centrifuged so that the w/o droplets cross the interface becoming covered by the second lipid monolayer, forming a lipid bilayer, as schematically shown in the zoom. Consequently, w/o droplets become completely covered by a lipid bilayer when successful transferred in the O-solution (not shown). (d) Schematic representation of bc1 reconstitution in the lipid membrane.

3 Results

Enzyme Activity as Micellar Suspension

We next investigated the retained activity of the bc1 after the extraction and purification. bc1@micelles in Tris–HCl buffer (pH = 7.4), in the presence of 0.05% DM as stabilizer, were added to a solution containing the substrates of the enzymes, namely cyt^{3+}, QH_2 and Q. The enzyme-catalysed reaction is the following:

$$QH_2 + 2\,cyt^{3+} \rightarrow Q + 2\,cyt^{2+} + 2H^+$$

The time course of the reaction is monitored by following the absorbance at 550 nm exploiting the difference between cyt^{2+} and cyt^{3+} molar extinction coefficients (namely, $\varepsilon_{cyt2+} = 28$ mM/cm and $\varepsilon_{cyt3+} = 9$ mM/cm) (Fig. 3). The grey curve is referred to the cyt^{3+} enzymatic reduction in a solution containing: [bc1] = 210 nM, $[cyt^{3+}]_0 = 20$ μM, $[cyt^{2+}]_0 = 0$ μM, $[Q]_0 = 50$ μM, $[QH_2]_0 = 150$ μM. This curve clearly shows that bc1 in form of a micellar suspension keeps its enzymatic activity if compared with the orange curve that reproduces the spontaneous reduction of cyt^{3+} when the bc1 is not present in solution. Moreover, control experiments are referred to the spontaneous ([bc1] = 0 nM, blue curve) and catalysed ([bc1] = 210 nM, yellow curve) oxidation of cyt^{2+} when $[cyt^{2+}]_0 = 20$ μM, $[cyt^{3+}]_0 = 0$ μM, with the same concentrations of Q and

QH_2, as before. It is important to mention that the cyt^{3+} reduction is the thermodynamically favoured process, as it can be rationalized by looking the standard reduction potentials, which favour the cyt^{2+}/Q state ($E'^{\circ} cyt^{3+}/cyt^{2+} = +0.22$ V; E'° $Q/QH_2 = +0.10$ V [35].

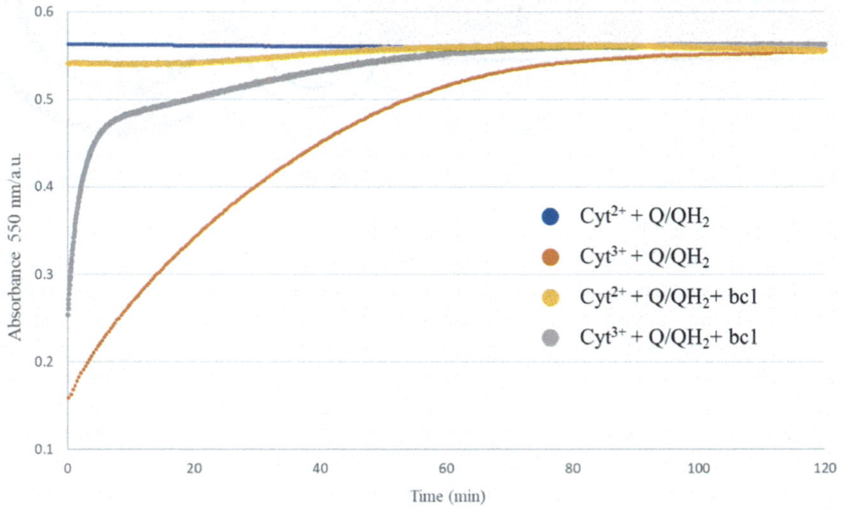

Fig. 3. Kinetics of redox reactions involving cytochrome c_2 redox species, monitored following the absorbance at 550 nm ($\varepsilon_{cyt2+} = 28$ mM/cm and $\varepsilon_{cyt3+} = 9$ mM/cm). The grey curve is obtained for $[cyt^{3+}]_0 = 20$ μM, $[cyt^{2+}]_0 = 0$ μM, $[Q]_0 = 50$ μM, $[QH_2]_0 = 150$ μM. The control experiments are performed in the same conditions except for $[cyt^{2+}]_0 = 20$ μM and $[cyt^{3+}]_0 = 0$ μM for the spontaneous oxidation of cyt^{2+} in the presence of Q/QH_2 (blue trace) and bc1 (yellow curve). In orange, the spontaneous reduction of cyt^{3+} in the presence of Q and QH_2.

Photo-Activity in Lipid Membrane

GVs have been prepared as previously described (Methods section) adding in the so-called I-solution also 5 μM bc1 as micellar suspension, 20 μM of the pH sensitive dye pyranine and 20 μM of cyt^{3+}. The phosphate buffer (10 μM, pH 7.5) is added at low concentration to set the initial pH value around 7.5, letting possible the fluorometric detection of the pH time change. On the left of the Fig. 4 is shown the scheme of the experiment in which a GV is represented functionalized with several reconstituted bc1 (two of them highlighted with violet dashed ovals) and pyranine and cyt^{3+} molecules confined in its lumen. After the addition of the mixture Q/QH_2 in the bulk environment, bc1 can catalyse the redox reaction across the membrane consisting of the reduction of cyt^{3+} to cyt^{2+} and the oxidation of QH_2 to Q with a proton translocation in the GV lumen. This process causes an acidification of the aqueous core of the compartments and consequently the decrease of the pyranine green fluorescence after 15 min of incubation (Fig. 4 green box). Due to the vectorial properties of the protein complex we can assume that only bc1 oriented with the hydrophilic domain towards the internal aqueous phase of the GV can contribute to the acidification process. The bc1 oriented in the opposite way cannot catalyse any reaction since its active site is not in contact with cyt^{3+} molecules.

Control experiment (Fig. 4 yellow box on the right) confirms that in absence of QH_2 there is no fluorescence change after 15 min of incubation.

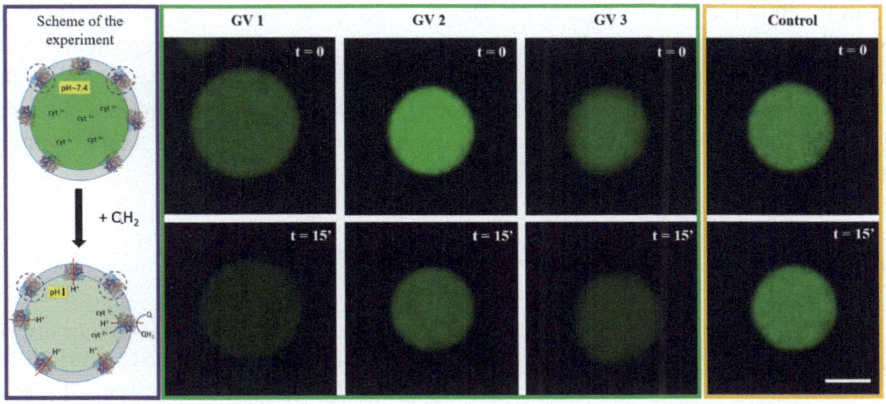

Fig. 4. On the left, cartoon representation of the experiment: GV containing pyranine and cyt^{3+} with bc1 molecule reconstituted in the lipid membrane. Two of them marked with dash ovals are reconstituted with opposite orientation. After the addition of QH_2 only bc1 oriented with the hydrophilic domain exposed towards the aqueous lumen, catalyze the redox reaction $2cyt^{3+} + QH_2 \rightarrow 2cyt^{2+} + Q + 2H^+$ with a net transport of protons inside the GV. The decrease of the green fluorescence confirms the acidification process. In the green box are shown confocal images of three different GVs as soon as QH_2 is added and after 15 min of incubation. The evident fluorescence reduction proves the pH decrease inside the GVs. The yellow box on the right refers to the control experiment of a GV analyzed without the addition of QH_2. As expected, the internal fluorescence results unaltered if one of the redox species is absent. Scale bar is 10 μm.

4 Conclusion

The most ambitious goal in bottom-up synthetic biology is the construction from scratch of synthetic cells capable of displaying characteristics and behavior of living cells. A crucial and challenging requirement for these protocells is to be energetically autonomous, i.e. the capability of taking up energy from the external environment to sustain an internal metabolism. In this contribution, we have recalled some results recently achieved by our group in this field [24, 27, 28, 30] and some new outcomes. In particular, we have shown as, being inspired by photosynthetic bacteria, it could be possible to extract and reconstitute in the lipid membrane of giant vesicles the protein complexes involved in the bacterial photosynthesis. With this approach, that we called the Single Compartment Approach, it has been possible to prepare transducing light energy vesicles, i.e. protocells able to convert light energy into chemical energy in form of a transmembrane proton gradient [28]. This has been achieved by reconstituting in the GV lipid membrane only one photosynthetic protein: the reaction center RC. However, although the RC embedding giant vesicles behave like light-chemical energy transducers, they can work only in an excess of substrates: cyt^{2+} and Q. Thus, to be energetically autonomous, at least the second enzyme bc1 is required, to implement a phot-redox cycle that continuously exploits light energy to pump protons across the lipid membrane. In this contribution, the first results of the reconstitution of bacterial bc1 in the lipid membrane of giant vesicles has been presented and discussed, showing how bc1 keeps its enzymatic activity also in synthetic protocells. This represents a step forward the reconstitution of the complete bacterial photosynthetic machinery in the vesicle membrane and allows preparing protocells able to transduce light into chemical energy in form of ATP molecules by the enzymatic phosphorylation of ADP driven by a transmembrane pH gradient.

References

1. Fujii, S., Matsuura, T., Sunami, T., Nishikawa, T., Kazuta, Y., Yomo, T.: Liposome display for in vitro selection and evolution of membrane proteins. Nat. Protoc. **9**, 1578–1591 (2014)
2. Kurihara, K., Tamura, M., Shohda, K.-I., Toyota, T., Suzuki, K., Sugawara, T.: Self-reproduction of supramolecular giant vesicles combined with the amplification of encapsulated DNA. Nat. Chem. **3**, 775–781 (2011)
3. Kuruma, Y., Stano, P., Ueda, T., Luisi, P.L.: A synthetic biology approach to the construction of membrane proteins in semi-synthetic minimal cells. Biochim. Biophys. Acta **1788**, 567–574 (2009)
4. Lentini, R., Santero, S.P., Chizzolini, F., Cecchi, D., Fontana, J., Marchioretto, M., Del Bianco, C., Terrell, J.L., Spencer, A.C., Martini, L., Forlin, M., Assfalg, M., Dalla Serra, M., Bentley, W.E., Mansy, S.S.: Integrating artificial with natural cells to translate chemical messages that direct E. coli behaviour. Nat. Commun. **5**, 4012 (2014)
5. Noireaux, V., Libchaber, A.: A vesicle bioreactor as a step toward an artificial cell assembly. Proc. Natl. Acad. Sci. U.S.A. **101**, 17669–17674 (2004)
6. Pohorille, A., Deamer, D.: Artificial cells: prospects for biotechnology. Trends Biotechnol. **20**, 123–128 (2002)
7. Chakrabarti, A., Breaker, R., Joyce, G., Deamer, D.: Production of RNA by a polymerase protein encapsulated within phospholipid-vesicles. J. Mol. Evol. **39**, 555–559 (1994)
8. Walde, P., Goto, A., Monnard, P., Wessicken, M., Luisi, P.: Oparins reactions revisited—enzymatic-synthesis of poly (adenylic acid). J. Am. Chem. Soc. **116**, 7541–7547 (1994)
9. Oberholzer, T., Wick, R., Luisi, P.L., Biebricher, C.K.: Enzymatic RNA replication in self-reproducing vesicles: an approach to a minimal cell. Biochem. Biophys. Res. Commun. **207**, 250–257 (1995)
10. Oberholzer, T., Albrizio, M., Luisi, P.L.: Polymerase chain-reaction in liposomes. Chem. Biol. **2**, 677–682 (1995)
11. Oberholzer, T., Nierhaus, K.H., Luisi, P.L.: Protein expression in liposomes. Biochem. Biophys. Res. Commun. **261**, 238–241 (1999)
12. Yu, W., Sato, K., Wakabayashi, M., Nakaishi, T., Ko-Mitamura, E.P., Shima, Y., Urabe, I., Yomo, T.: Synthesis of functional protein in liposome. J. Biosci. Bioeng. **92**, 590–593 (2001)

13. Stano, P., Carrara, P., Kuruma, Y., de Souza, T.P., Luisi, P.L.: Compartmentalized reactions as a case of soft-matter biotechnology: synthesis of proteins and nucleic acids inside lipid vesicles. J. Mater. Chem. **21**, 18887–18902 (2011)
14. Varela, F.G., Maturana, H.R., Uribe, R.: Autopoiesis: the organization of living systems, its characterization and a model. Biosystems **5**, 187–196 (1974)
15. Luisi, P.L.: Autopoiesis: a review and a reappraisal. Naturwissenschaften **90**, 49–59 (2003)
16. Shirt-Ediss, B., Ruiz-Mirazo, K., Mavelli, F., Solé, R.V.: Modelling lipid competition dynamics in heterogeneous protocell populations. Sci. Rep. **4**, 5675 (2014)
17. Mavelli, F., Altamura, E., Cassidei, L., Stano, P.: Recent theoretical approaches to minimal artificial cells. Entropy **16**(5), 2488–2511 (2014)
18. Mavelli, F.: Stochastic simulations of minimal cells: the Ribocell model. BMC Bioinf. **13** (Suppl 4), S10 (2012)
19. Mavelli, F., Stano, P.: Kinetic models for autopoietic chemical systems: the role of fluctuations in a homeostatic regime. Phys. Biol. **7**, 016010 (2010)
20. Mavelli, F., Ruiz-Mirazo, K.: ENVIRONMENT: a computational platform to stochastically simulate reacting and self-reproducing lipid compartments. Phys. Biol. **7**, 036002 (2010)
21. Mavelli, F., Ruiz-Mirazo, K.: Stochastic simulations of minimal self-reproducing cellular systems. Philos. Trans. R. Soc. Lond. B Biol. Sci. **362**, 1789–1802 (2007)
22. Carrara, P., Altamura, E., D'Angelo, F., et al.: Measurement and numerical modelling of cell-free protein synthesis: combinatorial block-variants of the PURE system. Data **3**(4), 41 (2018)
23. Mavelli, F., Albanese, P., Altamura, E.: Modelling a light transducing protocell population. In: Piotto, S., Rossi, F., Concilio, S., et al. (eds.) Advances in Bionanomaterials. Lecture Notes in Bioengineering. Springer, Cham (2020)
24. Altamura, E., Stano, P., Walde, P., et al.: Giant vesicles as micro-sized enzymatic reactors: perspectives and recent experimental advancements. Int. J. Unconv. Comput. **11**(1), 5–21 (2015)
25. Fayolle, D., Altamura, E., D'Onofrio, A., et al.: Crude phosphorylation mixtures containing racemic lipid amphiphiles self-assemble to give stable primitive compartments. Sci. Rep. **7**, 18106 (2017)
26. de Souza, T.P., Stano, P., Steiniger, F., et al.: Encapsulation of ferritin, ribosomes, and ribo-peptidic complexes inside liposomes: insights into the origin of metabolism. Orig. Life Evol. Biosph. **42**(5), 421–428 (2012)
27. Albanese, P., Altamura, E., Milano, F., et al.: Transducing light bacterial vesicles as potential organelles in artificial compartments. In: Piotto, S., Rossi, F., Concilio, S., et al. (eds.) Advances in Bionanomaterials. Lecture Notes in Bioengineering. Springer, Cham (2020)
28. Altamura, E., Milano, F., Tangorra, R.R., Trotta, M., Omar, O.H., Stano, P., Mavelli, F.: Highly oriented photosynthetic reaction centers generate a proton gradient in synthetic protocells. Proc. Natl. Acad. Sci. U.S.A. **114**(15), 3837–3842 (2017)
29. Esser, L., Elberry, M., Zhou, F., Yu, C.-A., Yu, L., Xia, D.: Inhibitor-complexed structures of the cytochrome bc1 from the photosynthetic bacterium Rhodobacter sphaeroides. J. Biol. Chem. **283**, 2846–2857 (2008)
30. Altamura, E., Fiorentino, R., Milano, F., et al.: First moves towards photoautotrophic synthetic cells: in vitro study of photosynthetic reaction centre and cytochrome bc1 complex interactions. Biophys. Chem. **229**, 46–56 (2017)
31. Guergova-Kuras, M., Salcedo-Hernandez, R., Bechmann, G., Kuras, R., Gennis, R.B., Crofts, A.R.: Expression and one-step purification of a fully active polyhistidine-tagged cytochrome bc1 complex from Rhodobacter sphaeroides. Protein Expr. Purif. **15**, 370–380 (1999)

32. Capaldi, R.A., Schulenberg, B., Murray, J., Aggeler, R.: Cross-linking and electron microscopy studies of the structure and functioning of the Escherichia coli ATP synthase. J. Exp. Biol. **203**, 29–33 (2000)
33. Pautot, S., Frisken, B.J., Weitz, D.A.: Production of unilamellar vesicles using an inverted emulsion. Langmuir **19**, 2870–2879 (2003)
34. Rampioni, G., et al.: Synthetic cells produce a quorum sensing chemical signal perceived by Pseudomonas aeruginosa. Chem. Commun. **54**, 2090–2093 (2018)
35. Berg, J.M., Tymoczko, J.L., Stryer, L.: Biochemistry, 5th edn. W. H. Freeman, New York (2002)

Image-Guided Mini-Invasive Treatments for Vascular and Oncologic Diseases: Embolization Therapy

Jacopo Santoro[1], Miriam Di Martino[1], Stefano Piotto[1], Simona Concilio[2], and Marco Midulla[3(✉)]

[1] Department of Pharmacy, University of Salerno, via Giovanni Paolo II 132, Fisciano, SA, Italy
[2] Department of Industrial Engineering, University of Salerno, via Giovanni Paolo II 132, Fisciano, SA, Italy
[3] Department of Diagnostic and Therapeutic Radiology, Center for Mini-Invasive Image-Guided Therapies, Université de Bourgogne Franche-Compté, Centre Hospitalier Universitaire de Dijon, Dijon, France
marcomidu@gmail.com

Abstract. Transcatheter Embolization, also called Embolotherapy, is a mini-invasive, non-surgical therapeutic solution used in Interventional Radiology to close blood vessels deliberately. A wide range of embolic agents is available in clinical practice, including metal coils and liquid agents. More recent advances in biomaterials such as shape-memory foam and in-situ gelling solutions have led to the development of new pre-clinical embolic agents. This review offers a brief overview of current and emerging technologies in the field of endovascular embolization. The focus is on devices, materials and techniques.

Keywords: Interventional Radiology · Embolic agents · Endovascular treatment

1 Introduction

Recent decades have seen a gradual shift from complex open surgery to new image-guided treatments providing mini-invasive patient care with fewer complications, comparable or sometimes better clinical outcomes, and eventually lower costs. The endovascular approach uses the vascular system, either arteries or veins, to get direct access to the different organs by navigation into the vessels from a peripheral puncture site. Embolization is defined as the intentional occlusion of a vessel or a whole vascular territory utilizing different agents injected from an intravascular catheter [1]. Embolization can be used to treat a wide range of pathologies. For instance, it can be used to treat life-threatening bleeding after trauma, different vascular lesions such as aneurysms (Fig. 1A), and malformations like arteriovenous malformation (AVM) (Fig. 1B), and eventually tumors in combination with other treatment modalities.

S. Piotto et al. (Eds.): BIONAM 2019, LNBE, pp. 233–242, 2020.
https://doi.org/10.1007/978-3-030-47705-9_20

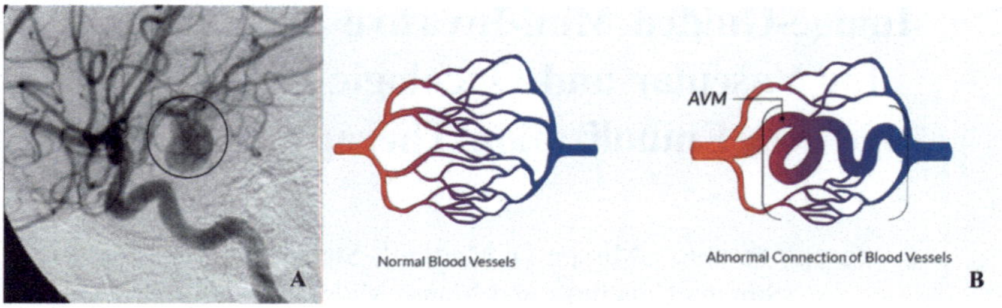

Fig. 1. A) Angiography of an aneurysm in a cerebral artery [2]; B) Schematic representation of an arteriovenous malformation (AVM) [3].

The first embolization procedures were performed in the late 70 s using muscle fragments, blood clots, and stainless-steel granules injected through the catheters [4]. Since then, significative advances in image-guidance technology and endovascular materials allowed a considerable development of the embolization field in a wide range of different therapeutic techniques, so-called Embolotherapy [5].

Several embolic agents have been developed to adapt to various clinical conditions, and they can be sorted into three main categories: metallic implants (coils and plugs), particle agents, and liquid agents.

Coils and plugs allow complete occlusion of a large or small vessel in a specific site. They are commonly used to treat bleeding conditions such as vessel rupture after trauma (visceral organs: spleen, kidneys, liver; upper and lower extremities; neck) by stopping the flow towards the damaged territory. Embolization is nowadays considered as the first-line treatment, if feasible, before surgery.

Coils and plugs can also be used to fill aneurysms, which are a focal dilatation of a cerebral of peripheral artery potentially at risk of rupture.

Particle agents are more adopted to close small vessels, such as the capillary bed of tumors. They are injected through the catheter and vehiculated through the blood flow into the pathologic territory (flow-directed embolization). This provides ischemic effects and, hopefully, necrosis of the pathologic lesion.

Liquid agents such as glue and polymers are the most recent and useful to manage complex conditions such as arteriovenous malformations (AVM). In these congenital pathological entities, the physiologic capillary bed is replaced by direct communication between the arterial and venous system in a specific region (brain; extremities; organs), so inducing different clinical consequences (local tissue alterations; heart insufficiency; cerebral stroke). The design of polymers is particularly delicate since their physical-chemical properties, and their phase transition must be considered. The design of a new polymer can be performed by traditional computational methods [6–8] or by artificial intelligence techniques. The design of new polymers for embolization therapy is particularly challenging because of the spatial confinement effect [9] exerted by the walls of veins and arteries.

2 Embolization Materials

2.1 Mechanical Occlusion Devices

2.1.1 Coils

A coil is pushed through the catheter by pulsed manual injection or using a pushing wire so that once into the vessel, it will shape and block the flow. Two main types of coils exist: pushable and removable. The second one allows better control as the device can be removed in case of misplacement. Coils consist mainly of a bioinert metal core (stainless steel, platinum or nitinol) with, eventually, an external coating. Metals are radio-opaque, thus allowing visualization under fluoroscopy. The embolization mechanism of the metal coil is mainly a mechanical occlusion [10]. Therefore, bioactive coatings have been developed to improve clot formation and thrombogenicity of the device [11]. This aspect might become a critical feature for specific conditions.

Polymer-Coated Coils

Biodegradable copolymers of polyglycol/polylactic acid copolymers (PGLA) or poly (glycolic acid) polymers (PGA) have been used as a coating for platinum coils to stimulate tissue reaction and clot formation [12]. It has been reported that the use of platinum coils coated with PGLA accelerates the healing of aneurysms with increased neointimal formation and pronounced fibrosis compared with bare platinum coils [13]. Unfortunately, PGLA coated coils were also more difficult to transport through the catheter due to high stiffness and to the friction against the catheter wall [14]. An alternative was found by Ahuja et al. [15] with a polyurethane coating. In their work [15], the thrombogenicity of the polyurethane-coated platinum coil was tested against a bare platinum one in rabbit aorta. The bare platinum coil in the aorta resulted in a 40% embolization in 30 min, while polyurethane-coated coil resulted in a 100% occlusion in 30 min [15]. SEM images of the coils obtained after implantation confirmed that the most thrombogenic polyurethane-coated platinum coils were covered with the most massive amount of thrombus.

Protein-Coated Coils

Fibroblast invasion and thrombus formation are critical steps in healing an aneurysm [16]. The essential proteins involved in the process of healing and recanalization are monocyte chemoattractant protein 1 (ICB-1), [17] osteopontin (OPN), interleukin-10 (IL-10) and matrix metallopeptidase 9 (MMP-9) [16]. In particular, a platinum coil coated with a human vascular endothelial growth factor promotes endothelialization, blood clotting, and tissue/coil integration to accelerate aneurysm closure [18].

Shape-Memory Polymeric Coils

While embolic coils are considered primarily for permanent occlusion, a bioabsorbable water-induced shape-memory plug has recently been introduced for temporary occlusion of vessels [19]. Radio opaque substances (e.g., barium sulfate, tantalum, and bismuth oxychloride) are used as a doping agent for the coil core. The coil is coated with cross-linked polyethylene glycol diacrylate (PEGDA) to enhance vessel occlusion. Shape-memory behavior is caused by phase transformations between the crystalline and amorphous states of PLGA and PEGDA polymers in response to both physiological infusion and thermal activation. At high temperatures (70 °C), both polymers are in amorphous phase and can be temporarily deformed. The temporary configuration can then be fixed at temperatures below the glass transition of PEG by immobilizing the crystalline phase of PEG. The PEGDA coating is expanded to improve occlusion in combination with the activated crystalline phase of PEG. In particular, the fold begins in the PLGA core, which gradually overcomes the resistance of the PEGDA coating, and let the coil to regain its programmed shape. In Fig. 2, a schematic representation of the coil formation, as proposed in the work of Zhang et al. [20], is shown. The composite polymeric plug has been investigated in several arteries in rabbits and showed a total occlusion time from 38 s to 2 min [19].

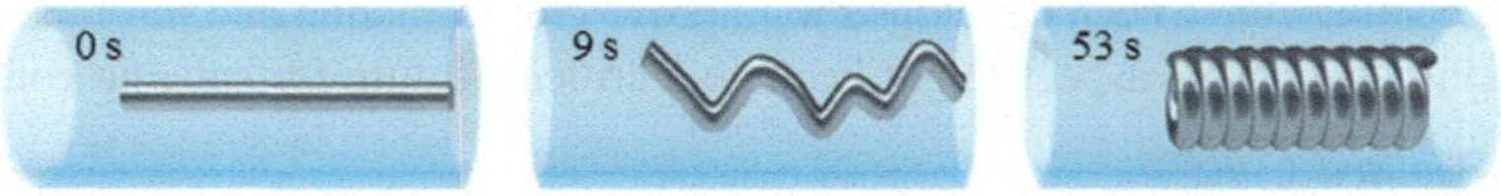

Fig. 2. The shape-memory-induced embolism overtime at 37 °C [20]. The blood vessel is represented as a cyan tube; the coil is in gray at 0 s, 9 s, and 53 s.

2.2 Particles Agents

Particles were the first embolic agent to be developed and are currently the most widely used due to their versatility [21, 22]. Although noncalibrated particulates were commonly used in the past and are still used today in hospitals, in vivo trajectories and degree of occlusion are generally unpredictable. Nowadays, particles with a controlled and narrow distribution, and spherical shape, are primarily preferred. There is a limit to how small particles can be used for embolization. It has been reported that particles with an average diameter of 9 μm can cause pulmonary embolism in rabbits [23]. 40 μm is the accepted lower limit for particulate embolism agents. Poly(vinyl alcohol) particles (PVA) have been used since the 1970s. They are produced by mechanical fragmentation of the PVA polymer block and subsequent sieving. However, PVA particles tend to aggregate due to surface charges and surface hydrophobicity in physiological solutions. This can lead to unintentional occlusion of larger proximal vessels [24]. The main problems of PVA particles are associated with their irregular shape and lack of size accuracy. Noncalibrated PVA particles cause immediate mechanical occlusion of the blood vessels, followed by thrombus formation [25]. The exact size of the particles is crucial for local embolization, as the blood flow drives the supply of microspheres and their accumulation in vivo is size-dependent. For example, distal microvascular embolization is often desired in tumor embolization [10]. Once distal embolization is complete, more proximal vessels can be embolized with larger particles.

2.3 Liquid/Gel Embolic Agents

There is a growing interest in liquid-gel embolic agents for peripheral interventions. These agents can generally diffuse to targets that catheters and coils cannot reach.

Liquid/gel embolisms block blood flow, creating permanent occlusions that are not dependent on the patient's coagulation system and lead to the formation of thrombosis [26]. Liquid/gel embolisms usually require a phase transformation of the material between administration (i.e., in syringe and catheter) and the body. Low viscosity is desired to pass through the delivery system. At the same time, the formation of a more rigid and stable structure is required after the catheter is removed to allow the vessel to be occluded. The phase transition can be achieved by physical reticulation (e.g., precipitation and ionic reticulation) or chemical reticulation (e.g., the formation of a stable polymer network) [27].

2.3.1 Sclerosing Agents

Sclerosing agents are, for example, detergents, osmotic agents, and chemical irritants that induce blood vessel sclerosis. Ethanol is one of the most widely used sclerosing agents for embolization [28, 29]. It leads to permanent damage to endothelial cells and denaturation of blood proteins; in addition, it causes necrosis of the vascular wall. Ethanol is affordable and accessible, although its low viscosity limits its therapeutic window, given the risk of diffusion beyond the target. A more viscous ethanol-based emulsion ($\approx$0.55 Pa s at 20 °C) is commercially available as Ethibloc (Ethnor Laboratories/Ethicon, Norderstedt, Germany), which includes corn protein zein, sodium diatrizoate, absolute ethanol and oleum papaverius [28]. It is a radiopaque agent for the treatment of AVM, although its use has declined over the past ten years [28].

2.3.2 In Situ Polymerization

N-Butyl-2-cyanoacrylate (nBCA) (e.g. TruFill), commonly known as "glue", is a transparent liquid used for embolization since the mid-1980s [30]. The low viscosity nBCA adheres to the tissue through in situ polymerization, initiated by a radical activator in contact with the tissue fluid. Due to rapid polymerization [31], the nBCA cannot penetrate sufficiently into the tumor bed or reach distal vessels, but it is excellent for embolizing branches that are 500 μm or larger. Clinically, the nBCA has been used primarily to treat high flow vascular malformations [32]. Its high thrombogenicity and excellent hemostatic properties require only minimal injected volume for occlusion, reducing embolization/fluoroscopy time and radiation exposure [32].

2.3.3 In Situ Precipitating Fluids

In-situ precipitation gelling solutions are composed of a polymer in a solvent. After replacing the solvent with body fluids, the polymer deposits and forms a barrier [28, 33]. One of the first gelling solutions ever made consisted of ethylene-vinyl alcohol (EVOH) and dimethyl sulfoxide solvent (DMSO) [34]. The main limitation of the EVOH system is associated with the use of an organic solvent. DMSO showed both local and systemic cardiovascular toxicity. In addition, since DMSO is a powerful plastic polymer solvent, specialized and expensive catheters compatible with DMSO are required. It is important to notice that the rapid injection of gelling solutions containing DMSO induces vasospasm and can cause vascular toxicity [35]. Therefore, a slow injection rate (2–4 mL min^{-1} in peripheral circulation and 0.6 mL min^{-1} in cerebral circulation) is critical and necessary [36]. An alternative organic solvent to replace DMSO is N-methyl-2-pyrrolidone (NMP), which produces less vasospasm than

DMSO when used as a vector for precipitation of liquid embolisms. Iodinated PVA liquid agents dissolved with NMP can effectively embolize aneurysms [37]. The composition of the gel can be changed to reach good injectability and good mechanical properties, without the use of organic solvents [38].

A commercially available product based on EVOH gelation solutions is Onyx (Onyx, Micro Therapeutics, Inc., Irvine, CA, USA), used primarily in cerebral and pulmonary AVMs, as well as in pseudoaneurysms [39]. Onyx has been approved by the FDA for the treatment of cerebral AVMs in 2005 [32]. It has been shown to block vessels up to 5 μm, demonstrating Onyx's ability to penetrate deeper than nBCA [40]. Onyx is also easier to manipulate than nBCA [41]. Both agents cause inflammatory reactions [40], but Onyx gives rise to smaller inflammatory areas, increasing tolerability for patients [42].

3 New Frontiers of the Embolotherapy and Mini-Invasive Treatments

3.1 Memory Polymer Foam

Due to the inherent risks of metal-based coils, including inflammation and recanalization, shape-memory polymer (SMP) foam occlusion devices have been proposed as a potential alternative [43]. They have many favorable properties that make them suitable as aneurysm filler materials [44]. The main advantages include configurable pore sizes that can be adapted for optimal cell infiltration and stable tissue integration into the foam (Fig. 3).

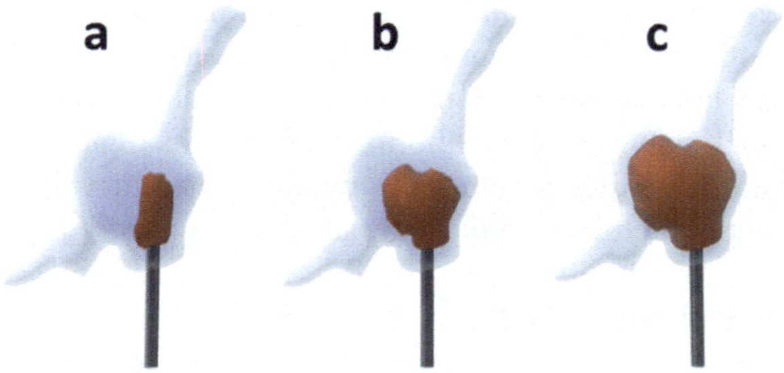

Fig. 3. The shape of an SMP foam for catheter delivery into an aneurysm [44, 45]. From left to right is represented (a) the insertion of the polymer foam (in brown), (b) the expansion of the foam, (c) the total recovery of the shape of the SMP foam until the aneurysm (in blue) is filled.

Compared to the extensive coagulation that occurs in conventional metal coils, the porous polymeric foam allows the formation and binding of multiple fine clots, resulting in faster clotting and healing. Porosity also allows the growth of granulation tissue and the formation of neointima to seal the aneurysm neck. These properties effectively reduce the chances of recanalization [46]. Various SMP materials, including polyurethanes and epoxies, have been designed for a wide range of applications, and most of them are thermo-sensitive. Compared to shape-memory metal alloys, like Nitinol that is also used for some endovascular stents, polyurethane-based SMP foams offer the advantages of lower density, higher recovery in elasticity (after plastic deformation), and lower cost. In addition, the mechanical properties and transition temperature of SMP foams can be adapted over a wide range with minimal changes in chemical structure and composition [46]. The polyurethane, besides its biocompatibility and non-mutagenicity, can also serve as a carrier of drugs and hemostatic agent [47]. SMP materials can be temporarily programmed to a secondary compressed form, and to maintain that form until the temperature of the material is raised above the T_g. In general, these polymers often have cross-linked structures (chemical or physical cross-linking) that determine their permanent form. To achieve high efficiency in medical embolic applications, such as aneurysm treatment, these materials must have very low density to allow transcatheter delivery and a high expansion volume. Still, at the same time, they should maintain good mechanical properties and excellent shape recovery. In recent years, extensive investigations have been conducted on these materials regarding their application for endovascular treatment of aneurysms. Most of the shape memory foams reported in the literature for biomedical applications are based on thermoplastic polyurethanes developed by Mitsubishi Heavy Industries, mainly MF5520, MF6020, and MF21 [48]. Singhal et al. [49] have reported the development of new polyurethane foams of different compositions with high recovery force, controllable drive temperatures, excellent biocompatibility, and open-cell morphology. Wilson et al. [50] have synthesized a polymer foam composed of N,N,N',N'-Tetrakis(2-hydroxypropyl) ethylenediamine (HPED), 2,2',2''-Nitrilotriethanol (TEA) and 1,6 diisocyanatohexane (HDI), this foam has a highly chemically crosslinked structure [50], high modulus, high recovery stress, finely controllable activation temperature, and excellent shape memory behavior.

The activation temperature of the foams can be customized in the range of 45–70 °C by modifying the composition of the constituent monomers, very low densities (0.015-0.021 g/cm^3), and excellent shape recovery of 97–98% can be achieved [51]. However, a premature activation may cause a shape recovery of the polymer inside the catheter [52], and this event can dramatically increase friction inside the catheter and potentially inhibit the delivery of the device. A new series of low-density SMP foams have been synthesized by modulating the hydrophobicity of the diisocyanate monomer, in order to facilitate transcatheter delivery of the device [52].

In a study by Rodriguez et al. [53], SMP foams have been implanted in a porcine aneurysm, and the devices were monitored for up to 90 days in order to assess their biocompatibility. Rodriguez et al. [53] have demonstrated that SMP foam has a reduced inflammatory response compared to conventional suture materials (polypropylene monofilament, silk). Subsequent histological examinations have verified that SMP foams are biocompatible and effective in providing a biological scaffold that can improve the healing response.

These studies emphasize the potential of SMP foams as viable mechanical occlusion devices for the treatment of aneurysms with efficiency and long-term occlusion benefits.

3.2 Conclusions and Future Outcomes

Since the early years of Interventional Radiology, the creative approach of the physicians to find alternative solutions to the classic surgery for different diseases, mainly vascular and oncologic, has been supported by researchers and industry involved in the field of biomaterials. The area of embolotherapy had a tremendous expansion in virtue of a wide range of biomaterials available to cope with different clinical situations. The use of liquid agents to occlude complex vascular lesions such as AVM and aneurysms is a promising approach. However, the number of parameters that control polymerization makes this approach still not optimal. The liquid agent approach in embolotherapy is still in its infancy. Physical (UVA-induced) or chemical controlled polymerization under image-guidance (fluoroscopy, ultrasonography, Computed Tomography, Magnetic Resonance) could offer interesting alternatives for controlled delivery. Image-guided nanoparticles can vehicle drugs to a specific target such as an organ, a muscle, or even a single lesion to maximize the efficacy and reduce the systemic collateral effects.

Close interaction with biochemists, biologists, and medical researchers is mandatory to support the emerged fields of molecular and cellular treatments, such as the gene- and immuno-therapies.

References

1. Kessel, D., Ray, C.: Transcatheter Embolization and Therapy. Springer, Heidelberg (2010)
2. Wikimedia Commons. https://commons.wikimedia.org/w/index.php?title=File:Aneurysem.jpg&oldid=359841565
3. Association, A.H. (2018). https://www.stroke.org/en/about-stroke/types-of-stroke/hemorrhagic-strokes-bleeds/what-is-an-arteriovenous-malformation
4. Poursaid, A., et al.: Polymeric materials for embolic and chemoembolic applications. J. Controlled Release **240**, 414–433 (2016)
5. Lubarsky, M., Ray, C.E., Funaki, B.: Embolization agents-which one should be used when? Part 1: large-vessel embolization. Semin. Interv. Radiol. **26**(4), 352–357 (2009)
6. Caracciolo, G., et al.: Segregation and phase transition in mixed lipid films. Langmuir **21** (20), 9137–9142 (2005)
7. Bianchino, E., et al.: DPD simulations of PMMA-oleic acid mixture behaviour in organic capped nanoparticle based polymer nanocomposite. Macromol. Symp. **286**, 156–163 (2009). WILEY-VCH Verlag, Weinheim
8. Piotto, S., et al.: Dissipative particle dynamics study of alginate/gelatin aerogels obtained by supercritical drying. In: Advances in Bionanomaterials, pp. 75–84. Springer, Cham (2018)
9. Piotto, S., et al.: Computer simulations of natural and synthetic polymers in confined systems. Macromol. Symp. **286**, 25–33 (2009). WILEY-VCH Verlag, Weinheim
10. Lazzaro, M.A., et al.: Endovascular embolization of head and neck tumors. Front. Neurol. **2**, 64 (2011)
11. Reinges, M.H.T., et al.: Bare, bio-active and hydrogel-coated coils for endovascular treatment of experimentally induced aneurysms: long-term histological and scanning electron microscopy results. Interv. Neuroradiol. **16**(2), 139–150 (2010)
12. Fiorella, D., Albuquerque, F.C., McDougall, C.G.: Durability of aneurysm embolization with matrix detachable coils. Neurosurgery **58**(1), 51–58 (2006)

13. Murayama, Y., et al.: Bioabsorbable polymeric material coils for embolization of intracranial aneurysms: a preliminary experimental study. J. Neurosurg. **94**(3), 454–463 (2001)
14. Taschner, C.A., et al.: Matrix detachable coils for the endovascular treatment of intracranial aneurysms: analysis of early angiographic and clinical outcomes. Stroke **36**(10), 2176–2180 (2005)
15. Ahuja, A.A., et al.: Platinum coil coatings to increase thrombogenicity: a preliminary study in rabbits. Am. J. Neuroradiol. **14**(4), 794–798 (1993)
16. Chen, J., et al.: Controlled release of osteopontin and interleukin-10 from modified endovascular coil promote cerebral aneurysm healing. J. Neurol. Sci. **360**, 13–17 (2016)
17. Hoh, B.L., et al.: Monocyte chemotactic protein-1 promotes inflammatory vascular repair of murine carotid aneurysms via a macrophage inflammatory protein-1α and macrophage inflammatory protein-2-dependent pathway. Circulation **124**(20), 2243–2252 (2011)
18. Abrahams, J.M., et al.: Delivery of human vascular endothelial growth factor with platinum coils enhances wall thickening and coil impregnation in a rat aneurysm model. Am. J. Neuroradiol. **22**(7), 1410–1417 (2001)
19. Wong, Y.S., et al.: Bioabsorbable radiopaque water-responsive shape memory embolization plug for temporary vascular occlusion. Biomaterials **102**, 98–106 (2016)
20. Zhang, Y.Y., et al.: Adv. Funct. Mater. **28** (2018)
21. Sheth, R.A., et al.: Endovascular embolization by transcatheter delivery of particles: past, present, and future. J. Funct. Biomater. **8**(2), 12 (2017)
22. Lubarsky, M., Ray, C., Funaki, B.: Embolization agents—which one should be used when? Part 2: small-vessel embolization. Semin. Interv. Radiol. **27**(01), 099–104 (2010)
23. Bannerman, D., Wan, W.: Multifunctional microbeads for drug delivery in TACE. Expert Opin. Drug Deliv. **13**(9), 1289–1300 (2016)
24. Bendszus, M., et al.: Efficacy of trisacryl gelatin microspheres versus polyvinyl alcohol particles in the preoperative embolization of meningiomas. Am. J. Neuroradiol. **21**(2), 255–261 (2000)
25. Davidson, G.S., Terbrugge, K.G.: Histologic long-term follow-up after embolization with polyvinyl alcohol particles. Am. J. Neuroradiol. **16**(Suppl.), 843–846 (1995)
26. Medsinge, A., et al.: A case-based approach to common embolization agents used in vascular interventional radiology. Am. J. Roentgenol. **203**(4), 700–708 (2014)
27. Zhou, F., et al.: Novel hydrogel material as a potential embolic agent in embolization treatments. Sci. Rep. **6**, 32145 (2016)
28. Loffroy, R., et al.: Endovascular therapeutic embolisation: an overview of occluding agents and their effects on embolised tissues. Curr. Vasc. Pharmacol. **7**(2), 250–263 (2009)
29. Albanese, G., Kondo, K.L.: Pharmacology of sclerotherapy. Semin. Interv. Radiol. **27**(4), 391–399 (2010)
30. Tamatani, S., et al.: Embolization of arteriovenous malformation with diluted mixture of NBCA. Interv. Neuroradiol. **6**(SUPPL. 1), 187–190 (2000)
31. Li, Y.J., Barthès-Biesel, D., Salsac, A.V.: Polymerization kinetics of n-butyl cyanoacrylate glues used for vascular embolization. J. Mech. Behav. Biomed. Mater. **69**, 307–317 (2017)
32. Rosen, R.J., Contractor, S.: The use of cyanoacrylate adhesives in the management of congenital vascular malformations. Semin. Interv. Radiol. **21**(1), 59–66 (2004)
33. Blakely, B., et al.: Formulation and characterization of radio-opaque conjugated in situ gelling materials. J. Biomed. Mater. Res.-Part B Appl. Biomater. **93**(1), 9–17 (2010)
34. Taki, W., et al.: A new liquid material for embolization of arteriovenous malformations. Am. J. Neuroradiol. **11**(1), 163–168 (1990)
35. Grasso, R.F., et al.: Pancreatic arteriovenous malformation involving the duodenum embolized with ethylene-vinyl alcohol copolymer (onyx). Cardiovasc. Interv. Radiol. **35**(4), 958–962 (2012)

36. Regine, R., et al.: Embolization of traumatic and non-traumatic peripheral vascular lesions with Onyx. Interv. Med. Appl. Sci. **7**, 22–29 (2015)
37. Dudeck, O., et al.: Embolization of experimental wide-necked aneurysms with iodine-containing polyvinyl alcohol solubilized in a low-angiotoxicity solvent. Am. J. Neuroradiol. **27**(9), 1849–1855 (2006)
38. Ahmed, E.M.: Hydrogel: preparation, characterization, and applications: a review. J. Adv. Res. **6**(2), 105–121 (2015)
39. Vanninen, R.L., Manninen, I.: Onyx, a new liquid embolic material for peripheral interventions: preliminary experience in aneurysm, pseudoaneurysm, and pulmonary arteriovenous malformation embolization. Cardiovasc. Interv. Radiol. **30**(2), 196–200 (2007)
40. Sabareesh Kumar, N., et al.: Histopathological changes in brain arteriovenous malformations after embolization using Onyx or N-butyl cyanoacrylate. J. Neurosurg. JNS **111**(1), 105–113 (2009)
41. Leyon, J.J., et al.: Preliminary experience with the liquid embolic material agent PHIL (Precipitating Hydrophobic Injectable Liquid) in treating cranial and spinal dural arteriovenous fistulas: technical note. J. NeuroInterv. Surg. **8**(6), 596 (2016)
42. Duffner, F., et al.: Combined therapy of cerebral arteriovenous malformations: histological differences between a non-adhesive liquid embolic agent and n-butyl 2-cyanoacrylate (NBCA). Clin. Neuropathol. **21**(1), 13–17 (2002)
43. Small Iv, W., et al.: Shape memory polymer stent with expandable foam: a new concept for endovascular embolization of fusiform aneurysms. IEEE Trans. Biomed. Eng. **54**(6), 1157–1160 (2007)
44. Landsman, T.L., et al.: Design and verification of a shape memory polymer peripheral occlusion device. J. Mech. Behav. Biomed. Mater. **63**, 195–206 (2016)
45. Rodriguez, J., et al.: In vivo response to an implanted shape memory polyurethane foam in a porcine aneurysm model: implantation of shape memory polyurethane foam. J. Biomed. Mater. Res., Part A **102**, 1231–1242 (2014)
46. Metcalfe, A., et al.: Cold hibernated elastic memory foams for endovascular interventions. Biomaterials **24**(3), 491–497 (2003)
47. Cherng, J.Y., et al.: Polyurethane-based drug delivery systems. Int. J. Pharm. **450**(1–2), 145–162 (2013)
48. Tobushi, H., et al.: Thermomechanical properties of polyurethane-shape memory polymer foam. J. Intell. Mater. Syst. Struct. **12**(4 SPEC. ISS.), 283–287 (2001)
49. Singhal, P., et al.: Ultra low density and highly crosslinked biocompatible shape memory polyurethane foams. J. Polym. Sci., Part B: Polym. Phys. **50**(10), 724–737 (2012)
50. Wilson, T.S., et al.: Shape memory polymers based on uniform aliphatic urethane networks. J. Appl. Polym. Sci. **106**(1), 540–551 (2007)
51. Singhal, P., et al.: Low density biodegradable shape memory polyurethane foams for embolic biomedical applications. Acta Biomater. **10**, 67–76 (2013)
52. Singhal, P., et al.: Controlling the actuation rate of low-density shape-memory polymer foams in water. Macromol. Chem. Phys. **214**(11), 1204–1214 (2013)
53. Rodriguez, J.N., et al.: In vivo response to an implanted shape memory polyurethane foam in a porcine aneurysm model. J. Biomed. Mater. Res., Part A **102**(5), 1231–1242 (2014)

Fluorescent Probes for Applications in Bioimaging

Miriam Di Martino[1], Francesco Marrafino[1], Rosita Diana[2], Pio Iannelli[1], and Simona Concilio[3(✉)]

[1] Department of Pharmacy, University of Salerno, via Giovanni Paolo II 132, Fisciano, SA, Italy
[2] Department of Agriculture, University of Napoli Federico II, Via Università, 100, Portici, NA, Italy
[3] Department of Industrial Engineering, University of Salerno, via Giovanni Paolo II 132, Fisciano, SA, Italy
sconcilio@unisa.it

Abstract. Optical bioimaging has played a central role in fundamental research and clinical practice. The signals emitted by biological tissues can provide molecular information about various physiological and pathophysiological processes. NIR light (650–1700 nm) can penetrate the blood and biological tissues more profoundly and effectively because, at longer wavelengths, less light is diffused and absorbed. Therefore, many probes have been developed for bioimaging in the NIR window for real-time, high-sensitivity deep tissue imaging. The library of optical probes has been expanded in recent years to include a wide range of probes with emission in the Red-NIR window. The emergence of these new contrast media has provided an essential alternative for realizing the full potential of bioimaging. The most recent advances in small molecule potential probes for detection and imaging in biological systems are examined below.

Keywords: Fluorescence probes · NIR · Optical bioimaging · Small-molecule

1 Introduction

Biological imaging (bioimaging) is a powerful tool in biological research nowadays because it offers a unique approach to visualize the morphological details of cells.

Living organisms include several cells, which in turn, contain a tremendous variety of biomolecules [1]. Very often, in biological systems, a small difference between remarkably similar biomolecules can lead to different functions [2, 3]. For example, cysteine (Cys), homocysteine (Hcy), and N-acetyl-L-cysteine (NAC), three analog amino acids with the only difference of one methylene or acetyl group in their residues, have been shown to play a different role in cellular processes. Some diseases, such as loss of leukocytes and psoriasis, are associated with Cys deficiency, while excess Hcy is a risk for cardiovascular disease and Alzheimer's. As a result, it is of crucial importance to probe Cys, Hcy, and NAC separately. However, their structural similarity makes it difficult to discriminate one from the other [4]. To this end, several

S. Piotto et al. (Eds.): BIONAM 2019, LNBE, pp. 243–258, 2020.
https://doi.org/10.1007/978-3-030-47705-9_21

technologies have been developed over the years to study the interactions of these molecules within the cells. The technologies used must respond to several conditions: they must be sensitive enough to visualize biological compounds and physiological concentrations (usually from nanomolar to micromolar), they must have a sufficient spatial-temporal resolution to analyze dynamic cell processes, not invasive and above all not expensive. Among the technologies used over the years, radioisotope labeling, magnetic resonance imaging, electrochemical detection and fluorescence bioimaging, fluorescence-based technique has proven to be the most powerful as it allows a highly sensitive, convenient, non-invasive and safe detection [3, 5]. This technique requires the use of “fluorescent probes” that, reacting specifically with biological molecules, present changes in their photochemical properties (fluorescence intensity, excitation, and emission wavelength). In the last years, there has been an increasing development of new probes, and some of them are already used in *in vivo* imaging as probes based on fluorescent proteins, inorganic nanoparticles and conjugated polymers [6–9].

The basic principle of fluorescence imaging is similar to that used in fluorescence microscopy techniques, which is based on the spectral range of visible light between 400–650 nm. However, when we consider biological tissue imaging, we must consider that microscopic components such as sugars, amino acids, nucleotides, and macromolecules such as proteins, phospholipids, and polysaccharides can absorb light and can limit light penetration to a few hundred microns in the tissue. Also, the autoflorescence of some of these molecules is located in the visible area. Higher depths can be achieved by using light in the far-red or near-infrared (NIR) range (660–1700 nm). In this spectral region, the absorption by the cellular components is lower and provides penetration through several centimeters of tissue [10]. Therefore, bio-imaging in the window of the red/NIR required the development of a considerable number of NIR fluorescent probes, such as those based on small organic molecules, designed to detect various biologically essential species, including ROS/RNS, metal ions, anions, enzymes and other related species, as well as intracellular pH variations [11, 12].

Typically, fluorescent probes are exploited to label the target with specific chemical structures and thus to generate fluorescent signals during the fluorescence-based bioimaging. Also, nanostructure-based detection platforms can provide many advantages over traditional approaches in terms of sensitivity, signal stability, and multiplexing capability so that growing interest has been shown recently in the design of different fluorescent nanostructured probes for bioimaging. The emergence of these new contrast agents has provided a valuable alternative to realize the full potential of NIR bioimaging.

The optical properties of probes can be modulated through structural modifications, for example, the absorption and emission wavelengths of small-molecule probes can be extended from the visible-light region to the NIR region by enhancing intramolecular charge transfer and expanding the π - conjugated system. This allows the design of probes that present significant changes in their spectroscopic properties as a result of interaction with specific biomolecules or even that can be anchored to specific sites based on modifications with target groups [13, 14]. In the last decade, a variety of probes have been developed, exploiting one of the mechanisms of modulation of the fluorescence properties listed below: Photoinduced electron Transfer (PET), Förster Resonance Energy Transfer (FRET), Intramolecular Charge Transfer (ICT) [9, 15–21].

This flexibility enables the achievement of long-term and in situ bioimaging and raises the possibility of generating small-molecule NIR probes for deep-tissue and high-contrast bioimaging.

In this mini-review, we will focus on recent advances in small-molecule probes in selected categories based on fluorescent nuclei. For each category, we will see how, depending on small changes in chemical structure, probe applications may vary from metal ion recognition to pH changes.

Finally, in the last paragraphs, a selection of active molecules in the NIR II region is reported, and some promising solid-phase fluorescent probes.

2 Organic Small Molecule Fluorescent Probes

To reach red emission, dye molecules generally have large planar rings with extended conjugation or strong π-conjugated electron-donating and accepting groups. Although these red fluorophores are strongly emissive when dissolved in solution, the emission signals are often weakened or even annihilated in aggregates, due to what is known as the aggregation-caused quenching (ACQ) effect. Conventional organic dyes tend to aggregate, which is highly dependent on their intrinsic hydrophobicity, and the excited state often decays via non-radiative channels, which reduces the brightness and sensitivity in biological applications. Many chemical and physical approaches have been employed to prevent fluorophore aggregation, but they have met with only limited success.

Interestingly, some propeller-like organic dyes show unique fluorescence phenomena with aggregation-induced emission (AIE) characteristics. These fluorogens generally contain rotor structures and exhibit low-frequency vibration motions in dilute solutions. However, in aggregated states, instead of quenching, AIEgens emit efficiently due to the restriction of intramolecular rotations (RIR) and the lack of energy dissipation via non-radiative channels. Since the first discoveries by Oelkrug and Hanack in 1998 [22] and then by Tang and coworkers in 2001 [23], luminescent materials with AIE features have attracted extensive research interest for applications such as electroluminescent devices, bio-sensing, and cell imaging. Until now, a wide variety of AIE dyes have been designed and synthesized based on the mechanism of RIR. Even so, synthesizing AIE-active probes with better biocompatibility, biodegradability, and red/near-infrared emission is essential for bioimaging applications.

Among traditional probes, cyanines, fluorescein, rhodamine, coumarins, and BODIPY (basic nuclei reported in Fig. 1) have shown to be the molecules most subject to structural modifications because they have favorable optical properties, a high absorption coefficient, biocompatibility, and low toxicity and are a valid choice for biomedical applications.

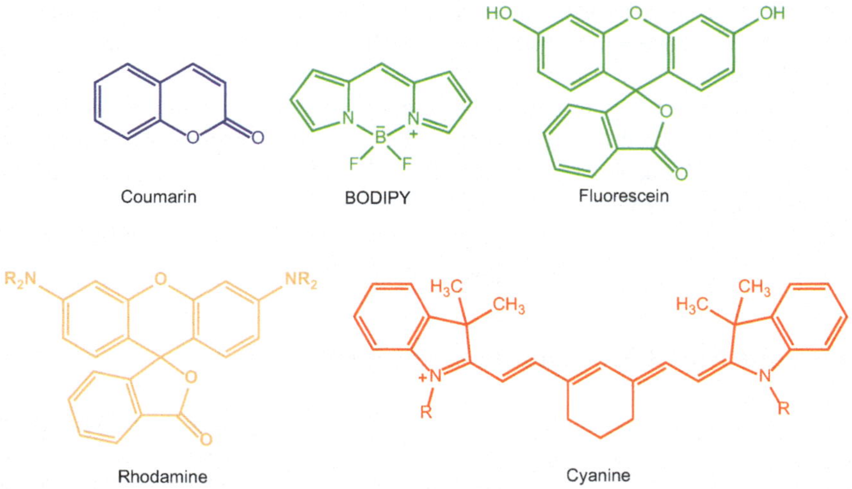

Fig. 1. A representative set of conventional dyes that can be transformed into fluorescent probes for bioimaging, sorted by structure and emission color. [2]

2.1 Cyanine-Derivates Probes

The development of NIR fluorescent probes for the study of metal ions has had a great focus on the role that ions have in biological environments [24].

In this context, probes based on modified cyanines are widely used for the detection of metal ions and [25–27]. For example, the introduction of a sulfur-based receptor, reported by Tasuku Hirayama et al. [28], shows selectivity for the Cu^+ ion in living cells. Copper is an essential element, and the maintenance of its correct homeostasis is essential for the growth and development of living organisms; consequently, the loss of copper homeostasis has severe consequences due to its redox activity. This fact results in diseases such as cancer, cardiovascular disease or Alzheimer's. The probe, Copper sensor CS790AM (Fig. 2), combines an infrared cyanine dye with a sulfur-rich receptor to provide a selective response to copper. The probe can detect increases in the level of copper in the case of Wilson's disease, a genetic disease characterized by the accumulation of copper. The probe also has no toxicity and is eliminated in a few days.

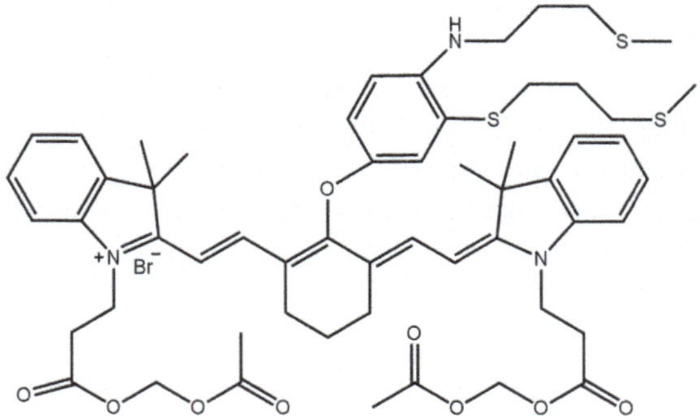

Fig. 2. Chemical structure of CS790AM copper-selective probe [28].

Similarly, selective probes have been designed for other ions such as Cu^{2+}, Ca^{2+}, Li^{+}, Zn^{2+}, Hg^{2+}, Cd^{2+}, and Ag^{+}, combining cyanine fluorophore with a selective metal ion receptor [27, 29, 30]. Tang group [31] developed the probe Cy–Cu, reported in Fig. 3, composed of tricarbocyanine (Cy) and a receptor for the Cu^{2+} ion, 2,2'-azanediyl bis(N-hydroxyacetamide). The probe is based on the photoinduced electronic transfer mechanism. When the Cu^{2+} ion is coordinated and binds to the receptor, PET is blocked, and an increase in fluorescence is observed (Fig. 3).

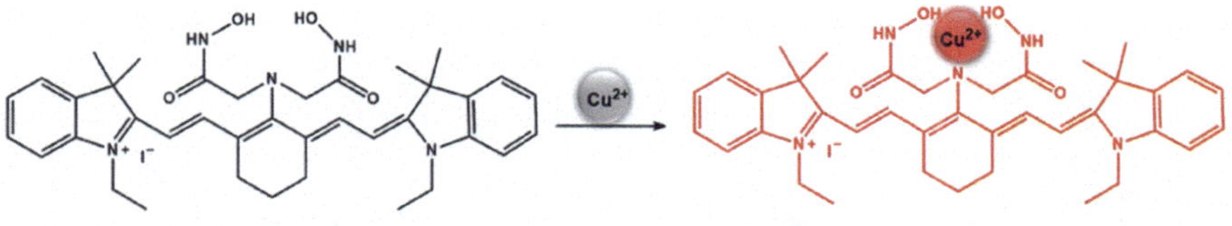

Fig. 3. Proposed binding mechanism of the probe Cy–Cu to Cu^{2+} [31].

This was the first NIR imaging probe for Cu^{2+} *in vitro* and *in vivo*. Fluorophores 2 and 3 show selectivity for Cd^{2+} by coordination of the tetramide receptor with the ion [32]. Alternatively, the incorporation of adenine as a substitute for the fluorescent nuclei of cyanine, is an active linker for Ag^{+} (probe 4 in Fig. 4) [33].

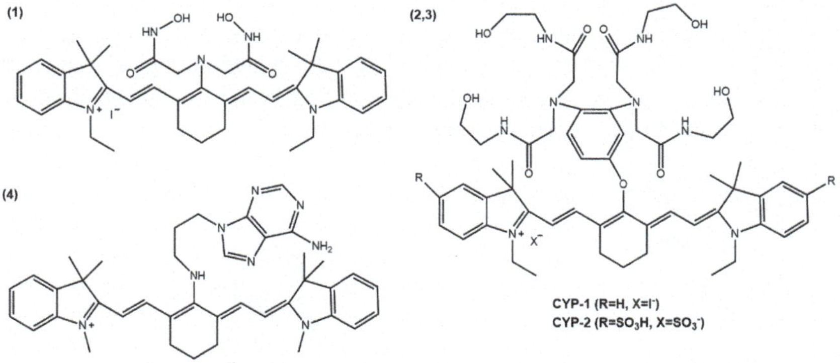

Fig. 4. Chemical structure of some cyanine probes selective for 1) Cu^{2+} [31]; 2,3) Cd^{2+} [32] e 4) Ag^{+} [33].

Many other cyanine-based probes have been developed to detect intracellular pH variation [34–36]. Intracellular pH changes lead to abnormal cell growth and division, resulting in inflammation or disease such as cancer or Alzheimer's disease. The pH-sensitive cyanine-derivate probes can be classified into two types: probes characterized by non-N-alkylated indolic group whose protonation or deprotonation depending on the protonic concentration; the second type of pH-sensitive probes are the tricarbocyanines that are based on the classical mechanisms ICT and FRET [37, 38]. In the first type of cyanine probes, the non-N-alkylated indole in the acid environment can be protonated inducing a strong fluorescence, in basic environment on the contrary, the deprotonation of the nitrogen atoms of the indole groups shows no fluorescence. Nagano and co-workers have designed several pH-sensitive probes based on tricarbocianines that have a diamino group. These probes have shown high stability under acidic conditions and a drastic shift in absorption towards red when the nitrogen is protonated. This behavior is due to the decrease of the electron-donor nature of the amine [39].

Some pH-sensitive tricarbonacyanine probes structures are shown in Fig. 5.

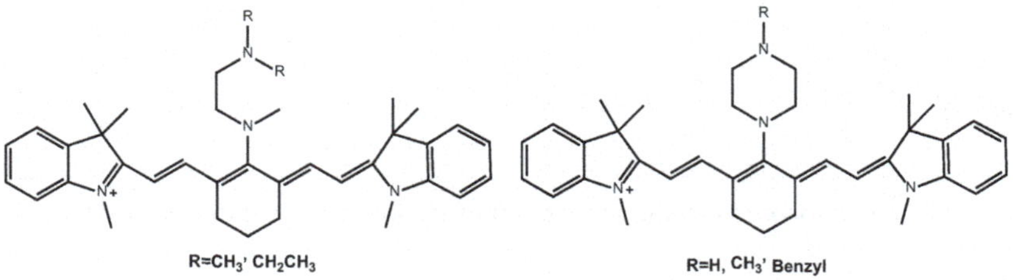

Fig. 5. Chemical structure of pH-sensitive tricarbocyanine probes [39].

2.2 Coumarin-Derivates Probes

Coumarin (2H-cromen-2-one) is a chemical compound of the benzopyrone class found in many natural species. Coumarin has a variety of biological activities and unique photophysical properties; among these, the fluorescent property has recently received attention because of its high quantum yield, high stability and biological compatibility [40–42]. Derivatives of benzocoumarin are classified based on the position of the aromatic ring fused to the nuclei, as shown in Fig. 6.

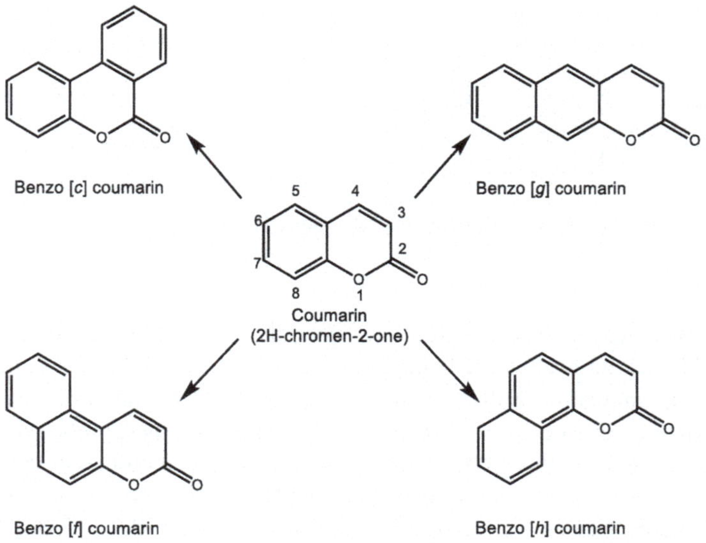

Fig. 6. Chemical structure of coumarin and benzo-derivates compounds [40–42].

Among the derivatives, benzo[g]coumarin exhibits better photophysical properties in bioimaging applications than other derivatives. Besides, a high two-photon absorption capacity, high photostability, and high chemical stability are the beneficial characteristics of benzo[g]coumarin derivatives. The photophysical properties of the analogs of benzocoumarin can also be predicted depending on the type and position of the substituents: it has been observed that the combination of electron-withdrawing and electron-donating substituents causes a significant shift of both absorption and emission spectra at longer wavelengths (red and NIR region) [42]. Cho et al. [43] reported a benzo[g]coumarin-derivative fluorescent probe for Cu^{2+} and quantitatively estimated ion concentrations in human tissues by two-photon microscopic imaging (Fig. 7). The probe, amide- and dimethylamino-substituted, to which is linked a benzo [h]coumarin analog as internal reference (sensitive to the environment or to the substrates, which maintains constant the fluorescence intensity) presents a decrease of the fluorescence intensity in the red, following chelation of the copper ion. It was thus possible to study the concentration of Cu^{2+} ions in live cells, rat brain tissue and human colon tissue samples using two-photon microscopy.

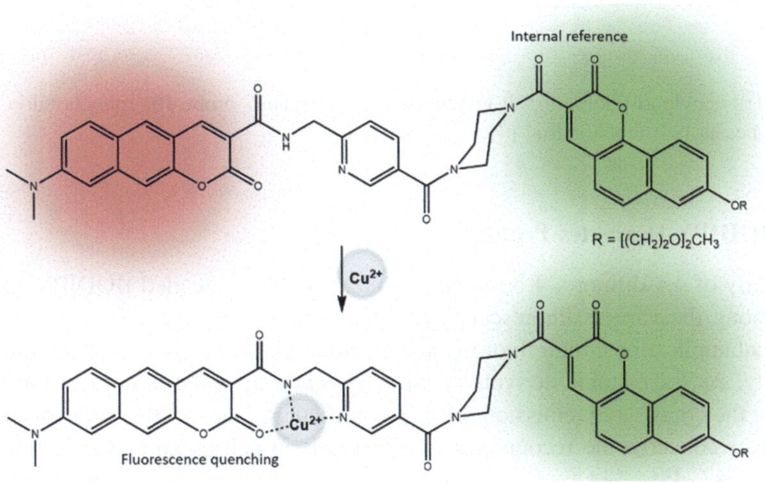

Fig. 7. Benzo[g]coumarin- and benzo[h]coumarin-based fluorescent probe for Cu^{2+} ions [43]

In 2014, a derivative of benzo[g]coumarin was reported by Kim et al. [44] for monitoring mitochondria, the malfunction of which could be related to many diseases (Fig. 8). The probe had a mitochondrial attack site consisting of the triphenylphosphonium salt bound to benzo[g]coumarin through an amide bond. The resulting probe showed maximum emission in red, without pH-sensitive variations. Mitochondria tracking ability has been verified on the T98G cell line with MitoTracker Green (MTG) as a known mitochondrial marker.

Subsequently, structural modifications resulted in a new probe that was sensitive to pH changes, through the introduction of an electron-donor hydroxyl group that can be protonated or deprotonated to pK_a near 8.0 (mitochondria pH). The shift of the absorption and emission peaks as a function of the pH variation indicates the capacity of the selective imaging probe for mitochondria [42, 45].

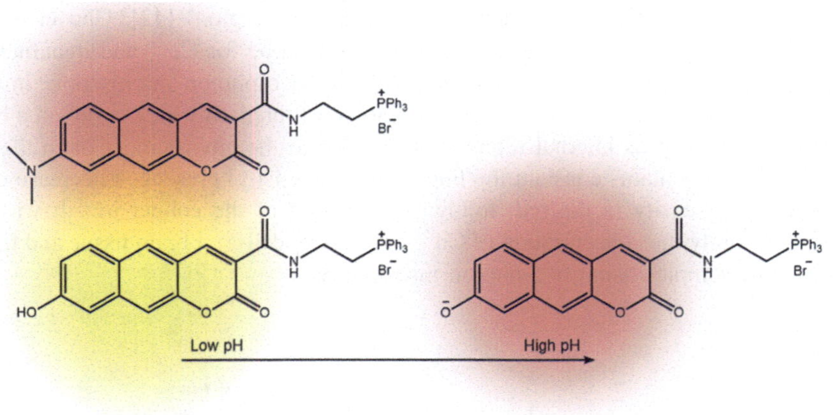

Fig. 8 Triphenylphosphonium salt-linked Benzo[g]coumarin probe for mitochondria tracking and for measuring mitochondrial pH values [44].

2.3 BODIPY-Derivates Probes

The family of 4,4-difluoro, 4-bora-3a, 4a-diaza-s-indacene (called BODIPY) is widely used among the various fluorescent dyes. These probes have excellent properties for biological applications: high fluorescence quantum yield, excellent photostability, high solubility. As for cyanine derivatives, also this class of probe is used in bio-imaging because they have red emission [46]. In particular, one of the uses of BODIPY derivatives concerns the recognition of metal ions in cells such as Cd^{2+}, Ca^{2+}, $Hg^{2+,}$ and Zn^{2+} [47, 48].

Akkaya and co-workers have developed a fluorescent NIR probe for Hg^{2+}, the bis (2-pyridyl)-replaced boratriazaindacene probe (AzaBODIPY) and found that the 2-pyridyl substituents create a metal ion pocket due to the rigid nature of the ligand and the donor's selectivity [49]. At a sufficiently high concentration of Hg^{2+}, there is a shift at higher wavelengths in the absorption and emission spectra (Fig. 9).

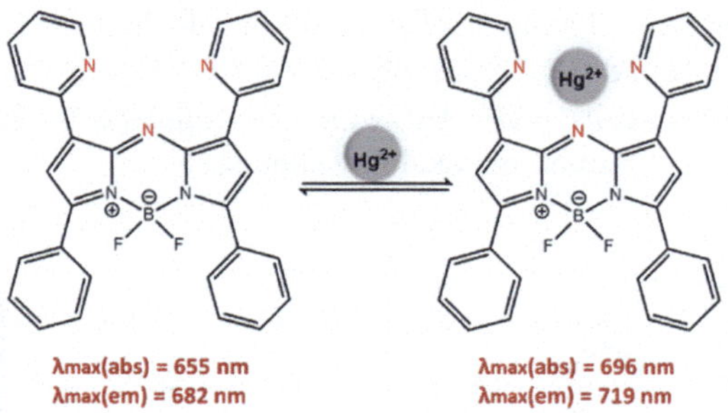

Fig. 9 Coordination mechanism of the Hg^{2+} to the AzaBODIPY probe [49].

An interesting pH-sensitive probe based on BODIPY has been developed by O'Shea et al. [50]. In the probe, two amine substitutes act as pH modulators leading to three-channel fluorescence (Fig. 10). Depending on whether the molecule was mono or di-protonate, a color change from red to purple, up to blue corresponding to the three forms: the non-protonate, mono-protonate, and di-protonate form in response to pH-induced change in the ICT (intramolecular charge transfer) properties of the system is observed.

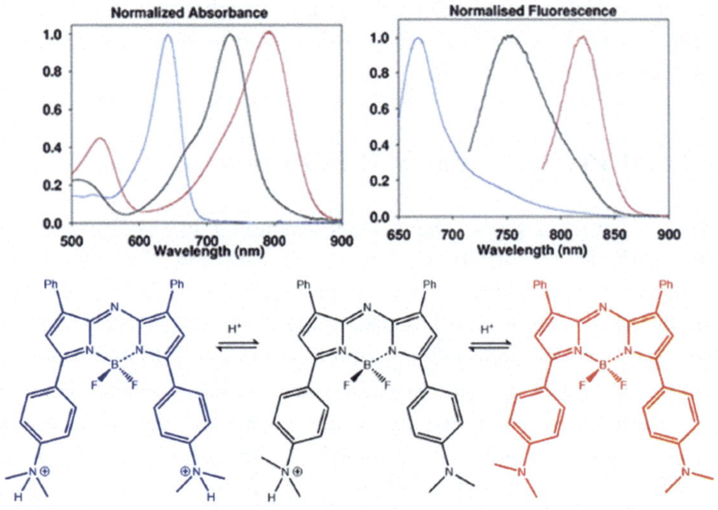

Fig. 10 Protonation and deprotonation of the pH-sensitive probe and related emission spectra [50].

As mentioned above, very often structurally similar biomolecules can have different functions, as in the case of cysteine, homocysteine, and N-acetyl-L-cysteine [4]. It is therefore essential to design probes that are selective for these biomolecules. Ravikanth et al. [51] have developed a red fluorescent probe for the specific detection of cysteine (Cys) and homocysteine (Hcy) in living cells based on BODIPY 3,5-bis(acrylaldehyde) (see Fig. 11). The probe, following cyclization of the aldehyde group with thiol, shows the formation of a hexaindro-1,4-thiazepine derivative, which results in a change in fluorescence properties (cells pretreated with Cys followed by incubation with the probe show bright yellow-green intracellular fluorescence).

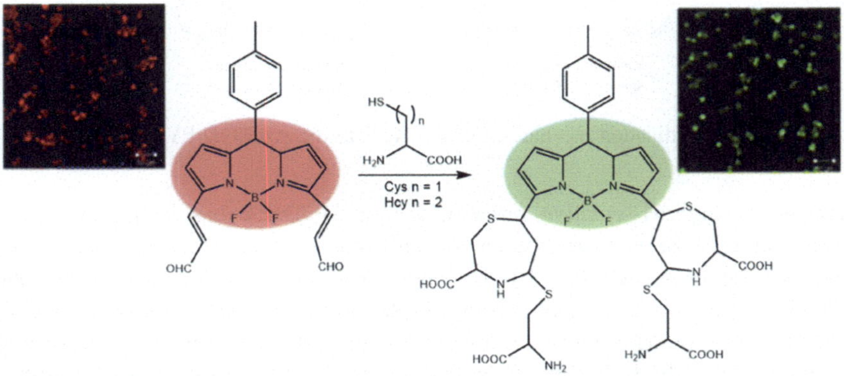

Fig. 11 The schematic reaction of BODIPY based probe with Cys or Hcy [51]

However, the NIR-I window (650–900 nm), in which almost all the above probes emissions fall, can present the problem of tissue autofluorescence, producing a considerable background noise [52, 53].

3 Fluorescent Probes in the NIR-II Region

Recently, new fluorescent probes have been attempted to operate at longer wavelengths, particularly in the NIR-II window (1000–1700 nm), in order to offer better penetration and overcome autofluorescence problems [54]. They meet these probes requirements based on Donor-Acceptor-Donor (D-A-D) systems that show emissions in the NIR-II zone as widely reported in the literature [55, 56]. In the work of L. Antaris et al. [17] a new probe based on the D-A-D system with bis-thiadiazole nuclei, called CH1055, is described. The reported probe has been modified with carboxylic acid groups to improve the solubility of the molecule; the carboxylic acid groups can be subjected to PEGylation to increase their solubility further, or even be conjugated to target ligands (Fig. 12).

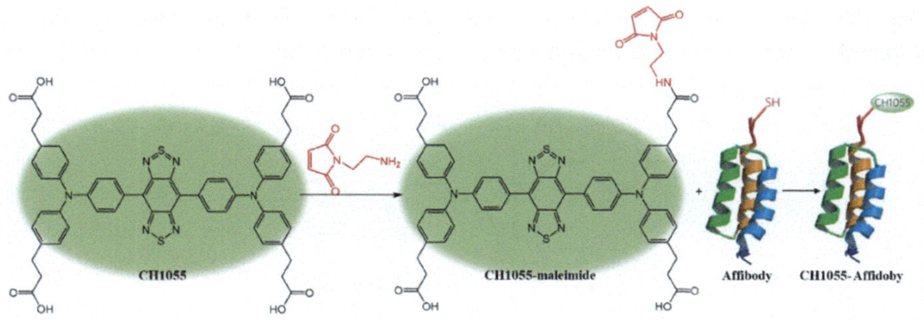

Fig. 12 A simplified reaction schematic illustrating the synthesis of the CH1055-a-body [17].

The CH1055-PEG probe has proven effective in mapping sentinel lymph nodes (SLN) and identifying brain tumors *in vivo*. It was noted that after intradermal injections of the CH1055-PEG probe used for imaging lymphatic vessels and lymph nodes, strong tumor fluorescence was observed, starting 57 h after injection.

The conjugation of the fluorescent molecule to a small protein such as anti-EGFR affibody (EGFR: receptor of epidermal growth factor) allows rapid and economical detection of early cancers and therefore represents a valuable aid for the removal of the imaging-guided tumor.

The probe showed rapid renal excretion (about 90% in 24 h) compared to other NIR-II probes, and PEGylation has become an accepted practice in the pharmaceutical industry as it offers many advantages such as longer probe circulation time, higher stability and protection against degradation. The probe also allowed the first guided NIR-II export surgery, simultaneously displaying the tumor excision under white light and the NIR-II fluorescence system.

4 Solid-State Fluorescent Probes

In the context of organic-core chromophores with emission properties in the near-infrared region, fluorophores using the effect of aggregation-induced emission have recently been of interest for their multiple applications such as dopants in the fabrication of OLED devices, in optical probes for bioimaging, and biomedical applications [57]. Unlike classic NIR fluorophores, which tend to form aggregates in highly concentrated solutions or solid-state, resulting in quenching (ACQ effect) of emission properties due to strong dipole-dipole interactions, these new types of fluorophores exploit the effect of emission induced by AIE aggregation. These molecules show weak emissions in solutions and strong emission in the aggregated state, high photoluminescence quantum efficiency, significant Stokes shifts (for eliminating self-absorption), excellent chemical and thermal stability, good solubility and processability [22, 58, 59]. In literature, there is a limited number of organic molecules with solid-state fluorescence.

Our research group has conducted several studies on the solid-state emission properties of chromophores, such as those of Phenylenevinylene (PV) derivatives based on a donor-acceptor system. In addition, the introduction of an electron-withdrawing group such as the cyan group leads to an increase in the energy of the π-occupied and π*-unoccupied states and causes a red shift in emission maxima compared to non-substituted phenylenevinylene systems [60–62] (see Fig. 13).

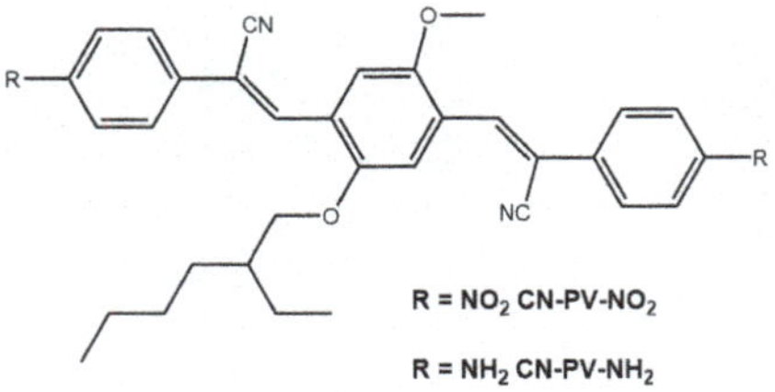

Fig. 13 Chemical structures of PV-based dyes [60]

All compounds show intense red photoluminescence with a pronounced shift of Stoke larger in the solid state than in solution.

This can be attributed to the presence of the cyano substitute that limits molecular movements and stabilizes the twisted conformation of the solid. This conformation may lead to intense aggregation-induced emission effect (AIE). Based on the above compounds, to improve photophysical performance, recently the same research group has synthesized new symmetric dyes with a modified skeleton phenylenevinylene (PV) and azobenzene (AB) adding groups such as N-N'-bis (salicylidene)ethylenediamine (salen) that would reduce molecular movements and improve their solid-state properties [63]. To this should be added the ESIPT process to which salen substitutes are subjected, which could potentially still improve the performance of the dyes. In the area of solid-state fluorescence, the same research group reported a study on the regulation of solid-state fluorescence properties of polymorphic compounds. In fact, polymorphism (different arrangements and conformations of the same molecule) can be used to produce materials with emissive properties in which thermal or mechanical stimuli can regulate the transition. The work [58] reports a derivative of N-salicylidene aniline containing the 2-methylbenzotriazole (BTz) showing solid-state ESIPT fluorescence, as shown in Fig. 14.

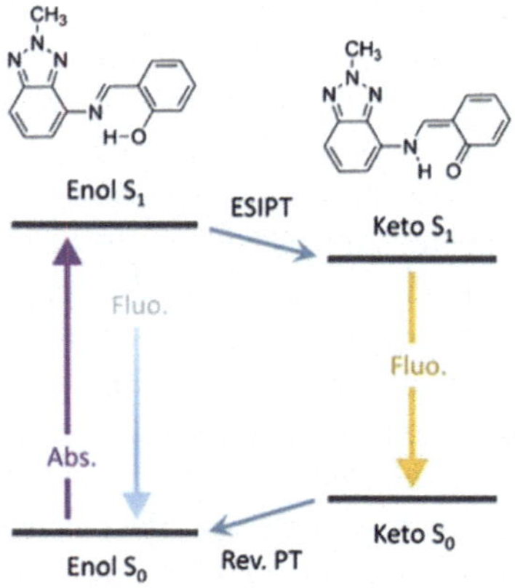

Fig. 14 ESIPT process in 2-methylbenzotriazole (BTz) chromophores [58].

The compound shows three phases: yellow (called 1-Y), orange (1-O), and red (1-R). The transformation from one shape to another is accompanied by a significant change in fluorescence intensity and color, as shown in Fig. 15.

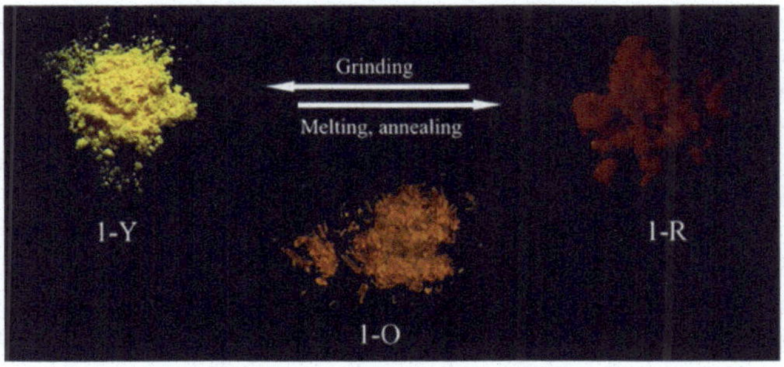

Fig. 15 Photograph of 1-Y, 1-O, and 1-R powders under a UV lamp and scheme of mechano-responsive behavior [58].

The 1-R form can be converted to 1-Y by grinding and returned to 1-R by melting and annealing, resulting in mechanical-responsive luminescence. The different photophysical properties have been explained by changes in molecular conformation in the three polymorphs. These findings offer an advantageous possibility to control the emission fluorescence for stimuli-response applications. The unique photochemical properties of these compounds make them potentially suitable for their encapsulation in nanoparticles (NPs) and use in bio-imaging. In bioimaging applications, AIE fluorophores can be immune to the limitations arising from the concentration of the dyes loaded into the nanoparticles (NPs), and AIE NPs are supposed to be more emissive, as well as more resistant to photobleaching [64].

References

1. Friedman, R., et al.: Understanding conformational dynamics of complex lipid mixtures relevant to biology. J. Membr. Biol. **251**(5–6), 609–631 (2018)
2. Chan, J., Dodani, S.C., Chang, C.J.: Reaction-based small-molecule fluorescent probes for chemoselective bioimaging. Nat. Chem. **4**(12), 973 (2012)
3. Terai, T., Nagano, T.: Fluorescent probes for bioimaging applications. Curr. Opin. Chem. Biol. **12**(5), 515–521 (2008)
4. Yang, Z., et al.: Highly selective red-and green-emitting two-photon fluorescent probes for cysteine detection and their bio-imaging in living cells. Chem. Commun. **48**(28), 3442–3444 (2012)
5. Terai, T., Nagano, T.: Small-molecule fluorophores and fluorescent probes for bioimaging. Pflügers Archiv-Eur. J. Physiol. **465**(3), 347–359 (2013)
6. Kiyose, K., Kojima, H., Nagano, T.: Functional near-infrared fluorescent probes. Chem.–Asian J. **3**(3), 506–515 (2008)
7. Miyawaki, A., Niino, Y.: Molecular spies for bioimaging—fluorescent protein-based probes. Mol. Cell **58**(4), 632–643 (2015)

8. Saito, K., Nagai, T.: Recent progress in luminescent proteins development. Curr. Opin. Chem. Biol. **27**, 46–51 (2015)
9. Panunzi, B., et al.: Photophysical properties of luminescent zinc(II)-pyridinyloxadiazole complexes and their glassy self-assembly networks. Eur. J. Inorg. Chem. **2018**(23), 2709–2716 (2018)
10. Leblond, F., et al.: Pre-clinical whole-body fluorescence imaging: Review of instruments, methods and applications. J. Photochem. Photobiol. B: Biol. **98**(1), 77–94 (2010)
11. Guo, Z., et al.: Recent progress in the development of near-infrared fluorescent probes for bioimaging applications. Chem. Soc. Rev. **43**(1), 16–29 (2014)
12. Concilio, S., et al.: Zn-complex based on oxadiazole/carbazole structure: Synthesis, optical and electric properties. Thin Solid Films **556**, 419–424 (2014)
13. Escobedo, J.O., et al.: NIR dyes for bioimaging applications. Curr. Opin. Chem. Biol. **14**(1), 64–70 (2010)
14. Li, J.-B., Liu, H.-W., Fu, T., Wang, R., Zhang, X.-B., Tan, W.: Recent progress in small-molecule near-IR probes for bioimaging. Trends Chem. **1**(2), 224–234 (2019)
15. Nagano, T.: Development of fluorescent probes for bioimaging applications. Proc. Jpn. Acad. Ser. B **86**(8), 837–847 (2010)
16. Yuan, L., et al.: FRET-based small-molecule fluorescent probes: rational design and bioimaging applications. Acc. Chem. Res. **46**(7), 1462–1473 (2013)
17. Antaris, A.L., et al.: A small-molecule dye for NIR-II imaging. Nat. Mater. **15**(2), 235 (2016)
18. Concilio, S., et al.: A novel fluorescent solvatochromic probe for lipid bilayers. Supramol. Chem. **29**(11), 887–895 (2017)
19. Diana, R., et al.: A real-time tripodal colorimetric/fluorescence sensor for multiple target metal ions. Dyes Pigm. **155**, 249–257 (2018)
20. Diana, R., Panunzi, B., Tuzi, A., Piotto, S., Concilio, S., Caruso, U.: An amphiphilic pyridinoyl-hydrazone probe for colorimetric and fluorescence pH sensing. Molecules **24**(21), 3833–3855 (2019)
21. Panunzi, B., et al.: Fluorescence pH-dependent sensing of Zn(II)by a tripodal ligand. A comparative X-ray and DFT study. J. Lumin. **212**, 200–206 (2019)
22. Oelkrug, D., et al.: Tuning of fluorescence in films and nanoparticles of oligophenylenevinylenes. J. Phys. Chem. B **102**(11), 1902–1907 (1998)
23. Luo, J., et al.: Aggregation-induced emission of 1-methyl-1,2,3,4,5-pentaphenylsilole. Chem. Commun. **18**, 1740–1741 (2001)
24. Yuan, L., et al.: Far-red to near infrared analyte-responsive fluorescent probes based on organic fluorophore platforms for fluorescence imaging. Chem. Soc. Rev. **42**(2), 622–661 (2013)
25. Kaur, M., Choi, D.H.: Diketopyrrolopyrrole: brilliant red pigment dye-based fluorescent probes and their applications. Chem. Soc. Rev. **44**(1), 58–77 (2015)
26. Yin, J., et al.: Cyanine-based fluorescent probe for highly selective detection of glutathione in cell cultures and live mouse tissues. J. Am. Chem. Soc. **136**(14), 5351–5358 (2014)
27. Oushiki, D., et al.: Development and application of a near-infrared fluorescence probe for oxidative stress based on differential reactivity of linked cyanine dyes. J. Am. Chem. Soc. **132**(8), 2795–2801 (2010)
28. Hirayama, T., et al.: Near-infrared fluorescent sensor for in vivo copper imaging in a murine Wilson disease model. Proc. Natl. Acad. Sci. **109**(7), 2228–2233 (2012)
29. Guo, Z., et al.: A cyanine-based fluorescent sensor for detecting endogenous zinc ions in live cells and organisms. Biomaterials **33**(31), 7818–7827 (2012)
30. Tang, B., et al.: A sensitive and selective near-infrared fluorescent probe for mercuric ions and its biological imaging applications. ChemBioChem **9**(7), 1159–1164 (2008)

31. Li, P., et al.: A near-infrared fluorescent probe for detecting copper (II) with high selectivity and sensitivity and its biological imaging applications. Chem. Commun. **47**(27), 7755–7757 (2011)
32. Yang, Y., et al.: Highly selective and sensitive near-infrared fluorescent sensors for cadmium in aqueous solution. Org. Lett. **13**(2), 264–267 (2010)
33. Zheng, H., et al.: A heptamethine cyanine-based colorimetric and ratiometric fluorescent chemosensor for the selective detection of Ag+ in an aqueous medium. Chem. Commun. **48** (16), 2243–2245 (2012)
34. Li, Y., et al.: Hemicyanine-based high resolution ratiometric near-infrared fluorescent probe for monitoring pH changes in vivo. Anal. Chem. **87**(4), 2495–2503 (2015)
35. He, L., et al.: A unique type of pyrrole-based cyanine fluorophores with turn-on and ratiometric fluorescence signals at different pH regions for sensing pH in enzymes and living cells. ACS Appl. Mater. Interfaces **6**(24), 22326–22333 (2014)
36. Fang, M., et al.: A cyanine-based fluorescent cassette with aggregation-induced emission for sensitive detection of pH changes in live cells. Chem. Commun. **54**(9), 1133–1136 (2018)
37. Han, J., Burgess, K.: Fluorescent indicators for intracellular pH. Chem. Rev. **110**(5), 2709–2728 (2009)
38. Hilderbrand, S.A., Weissleder, R.: Optimized pH-responsive cyanine fluorochromes for detection of acidic environments. Chem. Commun. **26**, 2747–2749 (2007)
39. Myochin, T., et al.: Rational design of ratiometric near-infrared fluorescent pH probes with various pKa values, based on aminocyanine. J. Am. Chem. Soc. **133**(10), 3401–3409 (2011)
40. Sethna, S.M., Shah, N.M.: The chemistry of coumarins. Chem. Rev. **36**(1), 1–62 (1945)
41. Thakur, A., Singla, R., Jaitak, V.: Coumarins as anticancer agents: a review on synthetic strategies, mechanism of action and SAR studies. Eur. J. Med. Chem. **101**, 476–495 (2015)
42. Jung, Y., et al.: Benzo[*g*]coumarin-based fluorescent probes for bioimaging applications. J. Anal. Methods Chem. **2018**, 11 (2018)
43. Kang, D.E., et al.: Two-photon probe for Cu^{2+} with an internal reference: quantitative estimation of Cu^{2+} in human tissues by two-photon microscopy. Anal. Chem. **86**(11), 5353–5359 (2014)
44. Sarkar, A.R., et al.: Red emissive two-photon probe for real-time imaging of mitochondria trafficking. Anal. Chem. **86**(12), 5638–5641 (2014)
45. Sarkar, A.R., et al.: A ratiometric two-photon probe for quantitative imaging of mitochondrial pH values. Chem. Sci. **7**(1), 766–773 (2016)
46. Ni, Y., Wu, J.: Far-red and near infrared BODIPY dyes: synthesis and applications for fluorescent pH probes and bio-imaging. Org. Biomol. Chem. **12**(23), 3774–3791 (2014)
47. Matsui, A., et al.: A near-infrared fluorescent calcium probe: a new tool for intracellular multicolour Ca^{2+} imaging. Chem. Commun. **47**(37), 10407–10409 (2011)
48. Cao, J., et al.: Target-triggered deprotonation of 6-hydroxyindole-based BODIPY: specially switch on NIR fluorescence upon selectively binding to Zn^{2+}. Chem. Commun. **48**(79), 9897–9899 (2012)
49. Coskun, A., Yilmaz, M.D., Akkaya, E.U.: Bis (2-pyridyl)-substituted boratriazaindacene as an NIR-emitting chemosensor for Hg (II). Org. Lett. **9**(4), 607–609 (2007)
50. McDonnell, S.O., O'Shea, D.F.: Near-infrared sensing properties of dimethlyamino-substituted BF_2–azadipyrromethenes. Org. Lett. **8**(16), 3493–3496 (2006)
51. Madhu, S., Gonnade, R., Ravikanth, M.: Synthesis of 3, 5-bis (acrylaldehyde) boron-dipyrromethene and application in detection of cysteine and homocysteine in living cells. J. Org. Chem. **78**(10), 5056–5060 (2013)
52. Zhao, J., Zhong, D., Zhou, S.: NIR-I-to-NIR-II fluorescent nanomaterials for biomedical imaging and cancer therapy. J. Mater. Chem. B **6**(3), 349–365 (2018)

53. Zhang, X., et al.: Near-infrared molecular probes for in vivo imaging. Curr. Protoc. Cytometry **60**(1), 12.27.1–12.27.20 (2012)
54. Zhu, S., et al.: Near-infrared-II (NIR-II) bioimaging via off-peak NIR-I fluorescence emission. Theranostics **8**(15), 4141 (2018)
55. Cui, M., et al.: Smart near-infrared fluorescence probes with donor–acceptor structure for in vivo detection of β-amyloid deposits. J. Am. Chem. Soc. **136**(9), 3388–3394 (2014)
56. Li, Y., et al.: Novel D–A–D based near-infrared probes for the detection of β-amyloid and Tau fibrils in Alzheimer's disease. Chem. Commun. **54**(63), 8717–8720 (2018)
57. Kim, M., et al.: A distyrylbenzene based highly efficient deep red/near-infrared emitting organic solid. J. Mater. Chem. C **3**(2), 231–234 (2015)
58. Borbone, F., et al.: On–off mechano-responsive switching of ESIPT luminescence in polymorphic N-salicylidene-4-amino-2-methylbenzotriazole. Cryst. Growth Des. **17**(10), 5517–5523 (2017)
59. Shi, J., et al.: Solid state luminescence enhancement in π-conjugated materials: unraveling the mechanism beyond the framework of AIE/AIEE. J. Phys. Chem. C **121**(41), 23166–23183 (2017)
60. Panunzi, B., et al.: Solid-state highly efficient DR mono and poly-dicyano-phenylenevinylene fluorophores. Molecules **23**(7), 1505 (2018)
61. Caruso, U., et al.: AIE/ACQ effects in two DR/NIR emitters: a structural and DFT comparative analysis. Molecules **23**(8), 1947 (2018)
62. Diana, R., et al.: Highly efficient dicyano-phenylenevinylene fluorophore as polymer dopant or zinc-driven self-assembling building block. Inorg. Chem. Commun. **104**, 145–149 (2019)
63. Diana, R., et al.: The effect of bulky substituents on two π-conjugated mesogenic fluorophores. Their organic polymers and zinc-bridged luminescent networks. Polymers **11**(9), 1379 (2019)
64. Lu, H., et al.: Highly efficient far red/near-infrared solid fluorophores: aggregation-induced emission, intramolecular charge transfer, twisted molecular conformation, and bioimaging applications. Angew. Chem. Int. Ed. **55**(1), 155–159 (2016)

Author Index

S. Piotto et al. (Eds.): BIONAM 2019, LNBE, pp. 259–260, 2020.
https://doi.org/10.1007/978-3-030-47705-9

The manufacturer's authorised representative in the EU is Springer Nature Customer Service Centre GmbH, Europaplatz 3, 69115 Heidelberg, Germany. If you have any concerns regarding our products, please contact ProductSafety@springernature.com

Printed and bound by CPI Group (UK) Ltd, Croydon, CR0 4YY
30/09/2026
02225557-0003